DATE DUE

DEMCO 38-297

England's Leonardo:
Robert Hooke
and the Seventeenth-Century
Scientific Revolution

England's Leonardo:
Robert Hooke
and the Seventeenth-Century
Scientific Revolution

Allan Chapman

IoP

Institute of Physics Publishing
Bristol and Philadelphia

British Library Cataloguing-in-Publication Data
A catalogue record for this book is available from the British Library.

ISBN 0 7503 0987 3

Library of Congress Cataloging-in-Publication Data are available

Commissioning Editor: Tom Spicer
Editorial Assistant: Leah Fielding
Production Editor: Simon Laurenson
Production Control: Sarah Plenty
Cover Design: Frédérique Swist
Marketing: Louise Higham and Ben Thomas

Published by Institute of Physics Publishing, wholly owned by The Institute of Physics, London

Institute of Physics Publishing, Dirac House, Temple Back, Bristol BS1 6BE, UK

US Office: Institute of Physics Publishing, Suite 929, The Public Ledger Building, 150 South Independence Mall West, Philadelphia, PA 19106, USA

Typeset by Academic + Technical Typesetting, Bristol
Index by Indexing Specialists (UK) Ltd, Hove, East Sussex
Printed in the UK by MPG Books Ltd, Bodmin, Cornwall

To Rachel, my wife, constructive critic,
and best friend

CONTENTS

PREFACE

It has taken almost three centuries to bring Dr Robert Hooke out of the shadows of the seventeenth-century Scientific Revolution, to stand at the very front of the stage and in the full beam of the historical spotlight. But there at last one hopes that he will remain: with the Hon. Robert Boyle, Sir Christopher Wren, Christiaan Huygens, Giovanni Domenico Cassini and, of course, Sir Isaac Newton. It is indeed astonishing that a man who, as he entered middle age in the mid-1670s, enjoyed a reputation as an astronomer, horologist, microscopist, physiologist, and as an out-standingly successful exponent of the 'experimental method' that resonated across Europe, should have thus fallen through the net of wider historical memory. And when one looks at the memory of Robert Hooke that did survive, it was never as a figure of merit in his own right, but as some sort of servant who did the bidding of others and who came a cropper when he had the temerity to contradict Sir Isaac Newton. Indeed, the Robert Hooke of folk myth is a mis-shapen, unkempt, ill-natured fellow, happiest working at his bench, filing away at a piece of metal, or else socialising with bricklayers and artisans; always being 'bidden' and 'ordered', and generally overworked by the fine gentlemen of science, whom he resented.

But it is hard to reconcile this image of Hooke with the clergyman's son who went to Westminster School and Christ Church, Oxford, and whose genius as a scientist was being noted both in London and in Oxford by the early 1660s: a man, indeed, who moved in the highest social circles, dined with Lord Mayors and Archbishops, and was even on easy social terms with His Majesty King Charles II himself. For the historical record — a record, moreover, that is immensely rich in primary sources — shows the kind of man that Hooke really was, and the company he kept.

Yet even by the mid eighteenth century, Hooke's historical standing was becoming obscured, and when Edmund Stone, in his English transla-tion of Nicholas Bion's *Construction of Mathematical Instruments* (1758), added a 'Supplement' which looked at the development of scientific instrumentation over the preceding century, Hooke scarcely got a mention. And in spite of his great success as an architect and town-planner after his appointment to one of the Surveyorships of the City of London after the Great Fire of 1666, even Hooke's reputation in that area was lost, and some of his surviving buildings came to be mistakenly attributed to his friend Sir Christopher Wren.

Robert Hooke's restoration to the historical canon, however, really began in 1891, when his manuscript Diary came on to the market, and passed into the possession of the City of London, and was deposited in the Guildhall Library. From those scholars who came to read it, the Diary demanded a drastic re-appraisal, giving as it does a spectacularly intimate portrait of Hooke's life and work, especially during his brilliant creative years of the 1670s. For far from being a shadowy recluse, this pre-Newton Hooke of the 1670s was a man about town, a smart dresser, and a lover of company, as well as a scientist of breathtaking ingenuity. The scholarly editing and publication of *The Diary of Robert Hooke, MA, MD, FRS, 1672-1680* by Henry W Robinson and Walter Adams in 1935 was in many ways the foundation for subsequent Hooke studies. And then, in the same year, Robert T Gunther in Oxford edited and published the shorter and more cryptic 'The Diary of Robert Hooke' covering the years 1688-90 and 1692-3. Between them, these scholars vastly amplified our knowledge of Hooke beyond those contemporary sources, the 'Life of Dr Hooke' published as part of Richard Waller's *Posthumous Works* of Hooke in 1705, the biographical entry in Anthony Wood's *Athenae Oxonienses* (1691-4; 1721), and the Bodleian Library manuscript 'Brief Lives' of John Aubrey, eventually edited and published by Andrew Clark in 1898.

The next milestone was Margaret Espinasse's *Robert Hooke*, published in 1956, which is the first modern scholarly study of Hooke's whole life and career, based on the Diaries, Hooke's own extensive body of publications, and the numerous references to Hooke in Thomas Birch's four-volume *History of the Royal Society* (1756-7), a compilation of the Royal Society's principal manuscript Registers covering the years 1660 to 1687, and a major primary source for any student of Hooke. Indeed, in the pages of Birch's *History* one can trace Robert Hooke's career through its most brilliant and creative years. Here, at the recorded weekly meetings of the Royal Society, one can follow through a multitude of reports of experiments and observations—on gravity, animal respiration, astronomy, and light—as well as find records of Hooke's salary and other details.

In 1960 and thereafter, when the Royal Society celebrated its tercentenary, Robert Hooke's reputation received further attention as part of a wider historical study of its foundation. And by the late 1960s and into the 1970s, a new breed of historical scholar started to work on Robert Hooke and his world. These scholars were not, in the main, scientists out to find illustrious professional ancestors, but rather academic historians looking at the Scientific Revolution and organizations like the Royal Society in their wider social and cultural context. Professor Michael Hunter's ground-breaking work on the early Fellowship of the Royal Society was an exemplar of this new approach. In consequence, Robert Hooke came to be caught up in scholarly studies of those contemporaries with whom he had either worked closely or else had fallen out, as in the books of R E W Maddison, Jim Bennett, and Richard S Westfall on Boyle, Wren and Newton respectively; while his letters to Boyle, Newton, Oldenburg, Flamsteed, Huygens and others came to be published as part of the works of these men.

This new scholarly approach, moreover, began to explore Hooke's extensive manuscript remains, such as those at Trinity College, Cambridge, the Guildhall Library, London (pertaining particularly to his City of London Surveyorship), and most of all, in the Royal Society itself. Modern scholars also made attempts to replicate particular Hooke devices described in his papers, starting back in the 1920s when working models of his airpump were built for the Museum of the History of Science, Oxford, and the Science Museum, London. Similarly, attempts were made to reconstruct his microscope in the 1920s (and in the late 1970s I also built a working replica of Hooke's *Micrographia* instrument to test its practical handling qualities), while Dr Michael Wright has reconstructed one of Hooke's early clock escapements, and in 2003 Dr Allan Mills his equatorial telescope clock drive mechanism. A major new monograph, *Robert Hooke: New Studies*, edited by Stephen Shapin and Simon Schaffer, came out in 1989, containing nine essays examining a wide range of modern researches into Hooke, and printing the hitherto unknown Inventory of Possessions found in Hooke's rooms at the time of his death in 1703, which had been discovered by Frank Kelsall in the Public Record Office. And in 1994, Richard Nichols published *The Diaries of Robert Hooke, the Leonardo of London, 1635–1703*, which looked at aspects of Hooke's work as recorded in his Diary entries.

By the 1990s, something of an international 'Hooke industry' had come into being, as scholars began to examine all accessible aspects of Hooke's life and work, from his extensive documentary remains. In 1994, I had the honour to be invited to give a 'Friday Evening Discourse' at the Royal Institution, where I looked at the life and experimental researches of Hooke, and this was followed in 1996 by the Henry Tizard Memorial Lecture at Westminster School, which I also devoted to the study of Hooke. But it was a broad wish felt across the scientific and historical communities to commemorate Robert Hooke's tercentenary year of 2003 which acted as a galvanizing force in Hooke studies. The Royal Society in conjunction with Gresham College, London, hosted a three-day international Conference at the rooms of the Royal Society in

July 2003, and this was followed in early October by a one-day Conference at Oxford University, under the aegis of Christ Church, in cooperation with the Royal Society, which filled the South Examination School with a capacity audience of scientists and academics, groups of sixth-formers from local schools, and interested members of the general public, some of whom had flown in to the event from overseas. And one of the upshots of the Oxford Conference was the consideration of a permanent memorial to Hooke, including proposals for a commemoration tablet in Christ Church Cathedral and the possible establishment of appropriate academic awards.

BBC television also commissioned a documentary programme dealing with Hooke's life, which was broadcast several times over 2003–4, though its particular focus was that subject of perennial fascination, Hooke's conflict with Newton. But Robert Hooke had made a previous BBC television appearance in 2002 in the *Great Britons* series, when he was depicted by an actor as Newton's obese, wine-swilling, *bête noire* in the programme advocating Newton's claim. Indeed, a more perfect example of grotesque historical mangling would be hard to find, pandering as the programme did to the 'Hooke the awkward, the jealous, and the obstructive' myth; though how this was supposed to relate to the emaciated, abstemious Hooke of historical record is anybody's guess. My letter of complaint to the BBC received neither an explanation nor an acknowledgement.

Then on the afternoon of 3 March 2003, a Memorial Service for and celebration of Robert Hooke was held at Willen Parish Church, Greater Milton Keynes, addressed by the Bishop of Buckingham and other dignitaries. Willen was, indeed, an ideal venue for such an event, for not only did Hooke design the Parish Church, but it is his only complete surviving building. This event, moreover, was in keeping with an earlier celebration of Robert Hooke held jointly between the Parish of St Mary Magdalene, Willen and Westminster School and Abbey, to mark the 300th anniversary of the Church's foundation, with a week of concerts and lectures in 1980.

The 2003 tercentenary has also acted as a focus for publication, seeing the appearance of four books dealing with Hooke's life and career. The first was Stephen Inwood's *The Man Who Knew Too Much* (2002), which, in accordance with its author's special research expertise, dealt especially with Hooke in the context of London history. Then came *London's Leonardo – The Life and Work of Robert Hooke* (2003), a collection of four separate essays written by Michael Cooper, Jim Bennett, Michael Hunter and Lisa Jardine respectively, and based on a series of lectures delivered at Gresham College, London. Lisa Jardine next published her *The Curious Life of Robert Hooke, The Man Who Measured London* (2003); and Michael Cooper, who has perhaps done more than any other scholar to bring to life Hooke's daily activities as Surveyor from his meticulous examination of City of London documents, his 'A *More Beautiful City', Robert Hooke and the Rebuilding of London after the Great Fire* (2003). All of these books are the work of scholars of the highest reputation and research expertise, as well as being beautifully and lucidly written. While I have myself been researching into Robert Hooke for over 25 years, I am indebted to these authors for a variety of insights, especially concerning Hooke's career as Surveyor and as a Londoner, as will be evident in my notes and references at the end of this volume. In addition to the above works, the 2003 Oxford Hooke Conference has resulted in *Robert Hooke and the English Renaissance*, a volume of papers edited by Paul Kent and myself, and based on the papers delivered.

While it is not the historian's task to heroise or write panegyric, it is his or her task to get at the truth as it can best be understood from the surviving evidences. And within that aspiration, it is hoped that the present book may play its part in the business of bringing Robert Hooke out of the shadows and into the light.

ACKNOWLEDGMENTS

I wish to acknowledge the help and inspiration of many people and institutions who have, in various ways, assisted with my work on Hooke over the years. Institutionally, these include the Royal Society (especially the Library Staff), Westminster School, Christ Church, Oxford, the Royal Astronomical Society Library, the Bodleian Library, the Radcliffe Science Library, Oxford, as well as the Museum of the History of Science, Oxford, Wadham College, Oxford and the Faculty of Modern History, Oxford, for granting me academic sanctuary. As regards individuals, I wish to thank the following: Professor Gerard L'E Turner; Dr Paul Kent and Sir Henry Harris of Christ Church; Cliff Davies; Rob Martin and members of the Isle of Wight Hooke historical research group; Dr Richard Whittington, Dr John Lester, the late Professor John Potter and other medical friends for kindly contributing their insights to my attempt to diagnose the cause of Hooke's death; Tony Morris and friends of the Mexborough and Swinton Astronomical Society, and Kevin Kilburn of the Manchester Astronomical Society, for their astronomical and computational help; Sarah Jane White, for all of her help with the graphics processing; Peter Hingley; Miss Mary Chibnall; Rob Marriott; Chris Mitchell; Rod Beavon; Tony Simcock; Sarah Jane White; Gain Lee; Kenneth Kennedy; Rita Greer; Nigel Frith; Peter Weldon and Monica Mears, of Willen; and Sandra Bates, née Kingsworth. And if I have forgotten anyone, then I crave their pardon. I am also very much indebted to Tom Spicer of the Institute of Physics Publishing for his encouragement and patience. My greatest single acknowledgment, however, must go to my wife Rachel, whose scholarship and proof-reading and word-processing skills have transformed my proudly dinosaurian fountain-pen-written manuscript into the book which you now have before you.

ACKNOWLEDGMENTS

CHAPTER 1

EARLY LIFE: THE PRODIGY FROM THE ISLE OF WIGHT

Getting to know Robert Hooke has been a fascinating process of discovery: full of twists and turns, with great volumes of data counterbalanced by maddening silences. Contrasts, indeed, that he himself must have experienced in the natural world when first he set out on that journey of scientific exploration as a schoolboy, and which were to drive him through 68 years of life. For while Hooke's scientific researches are documented with an astonishing thoroughness, and one can identify the origin and maturation of those great themes that run through his creative imagination, the inner life of the man himself is often elusive.

Nor is there any surviving portrait or contemporary visual likeness of Robert Hooke, and when the German antiquarian and scholar Zacharias Conrad von Uffenbach visited the Royal Society in 1710, and specifically mentioned being shown the portraits of 'Boyle and Hoock', which were said to be good likenesses, it is likely that the 'Hoock' in question was not Robert Hooke but Theodore Haak FRS, the German scholar who had settled in England, and whose name was pronounced 'Hawk' or 'Hark', and was perhaps misunderstood in the telling from the mouth of an Englishman to the ears of a German in a conversation probably conducted in Latin.[1] Boyle's and Haak's portraits survive in the collections of the Royal Society, but there has never been a record for one of Hooke. What is more, when Richard Waller, Secretary of the Royal Society, edited Hooke's *Posthumous Works* in 1705, the handsome and otherwise richly illustrated folio volume contained no frontispiece portrait: an omission, indeed, that was very peculiar for such a celebratory volume. Nor did the *Philosophical Experiments...of Robert Hooke* which William Derham edited in 1726 contain a portrait, and one suspects, quite simply, that there was no available oil painting for an engraver to copy. On the other hand, one cannot help but be tantalized by Hooke's cryptic Diary entry of 20 April 1674 which recorded 'sit at Mrs Beales. At Mrs Beales', for Mary Beale, who was three years older than Hooke, was a well-known London portrait painter who, it is thought, like the young Hooke himself, had passed through the studio of Sir Peter Lely.[2] But more will be said of this in the Appendix.

What we do possess for certain, however, are two detailed pen-portraits of Hooke written by men who knew him well. The first was that recorded by his friend John Aubrey, and describes Hooke in middle life and at the height of his creative powers:

> 'He is but of midling stature, something crooked, pale faced, and his face but little below, but his head is lardge, his eie full and popping, and not quick; a grey eie. He haz a delicate head of haire, browne, and of an excellent moist curle. He is and ever was temperate and moderate in dyet, etc.'[3]

The second is that by Richard Waller, who came to know Hooke well over the last decade or so of his life, and whose forthright account of the elderly Hooke can scarcely be said to err on the side of flattery:

> 'As to his Person he was but despicable, being very crooked, tho' I have heard from himself, and others, that he was strait till about 16 Years of Age when he first grew awry, by frequent practicing, with a Turn-Lath...He was always very pale and lean, and laterly nothing but Skin and Bone, with a Meagre Aspect, his Eyes grey and full, with a sharp ingenious Look whilst younger; his nose but thin, of a moderate height and length; his Mouth meanly wide, and upper lip thin; his Chin sharp, and Forehead large; his Head of a middle size. He wore his own Hair of a dark Brown colour, very long and hanging neglected over his Face uncut and lank...'[4]

Rachel E W Chapman's reconstructed portrait of Hooke is an attempt to give a visualization to these two descriptions.[5]

Yet in his day, Robert Hooke lived a very public life, and soon became one of the familiar figures of mid and late seventeenth-century London. He enjoyed a remarkable social mobility, his daily business bringing him into contact with all manner of men: fellow-scientists, craftsmen, builders, senior clergymen, members of the aristocracy, government officials, and even the King himself. The diarists Samuel Pepys and John Evelyn knew him well, while the early records of the Royal Society testify to his amazing versatility and creative powers.

In an age well known for its graphic and often blunt personal invective, Hooke knew how to hold his own. A highly-strung and sensitive man, he could often be snappish. He made enemies and became embroiled in a series of controversies of both a personal and professional character but, more important, he had the ability to make and retain loyal friends. Hooke had a deep and genuine affection for a wide range of people, encompassing at the one end his financially inept and ultimately suicidal grocer brother John and his family, and a string of natural and adopted dependents, servants and orphans, and at the other instrument makers and academic colleagues such as Sir Christopher Wren, Sir John Lawrence, Lord Mayor of London and the Honourable Robert Boyle.

One senses that the key to a good and lasting relationship with Hooke was never to patronize him, for though instinctively generous of himself and, in the days of his fame and prosperity, remarkably keen to help those who were still struggling up life's ladder, he had an intense pride in his achievements and a sense of the enduring value of what he had done. While Hooke never seems to have been a man who desired flattery, he always demanded respect and due acknowledgement, and when one finds him in various *imbroglios* at certain points in his life and career, they can usually be traced back to a real or perceived failure to give him his due.

For 40 years, and especially up to about 1685, Hooke produced a succession of ideas, inventions, books, buildings and discoveries that amazed even his enemies. His dedication to the New Science, and his ability to produce original creations in many of its branches, from astronomy through microscopy to physiology and geology, did much to guarantee the survival of that science in England. Samuel Pepys summed him up as 'mighty ingenious', while none other than his monarch, King Charles II, more familiar with rogues and pretenders than most, styled him 'a very able and honest man'.[6]

There are three primary sources for Hooke's early life. The most reliable is probably the short autobiographical draft which Hooke began to write in a pocket diary, beginning on 10 April 1697 but only covering his childhood; extracts from this 'Life' were printed by Richard Waller in 1705.[7] Then there is the 'Brief Life' written down by his friend John Aubrey probably some time around 1689, and clearly based on details passed from Hooke to Aubrey in various conversations, though containing several errors of fact, mainly of date, which become apparent when checked against other records. Finally, there is the substantial contemporary entry on Hooke contained in Antony à Wood's *Athenae Oxonienses*, and probably drawn from material supplied by Hooke himself, by Aubrey and by other fellow-Oxonians. But Wood, who himself was a notoriously difficult man, did not care for Hooke, and to Aubrey's chagrin had failed to include a biography and an account of Hooke's work in his 1674 History of Oxford University, for, chided Aubrey, 'England haz hardly produced a greater witt, viz for Mechaniqs...' and science than Hooke.[8]

But all sources agree that Robert Hooke was born at 12 noon on Saturday 18 July (Old Style) 1635 at Freshwater, Isle of Wight, where his father the Revd John Hooke was curate in charge of the parish, and that he was baptized on either 19 or 26 July.[9] According to what Robert later told Aubrey, he was descended 'of the family of the Hookes of Hooke in Hants. [Hampshire]', whose property had stood on the Salisbury to London road, they being 'a very ancient Family and in that place for many (3 or more) hundred yeares'.[10] One presumes that he meant the village of Hook, which nowadays stands on the A30 just east of Basingstoke. It seems, however, that the Hookes had declined in prestige by the seventeenth century, or else Robert was descended from a less powerful branch of the family, for his father John had been fulfilling a variety of curacies on the Isle of Wight since at least 1610, and had been at Freshwater since 1626. It would be nice to know more about the Revd John Hooke, and whether the Warwickshire lad of the same name—designated as 'cler. fil.' (son of clergyman)—who went up to Trinity College Oxford in 1599–1600, or else the 1602 Sizar (or poor scholar) of Emmanuel College, Cambridge, were he.[11] We do know from Isle of Wight records, however, that in 1615 the Revd John Hooke who became Robert's father was the Curate of Brading, an Island village to the south of Ryde, for there he had married Margaret Lawson who was herself a twice-widowed lady, though still young. But Margaret died only a few months after marrying John Hooke, in January 1616, and in 1622, still at Brading, John Hooke married Cecellie Gyles, the daughter of a Brading merchant. They had four children, Anne (?), Katherine (1628), John (1630) and Robert the scientist. The Isle of Wight must have been a quite healthy place upon which to bring up children in the seventeenth century, for of the four recorded children born to John and Cecellie Hooke, only Anne seems to have died young, with Katherine and John living into the 1670s, where their comings and goings were recorded in their younger brother Robert's Diary. Even so, Alice and Rebecca Lawson, the two children of Margaret Hooke with her previous husband, Edward Lawson, failed to get into their teens.[12]

As a boy Robert Hooke was not strong, being obliged to live on a milk diet and unable to eat meat, and in spite of an ability in running and leaping, his father, who was reluctant to subject him to the rigours of a boarding school in preparation for an intended career in the Church, educated him at home. That the Revd John Hooke was up to this task, however, is not to be doubted, for while curate of Brading, he had also worked as tutor to George, the son of Sir John Oglander, who was one of the leading gentlemen of the Island.[13] Yet no matter what gifts of intellect or training Robert may have imbibed from the Revd John Hooke, a scientific bent was not one of them for, as Aubrey recorded, 'his father was not Mathematicall at all'.[14] It was during these early years that, according to his contemporary biographers Aubrey, Wood and Waller, his talents began to manifest themselves. He was an extraordinarily quick learner, possessed a manual dexterity which enabled him to build an impressive array of mechanical devices, including clocks, sundials and model ships with firing cannon, which he sailed upon the river Yar, at Freshwater, while his untrained draughtsmanship so struck the visiting artist John Hoskins that he advised Mr Hooke to settle upon an artistic career for Robert. Indeed, one cannot help but notice the parallels between Hooke's early years and those of his subsequent arch-rival, Sir Isaac Newton, for Newton himself had been an isolated, sickly child in Lincolnshire, who similarly found release for his creative energies in devising a range of working models and instruments.

In 1648, the Reverend John Hooke died, and was buried in Freshwater churchyard on 17 October. John Aubrey in his 'Brief Life' of Robert Hooke recorded that 'His father died by suspending him selfe', though Hooke's own narrative which Waller used for his early life does not specify any cause of death, but simply states that 'for three or four years before his Death [the Revd John Hooke had] been much afflicted with a Cough, a Palsy, Jaundice and Dropsy'.[15] As John Hooke received Christian burial in the churchyard and had his Will proved in the normal way, however, it is highly unlikely that he died by his own hand. What is almost certainly the case is that Aubrey got his John Hookes mixed up, for while the father died naturally in 1648, his eldest son John, who was Robert's older brother, certainly died by hanging himself in 1678 at the age of 48. Though we do not know the cause of the Revd John Hooke's death, the presence of palsy, dropsy, jaundice and other prior symptoms suggests a degenerative syndrome in which cardiovascular failure and nephritis could have been significant contributory factors, and as we shall see in the final chapter of this book, a similar collection of symptoms was also recorded as being present prior to his son Robert's death in 1703. Mrs Cecellie Hooke survived her husband by 17 years, dying in the summer of 1665.

Yet while, according to Aubrey, Robert Hooke was said to have received a legacy of £100, the discovery of John Hooke's will by Hideto Nakajima in the Isle of Wight County Record Office in 1992 indicates that he got only £40, a wooden chest, and a supplementary payment of £10 from the estate of his deceased maternal grandmother, Ann Giles.[16] But once again, one senses a wider truth embedded within Aubrey's sometimes confused narrative, for although Robert only received £40 from his father, his elder brother John received £60, thereby giving some substantiation to Aubrey's talk of £100. In spite of the contradictions and inaccuracies of some of Aubrey's biographical writings on Hooke, we now know that Robert's elder brother John, who stayed on the Isle of Wight, became Mayor of Newport before his business as a grocer failed and he committed suicide. Indeed, one wonders if the 'melancholick' strain in Robert's personality to which Richard Waller referred was part of a wider depressive tendency which ran through the Hooke family.[17]

Even so, one should not forget that in 1648 a combined legacy of £50 was quite a respectable sum for a youngest child to receive, and argues against the suggestion that the Hookes were poor, especially when one remembers that John Hooke's other three children received a total of

£170 between them, while Cecellie would have received her widow's portion from John's estate, not to mention the very comfortable house (still standing) in which they lived in Freshwater. Indeed, when Robert became Gresham College Professor of Geometry in 1665, the salary carried by that prestigious City of London post was also £50 a year, while at the same time a Tutorial Fellow at Wadham College, Oxford, received £20 per annum plus allowances for meals.[18] In short, the Hookes were a comfortably-off, middle-class family who, as other records show, employed several servants.

In addition to the family, financial, or genetic factors which helped mould Robert Hooke, however, it is hard to discount the influence which the Isle of Wight itself — and especially the Island's extreme western tip around Freshwater where he was born and bred — had upon him. For this place was as rich and variegated as any land- and seascape in Britain, with its precipitous chalk cliffs, the strangely-coloured sands of Alum Bay, and its constantly changing panoply of wind, weather and light. Here, the young Hooke would have seen King Charles I's great gilded-sterned battleships riding into the Solent with fierce Atlantic westerlies billowing their sails, or else turning around their anchor cables, as they stood moored outside Portsmouth harbour, in response to the alternating tidal rips that passed around the island and the mainland.

One has only to stand on the cliffs overlooking the Needles, little more than a two-mile walk from Freshwater churchyard, on a November afternoon to sense the *genius* of the place. The island of Purbeck, some 15 miles down Channel to the west, seems as if it must once have been connected to the Isle of Wight and has a similar geology. The regular collapses of chalk cliff into the driving sea suggest that this is a landscape being gradually eroded away, while the constantly changing intensities and colours of the dappled light coming from the sky and scudding across the sea and the distant Hampshire and Dorset countryside impress upon one the grandeur and awesomeness of nature's forces.

Did the endless flux and reflux of the Island's complex tidal systems, and the sudden buffeting gales upon the great hulks and sails of passing ships lie, perhaps, at the heart of Hooke's lifelong scientific fascination with spring, compression, rebound, resistance, pressure and force? Did the Island's ever-changing light and weather conditions form the well-spring of his 40 years of inquiry into the nature of optics and meteorology? And did the visible erosion, and those beds of shells which he found in the chalk 60 feet above high water mark around nearby Hurst Castle, together with fossilized giant ammonites dug out of the quarries at Portland 40 miles down the English Channel,[19] constitute the genesis of his interest in geology, and of that concept of a dynamic earth history, which he developed in the long series of Earthquake Discourses delivered before the Royal Society between 1664 and 1700? He certainly made telling references to the Isle of Wight in these Discourses, as we shall see.

It is always hard to assess the influence of a particular place upon a subsequently famous individual's development, for one must never lose sight of all those other Islanders who saw the same things as Robert Hooke yet left no mark on history beyond the brief record of a parish clerk. But when a person possessed of a formidable latent talent witnesses them, perhaps they can help to lay the foundations of a rich and powerful creative imagination. For much of the stuff of history—and especially of biography, we must never forget—is human peculiarity, and the way in which particular circumstances mould and inspire individual people.

Though we have no documented record of who, following the death of his father, made the arrangements, the 13-year-old Robert went up to London for 'tryall' in the studio of Peter (later, Sir Peter) Lely, the leading portrait painter of the age. But according to the account that the middle-aged Hooke gave to John Aubrey, he 'quickly perceived what was to be done' in mastering the grounds of practical art and painting, and then, complaining that the oils and varnishes irritated his chest, he left the studio, with his indenture fee probably virtually intact, only to be enrolled at Westminster School under Dr Richard Busby.[20] One presumes that the 13-year-old must have had friends and a patron in London, for these adroit social moves would have been extraordinary even for a youth of Hooke's precocity unless he had received help. Though we have no details of who these helpers were, three obvious candidates spring to mind. Firstly, there is the Revd Cardell Goodman, the largely absentee Rector of Freshwater during the years when the Revd John Hooke was effectively running the parish in the 1640s. It is clear that John Hooke and Cardell Goodman got on well together, for in his 1648 will, John Hooke speaks of Goodman as one of 'my worthy and well-beloved friends' who was also entrusted with part of the task of ensuring that the will was properly and justly executed. Then there was the Revd Dr Samuel Fell, who had been Rector of Freshwater between 1615 and 1621, and must have known John Hooke, especially during Hooke's younger days. As will be noted shortly, both Goodman and Fell were old Westminster and Christ Church men. Thirdly, there was Sir John Oglander, the staunchly Royalist gentleman of Nunwell, near Brading, where John Hooke held his first recorded Island curacy and to whose family he had acted as tutor.[21]

At Westminster, Robert Hooke found his feet in the city that was to furnish the theatre of operations for the greater part of his career. The Revd Dr Richard Busby, who was still within the first decade of his 55-year reign as Head Master of what he was turning into the most intellectually distinguished public school of the seventeenth century, quickly recognized Hooke's genius. At Westminster, Hooke not only acquired a mastery of classical Greek and Latin, but also 'got some insight into the *Hebrew* and some other Oriental languages',[22] learned to play the organ, contrived 'thirty several ways of flying',[23] and mastered the first six books of Euclid's *Elements of Geometry* in a week. His exact status at Westminster is unclear, however, for he was neither rich, nor is there evidence that he was a King's Scholar, or its Cromwellian Republican equivalent. But instead of living in the School proper, he enjoyed the privilege of lodging in Dr Busby's own house, and as his Westminster contemporary Sir Richard Knight later told John Aubrey, 'he seldome saw him in the schoole'. Perhaps he received most of his teaching within Busby's household. If he did, the young Hooke was fortunate indeed, for Busby was one of the most learned men of the age. He remained on warm terms with Dr Busby for the rest of the Head Master's 89-year life, undertaking architectural design work for him and for Westminster Abbey, and at Willen, Buckinghamshire, where in 1680 Hooke built a new Parish Church that was paid for by Busby, and he mentions him in his Diary on several occasions during the period 1672–80, and even more so in his 1688–93 Diary.[24]

Yet when looking at his early life in terms of precocity and unfolding genius, it is all too easy to miss the fact that Hooke's years as a Westminster boy coincided with some of the most traumatic events of English history. When Hooke was up at Westminster, the School, which stands within the precincts of the Abbey, lay adjacent to the great Palace of Whitehall which, up until the outbreak of Civil War in 1642, had been the principal metropolitan residence of the King and the Royal Court. Only a few hundred yards away from the School gate in Whitehall stood King Charles I's great Banqueting House, built by Inigo Jones — one of the English classical architects whom the mature Hooke would greatly admire. But it had been in nearby Westminster Hall just after Christmas 1648 that a specially convened Court of Law sentenced King Charles to death, while the King's head was struck off on a scaffolding erected outside the Banqueting House on 30 January 1649.

We do not know whether Hooke was at Westminster School by January 1649, or whether he was still with Lely. But either way, the act of regicide shocked the nation, and ushered in a period of increasing Puritan rule. Being at Westminster School and possibly living in Busby's house in the early 1650s, when in his mid teens, he would have been all too aware of the revolution in politics and morals, and of the enforced Godliness which the self-appointed 'Saints', who now dominated the Parliament House only 200 yards down the road, were now trying to thrust upon England.

One should remember, moreover, that in Hooke's lifetime Westminster looked radically different from what it does now. For while the

Abbey, St Margaret's Church, Westminster Hall, Westminster School, and the King's Banqueting House all survive today, they do so only as separate components of the once vast enclosed and gated Whitehall Palace complex. That Palace was largely destroyed in the disastrous fire of 1698. The Westminster that Hooke would have known, both as boy and man, therefore, would not have been that of a broad Whitehall sweeping into an expansive Parliament Square, but a teeming warren of alleys, red-brick Tudor courts, the Gatehouse, towers, official residences, cramped government offices, great buildings of state, chapels, and cere- monial ways giving access to a variety of stairways that ran down to the principal highway of London—the Thames. For no bridge crossed the Thames at Westminster in his day, and the wide, tidal, un-embanked river, with its great mud-flats stretching out across Lambeth marshes— the flatness broken only by the towers of the Archbishop of Canterbury's Lambeth Palace—must have placed the countryside within a close walking distance. Close, that is, unless one walked to Charing Cross, continued eastwards, passed the great noble and episcopal residences on the Strand, went through the gates of Temple Bar, and then into that dense concourse of crowded humanity which was the City of London.

In his writings and subsequent correspondence, Hooke was to say astonishingly little about politics and affairs of state, though like so many other young intellectuals of his generation, he grew up with a hearty dislike of Puritanism and of the ecstatic religion from which it sprang. Even so, at Oxford he was to form enduring friendships with men who, while not Puritans, had nonetheless come to something of an accommodation with the new regime, and some of whom were even related by marriage to Oliver Cromwell himself.

In 1653 Robert Hooke left Westminster to take up a poor scholar's place at Christ Church, Oxford, where he was subsequently described as 'Servitor' to a Mr Goodman.[25] A Servitorship was the lowest type of undergraduate place in seventeenth-century Oxford, and often required the performance of menial duties, such as waiting at table and looking after the wealthier gentleman who had brought the Servitor up to the University with him. Who this individual could have been, however, is a mystery, for the Matriculation Registers of Oxford University show no one called Goodman coming up to the University, and certainly not to Christ Church, during the early 1650s. On the other hand, the Rector of Freshwater between 1641 and 1651 was the above-mentioned Revd Cardell Goodman, who had not only matriculated into Christ Church on 28 January 1626 at the age of 18, but matriculated at and took degrees from Cambridge University as well. It is true that the Revd Cardell Goodman did have a son, also-called Cardell, but he went up to Cambridge in November 1666 aged 13, by which time Robert Hooke was 31 years old and corresponding with scientists across Europe.[26]

9

The story about Hooke's Servitorship to Mr Goodman comes from Waller's 1705 narrative, and Waller may have got it from Hooke himself. On the other hand, one suspects that Waller was getting his wires crossed half a century after the event, especially if the information had been passed to him casually by Hooke in a coffee house conversation ten years before he wrote it down. It is likely, however, as indicated above, that the Revd Cardell Goodman, who came from a comfortably-off Hertfordshire family, took pity on his deceased curate's bright son in 1648, and helped to steer the 13-year-old Robert to Lely, to Westminster School, and then on to his own old Oxford College of Christ Church. Perhaps more archival research into Cardell Goodman could reveal something.

Another fruitful line of inquiry into Robert Hooke's early life and patronage could also lie with the Fell family, for the Revd Dr Samuel Fell, on his way up the ladder of ecclesiastical preferment after vacating the Rectory of Freshwater, eventually ascended to the dignity of the Deanery of Christ Church, before his staunch High-Church Royalism led to him and his wife being physically evicted from the Dean's House by Cromwell's soldiers at the behest of the Puritan Commissioners in 1647.[27]

There was, however, only one man from the Isle of Wight who seems to have come up to Christ Church during Hooke's time in the University, and he was Thomas Newnham, the son of a gentleman from the Isle of Wight. As a Gentleman Commoner, he could perhaps have been in a position to afford to bring up a Servitor from the Island with him, but the Registers are silent on any such matter, and there is no known connection between Newnham and Hooke, though in the close-knit society of an Oxford College one presumes that they would have known each other.[28]

In addition to any Goodman or Fell connection, Hooke is said to have had a Singing Man's place at Christ Church Cathedral, which was, and still is, an integral part of the collegiate foundation. But as the abolition of the Anglican Church and its sung services between 1646 and 1660 would have closed down the liturgical choirs, one presumes that Hooke must have received a Singing Man's or Lay Clerk's endowment, 'a pretty good maintenance',[29] as his friend John Aubrey later recorded, by way of a scholarship. For Robert Hooke was clearly a man who possessed musical abilities, as his subsequent interest in acoustics and musical vibrations suggests.

What probably happened, though in the absence of any known surviving record it is impossible to be sure, is that Hooke had a rather ambivalent status in Oxford between 1653 and 1658. Not until 1658, indeed, does he leave his first trace on the records of Oxford University, for in that year he was matriculated (or fully incorporated) into Christ

Church and designated 'fil. min.' or *filius ministri*, being the son of a clergyman; though the *ministri* ('of a minister') designation was used in the Puritan 1650s as what would turn out to be a temporary replacement for the traditional 'cler. fil.', or son of a clerk, or clergyman of the Church of England.[30] Even so, as Waller tells us, Hooke had been around Christ Church since 1653, though as a Singing Man or a Servitor, and relatively poor, he may have chosen not to go through the costly process of formal matriculation until he could afford it.

It is not known whether Hooke was especially inspired or encouraged by John Owen, the replacement Dean and Oliver Cromwell's ex-Chaplain, who had been imposed upon once-Royalist Christ Church by the Parliamentary Commissioners after the Civil War; and while there is no record of Hooke ever having taken his BA, he nonetheless received his MA degree in 1663. In the seventeenth century, however, it was not uncommon for poor undergraduates to delay the formal conferment of their degrees, which was an expensive procedure, involving tailor's bills for new gowns, along with hefty tips to various officials, until well after the completion of their course of studies, by which time they had hopefully become established in a career, as was clearly the case with Hooke. On the other hand, the awarding of Hooke's MA does not seem to have been routine, but may have been brought about by the influence of people in high places who wanted to ensure that the Royal Society's newly-appointed Curator of Experiments was a fully incorporated member of the learned establishment. For as Wood records, Hooke received his degree 'by the Favour of the Chancellor, nominated (among others) to have the degree of Master of Arts to be conferred upon him, but whether he was admitted or diplomated it appears not in the Register'.[31] This probably means that he was awarded his MA without the then usual exercises of Examination, and perhaps *in absentia*, possibly while living in London, though as he was still in the employ of Boyle at this time, he could have travelled quite regularly between Boyle's lodgings in Pall Mall, London, and those in High Street, Oxford. The Chancellor of the University in 1663, moreover, was Edward Hyde, the powerful Lord Clarendon, who was also Lord Chancellor of England, which suggests that by that date Robert Hooke was already a man whose services were coming to be highly valued by some of the most important men in England, and that he had influential friends. Indeed, there is ample evidence that this was the case, for around 1660 Sir Robert Moray, Lord Brouncker, and several early Royal Society dignitaries appear to have been negotiating a possible sea-clock or chronometer patent with Hooke, for if such a clock could be perfected it could perhaps have enabled the newly-restored King's ships to find their longitude at sea.

Also at some stage during his time at Christ Church as an undergraduate, according to John Aubrey, Hooke resided in the same set of

11

College rooms in which the late Robert Burton, author of *Anatomy of Melancholy*, had lived and hanged himself in 1640, though if he was only a Servitor or a Singing Man it is unlikely that he would have had a don's set of rooms all to himself.[32]

During the years that preceded Hooke's matriculation in 1658, and even in some of those that followed, however, it is quite possible that he did not live in Christ Church at all, but boarded out; and it is very likely that when he was working as an Assistant to Dr Thomas Willis around 1657, and even with Boyle after 1658, he lodged with them in their own houses, at nearby Beam Hall in present-day Merton Street, and Deep Hall in the High Street respectively. While one might think that living under one's employer's roof would cramp the style of a young man's student days, one must not forget that during the Puritan rule of Oxford, up to 1660, students residing in College were obliged to live under a strict religious discipline of sin-searching, sermon-attending and general moral control. But living with Boyle, who was a deeply devout but not fanatical Protestant, and Willis (who had married Samuel Fell's daughter, and like Robert's father John, Cardell Goodman, and the Fells, was a Royalist High Church Anglican who detested Puritanism), Robert Hooke could well have enjoyed an enviable personal freedom. Indeed, one cannot avoid the likelihood that Hooke's early educational patronage owed a very great deal to a Royalist High Church Anglican network that had been spiritually and politically estranged from national life by the new Puritan regime. This network, moreover, was replete with Westminster and Christ Church men such as Samuel Fell, Cardell Goodman, Richard Busby, Thomas Willis, the young Sir Christopher Wren, and others who had all been to one or both of these academic institutions, and one wonders what contacts and recommendations had been made and acted upon to safely guide the orphaned son of the curate of Freshwater in their own footsteps,

As to his studies, it must be remembered that in the seventeenth century no one went up to Oxford or Cambridge to read for a degree in a specific discipline, as one would today. Rather, the academic Statutes that set out the undergraduate curriculum specified a collection of texts and authors that were almost entirely classical, and a degree in Hooke's time would have demanded a proficiency in Greek and Latin (which were far from being 'dead' languages, for Latin in particular was to remain the formal spoken language of the universities until the nineteenth century), classical literature, classical philosophy and science, Protestant divinity and some civil law.[33] Relatively little of this curriculum would have derived from medieval Roman Catholic writers, however, and as is clear from some of his later remarks, Hooke, like most of his English contemporaries, would have held the Middle Ages in low esteem.

12

But in addition to polishing his Latin and refining that knowledge of classical literature on which he was frequently to draw for reference in his subsequent scientific publications, Oxford inspired Hooke in other, non-curricular, ways, for it was in this University city that he fell in with that group of men who within a decade would form the original Fellowship of the Royal Society. One of the first figures to initiate Oxford's scientific renaissance was Dr, later Sir William, Petty, the Hampshire-born, Leiden-trained medical doctor and experimental chemist, who brought new scientific techniques and methods into Oxford following his appointment to the Anatomy Lecturership in the wake of the Parliamentarian re-staffing of the University after 1648. It was in Oxford that Robert Hooke began his apprenticeship to science, and formed a collection of influential and creative friendships, some of which, as in the case of Sir Christopher Wren, would last to the end of his life.[34]

One of the most important of these friendships was with the Revd Dr John Wilkins, Warden of Wadham College, who, after Petty's departure to make his famous 'Survey' of Ireland, became the leader of the Oxford scientific 'Club'. Wilkins encouraged Hooke in astronomy, mathematics and mechanics, as did the newly-appointed Savilian Professor of Geometry, the Revd Dr John Wallis, as well as the young Christopher Wren. But the man whom he seemed to acknowledge as his first encourager in independent scientific research was the Revd Dr Seth Ward, Oxford's Savilian Professor of Astronomy and later Bishop of Exeter and then Salisbury. For not only did Ward teach him astronomy, but it was from Ward's promptings that 'I apply'd myself to the *Pendulum* for such [astronomical] Observations, and in the Year 1656, or 1657, I contriv'd a way to continue the motion of the *Pendulum*, so much commended by *Ricciolus* in his *Almagestum*, which Dr Ward had recommended to me to puruse'.[35]

Of central importance to Hooke's early and continuing career, however, was Dr Wilkins, who after the restoration of the monarchy and the Anglican hierarchy in 1660 acquired a succession of influential clerical appointments before becoming Lord Bishop of Chester in 1668. Wilkins's rooms or Warden's Lodgings in Wadham had become by 1654 not only one of Europe's leading scientific meeting places, but also, as the visiting John Evelyn recorded, a repository for a wide range of scientific instruments: for 'above in his Lodgings and gallery [was a] variety of shadows, dyals, perspectives and many other artificial, mathematical and magical curiosities, a way-wiser [land measurer], a thermometer, a monsterous magnet',[36] while the Warden's garden contained glass beehives and 'an hollow statue which gave a voice and utter'd words, by a long concealed pipe that went to its mouth'. One cannot help wondering to what extent Wilkins's own fascination with machines, instruments and automata, and especially his influential *Mathematical Magick* (1648), helped to frame Hooke's highly instrumental and mechanistic approach to science. One also wonders, though solid proof is hard to find, whether it was from his early friendship with Wilkins that Hooke later came to be acquainted with the staunchly anti-Papist John Tillotson, subsequently Dean of Canterbury and then of St. Paul's, London, and who on 7 December 1691, soon after becoming Archbishop of Canterbury, granted Hooke his Lambeth MD degree. For John Tillotson was not only intellectually and spiritually indebted to Wilkins, but had also married Wilkins's step-daughter Miss Elizabeth French, whose widowed mother had started life as Miss Robina Cromwell, the youngest sister of Lord Protector Oliver. According to the Diary which Hooke kept between 1672 and 1680, Robert Hooke was not only a visitor to the Tillotsons' London house, but was also on several occasions the recipient of Mrs Tillotson's home-made medical preparations.[37]

John Wilkins was not only an influential scientist, theologian and educator; he was also an astute politician and diplomat. During his tenure as Warden between 1648 and 1659, he had turned Wadham into a haven of toleration in 'troubled times', and during the 1650s had used the influence of his high office in Oxford to gradually untie the bonds which the Puritan government in Westminster had imposed upon the University (whch imposition, moreover, had included his own Wardenship on the reluctant and Royalist Wadham in 1648), and help restore self-government to the academic corporation. But Wilkins the Broad Church diplomat had established a network of important friendships across the intellectual 'middle England' of the 1640s and 1650s, which included members of the Cromwell family at one end and High Church Royalists like the Wrens at the other: friendships, indeed, which covered many political and religious tendencies, with the obvious exception of fanatical Royalists and fanatical Puritans—many of their sons came to Wadham.[38]

It was in Oxford, in the informal world of dedicated experimentalists that clustered in friendship and shared intellectual curiosity around John Wilkins and who met in his Wadham Lodgings, and no doubt at the less famous gathering—which also included Willis and other medical men—that met in Trinity College, Oxford, that Robert Hooke found his feet as a scientific apprentice, and where the talents of this extraordinary junior member of the University were first recognized and encouraged by men of already established reputation.[39] There was, however, one crucial practical problem to overcome; for while John Wilkins, John Wallis, Seth Ward and others were comfortably-off University dignitaries, Petty combined medicine with profitable political entrepreneurship, and the young Christopher Wren and the Honourable Robert Boyle were independently wealthy gentlemen, Hooke was a Christ

Church Servitor or Singing Man with only his genius to offer. Quite simply, he needed to earn his living, and this he did by acting as a paid assistant to the anatomist and chemist Dr Thomas Willis, and to the chemist Robert Boyle.

But none of these new scientific researches were undertaken in accordance with the formal academic or statutory requirement whereby these men held their Fellowships and Chairs in Oxford University. While sciences such as astronomy and medicine had been part of the curricula of the European universities since the Middle Ages, their mode of teaching had always been highly conservative and based upon classical models. The devising of complex experiments that were meant to open up hidden structures within nature was not, and never had been, part of their academic remit. By contrast, however, Wilkins's 'Philosophical Club' came into being for the very purpose of examining these hidden structures by the experimental method, but in the context of a private club of friends who met to pursue a common interest; and to see their enterprise as some kind of research institute headed by Wilkins, or as a forerunner of the German, French and American professorially-directed research institutes of the nineteenth century, is to entirely misunderstand its character and intention – for the Philosophical Club had no more official status in the University than would a private dining club, or a group of friends who came together to sing songs, pray in accordance with the Book of Common Prayer, or to discuss antiquities; activities for which, moreover, there were thriving clubs in Oxford in the 1650s. It seems to have given Robert Hooke a taste for gentlemanly clubs, fellowships and societies that lasted a lifetime.

Before we proceed to examine Hooke's informal apprenticeship and entry into the world of the wealthy Virtuosi and experimentalists, however, we must first note that such a career choice was very unusual, if not downright eccentric, for a hard-up yet clearly gifted graduate of 1660. A far more normal procedure, indeed, particularly if one had a circle of influential friends, would have been to have entered one of the established learned professions, especially the Church or the law, which were the golden roads whereby many a scholarship boy had risen to the episcopal or judicial benches in the seventeenth century, especially old Westminster boys. For entry into the priesthood of the Anglican Church, indeed, was virtually an expected path for a poor graduate to follow, especially if he had well-connected friends who could secure him a chaplaincy to a nobleman or a rich living on his way up to greater glories. Indeed, for a young man to actively choose to enter science before having first secured himself in a profession could have seemed, to an impartial observer at the time, like the squandering of opportunities. For in the infant state of science in 1660, what could a 'professional' career in science have foreseeably entailed?

15

A life of permanent dependency and 'client' status, doomed to a succession of insecure assistantships to wealthy researchers, just like those eternally hard-up curates who never stood a chance of getting their own benefices; a lack of control over one's own life which seemed to sit incongruously with having the letters MA after one's name; and an almost certain inability to secure a marriage to any girl whose father demanded that his daughter marry a gentleman possessing prospects.

How a man possessing the courage, fighting spirit, personal pride and love of independence that were such obvious components of Hooke's character eventually made this career choice, we will probably never know, although there is no real evidence to suggest that he ever considered other ways of earning his living than science. Perhaps what made the career gamble possible was the absence of any apparent romantic goal in his life, for while Hooke's Diary indicates that he had a number of relatively casual sexual encounters with servant girls between 1672 and 1680 — and no doubt had had similar ones before, especially when in his late teens and 20s — the driving passion of his life was scientific research, for as his biographer Richard Waller later recorded, 'Mechanicks' was his 'First and last Mistress'.[40] But if marriage never seriously figured amongst his life's goals, and he never felt the urgency of needing to win a bride and support a family in gentlemanly style, then Hooke's future plans would have possessed a simplicity which could have allowed that personal and financial leeway which was a necessary prerequisite for a scientific career for a man in his circumstances. In addition to apparently accepting the prospect of a bachelor life, Hooke never seems to have harboured ambitions of a political or administrative kind, such as might have been spurs towards more conventional lines of advancement and public reward. Even by 1690, when the many thousands of pounds of the 'Great Estate' which he had accumulated from his architectural work and City of London Surveyorship would have been sufficient to give him the standing to purchase a country estate, enter Parliament, or negotiate for a Knighthood, he showed no inclination towards becoming a figure of state.

By seeing Robert Hooke, therefore, as a man without any serious romantic, dynastic, or wider social ambitions, whose natural *milieu* was the all-male environment of the college, club, or coffee house, and whose intellectual curiosity was the driving force of his life, one can perhaps come some way towards understanding why he was so keen to take up the seemingly unconventional job opportunities which the Royal Society offered him. How the Royal Society would have developed in the longer term without a man of Hooke's peculiar genius, dedication and willingness to carry so many of the burdens, is in itself a matter of historical speculation.

CHAPTER 2

BREATHING, BURNING AND FLYING: HOOKE'S SCIENTIFIC APPRENTICESHIP

We do not know for certain when Hooke was first admitted to the Club of Virtuosi that met in Dr Wilkins's lodgings in Wadham College, Oxford, but according to Hooke's Obituarist, Richard Waller (who tried to be scrupulous about matters of historical record) it seems to have been around 1655. However, in September 1658 Oliver Cromwell, Lord Protector of England and Dr Wilkins's brother-in-law, died,[1] to be succeeded by his peaceable and politically unenthusiastic son, Richard. As a brother-in-law of the Lord Protector, and a person of some importance within the state, Wilkins was now regularly absent from Oxford and in London, as is clear from the Wadham College 'Bursar's Book', for Wilkins last signed the College accounts in midsummer 1658, just before Cromwell's sudden death on 3 September, but left the Christmas accounts to be signed by Walter Blandford, who was to succeed him five months later as Warden of Wadham.[2] Then in 1659 Wilkins left Oxford altogether, to take up the briefly-held Mastership of Trinity College, Cambridge, from which he would be reluctantly — as far as the Fellows of Trinity were concerned — expelled at the restoration of the Stuart monarchy in 1660.

Nearly a decade earlier, however, in the early 1650s, it is evident that the charismatic and congenial as well as intellectually far-sighted Wilkins had become the leader of the English Virtuosi. He had first entered, and had helped to develop, this group at Gresham College, London, where the earliest meetings took place around the mid 1640s, and then after his appointment to the Wardenship of Wadham College in April 1648 much of the activity was gradually transferred to Oxford. Although there had been English scientists of eminence before Wilkins's group, it is clear that the group's activities did presage a sea-change in English intellectual life, for as John Aubrey recorded, 'Till about the yeare 1649, when Experimental Philosophy was first cultivated at a Club at Oxford, 'twas held a strange presumption for a Man to attempt an Innovation in Learnings, and not to be good Manners, to be more knowing than his Neighbours and Forefathers'.[3]

John Wilkins was an English disciple of Galileo and Kepler, a Copernican, and an admirer of the English philosopher of science Sir Francis

Bacon, Lord Verulam, who had died in 1626. Wilkins was also deeply influenced by the published researches of Dr William Gilbert, the Elizabethan physician, whose *De Magnete* (1600) had introduced learned Europe to the scientific study of magnetism via the medium of a set of ingenious and rigorous experiments, and whose geomagnetic concepts helped to substantiate the Copernican theory. In an age when publication was not the scholarly necessity it later became, Wilkins produced his first book at the age of 24, in 1638, and though it was published anonymously his authorship became generally known. Wilkins's *Discovery of a New World in the Moone* argued for Copernicanism, and presented Galileo's ideas and discoveries in the vernacular to an English audience. His Chapter or 'Proposition' XIV, moreover, contained the first serious discussion in the English language about sustained flight and the feasibility of a journey to the moon in a mechanical 'Flying Chariot'. While Wilkins borrowed components of his thinking about the nature of other worlds and the means of flying to them from William Godwin, Kepler and other writers, their assemblage into a suggested coherent programme whereby mankind might be able to travel through space under mechanical power was very much his own.[4]

Then in 1648, the same year in which he became Warden of Wadham College, Wilkins published his influential *Mathematical Magick*, which was a visionary study of what might be achieved by the application of practical mathematics and mechanical technology to a range of problems. Here we find wind cars, self-acting domestic gadgets, labour-saving devices, hoists, cranes, and, of course, talk of — but no pictures of — flying machines. What runs through Wilkins's books is a wish to use invention and discovery to fully realize what Sir Francis Bacon had called mankind's 'right over nature',[5] to reduce our ancient burden of toil and drudgery by the ingenious use of those forces of weight, elasticity, mechanical action, wind, water and fire, when operating through gears, levers, sluices, and flues. For Bacon's visionary philosophy of science had inspired Wilkins and other men of his generation not only to conduct a 'very diligent dissection and anatomy of the world'[6] to see how nature worked, but also to harness its forces for our own benefit through mechanical inventions. For as Bacon had argued in the *Novum Organon* (1620), 'printing, gunpowder, and the magnet...have changed the whole face and state of things throughout the world', for 'no empire, no sect, no star seems to have exerted greater power and influence in human affairs than these mechanical discoveries'.[7]

When one combines this Bacon-derived visionary ingenuity with Wilkins's practical piety, gift for friendship, and the knack of knowing how to captivate people who mattered — not to mention his use of good, plain English and not Latin as a way of communicating his ideas — then one senses how Wilkins became the driving and organizing force that he was within the English scientific community by 1655. Being an adroitly social personage, moreover, with a clear vision of where science should go, he clearly head-hunted clever men to become part of his Oxford Philosophical, or scientific, 'Club', which met in Wadham. It was at Wilkins's invitation, for instance, that Robert Boyle moved to Oxford from London around 1654;[8] while we know that Wilkins was quietly active in softening the worst intolerances of Puritan zeal in Oxford and in freeing, wherever possible, its intellectual life from the stern grip of 'saintly' control.

If, as Richard Waller says, it was in 1655 that the Christ Church Servitor first made the acquaintance of the Warden of Wadham, there is no doubt that a good working relationship and warm friendship quickly blossomed thereafter between the 20- and the 40-year-old men. Sadly, hardly any of Wilkins's own papers survive, so we have no record from his side. However, from the encomium paid to Wilkins in the 'Preface' to Hooke's *Micrographia* (1665) and in his Diary, both before the Warden's death in 1672 and as reminiscence afterwards—not to mention in Samuel Pepys's Diary entries when Hooke, Wilkins and Pepys formed part of the same social groups—it is evident that John Wilkins played a crucial role in forming Hooke's scientific career.[9]

One could argue that several fundamental traits present in Hooke's mature scientific thought can be traced directly to Wilkins's influence. Hooke's lifelong fascination with mechanisms, inventions and self-acting devices is perhaps the most obvious of them, and it is hard to avoid recognizing the influence of *Mathematical Magick*. It is also likely, though no surviving documents can clinch the point, that Hooke's Baconianism and belief in the truth of the experimental method was imbibed from Wilkins (though we do know that it was Boyle who first instructed him in Cartesianism).[10] Though Sir Francis Bacon was the great prophet of empirical, experimentally-based science in the years after 1600, Bacon's science, nonetheless, hinged more on forensic inquisition and taxonomy than on practical laboratory technique, and Bacon seriously underestimated the power of instrumental investigation as a primary agent for the furthering of scientific knowledge. By Wilkins's time, however, working scientists were realizing that the new tools of science, be they devised to weigh, distil, observe or measure, were fundamental to their fresh understanding of the natural world. As is clear from John Evelyn's account of his visit to Wadham College in 1654,[11] and from his record of the large collection of instruments and scientific models owned by Wilkins, one can appreciate how the visiting Christ Church servitor would have quickly grasped their significance. The Rules for conducting inquiries into Natural Philosophy which Hooke drew up in his mature years, and which were published in his *Posthumous Works*, encapsulated the central Baconian tenets of investigation: taxonomy, experiment, analysis and induction, added to which he recognized the significance of instruments as the tools whereby nature's deepest secrets could be laid bare.

In addition to mechanism and instrumentation, one could say that other components of Hooke's subsequent scientific creativity had their roots in Wilkins's own vision of science. The most obvious of these was Hooke's own preference for good plain English as a mode of scientific communication, for Hooke, like Wilkins, knew how to describe things accurately and succinctly in his native tongue, in spite of his acknowledged fluency in the ancient learned languages of the universities. Like Wilkins, Hooke knew that a picture was worth a thousand words: for just as *Mathematical Magick* had beguiled its readers with engravings of real and suggested devices, so most of Hooke's own books and articles supplemented their texts with plates — plates which, in the case of *Micrographia* (1665), *Lampas* (1677) and *Cometa* (1678), were works of art in their own right.

On the other hand, it is obvious that Wilkins recognized an apt pupil right from the start, for in fostering Hooke's mechanical bent he was developing something which had already become evident at Westminster, for 'At schoole he was very mechanical', and prior to that, we must not forget, he had been a boyhood inventor back at Freshwater. But it was under Wilkins's influence that these 'mechanick' tendencies were graced with the laurels of academic respectability. Though nothing under Hooke's hand survives from the period when he was first working with Wilkins, perhaps around 1655 or 1656, he was nonetheless to refer to his own and the Warden's activities on several occasions over the next four decades of his life. His assistance to Wilkins in the devising of model flying machines which seem to have got off the ground, making 'artificial muscule' (muscle) whereby to obtain elastic tensions, as well as experiments with air and water pressure to produce novel fountains, all in the gardens of Wadham College, were mentioned in his subsequent writings, 'As he and Dr Wilkins of Wadham Coll: have reported'.[12]

We do not know whether Hooke's work with Wilkins involved a formal component, such as Hooke being a paid assistant, but it is clear that he did receive remuneration from Thomas Willis and Robert Boyle, for he was spoken of as working as Assistant to both men in succession.[13] Willis was, as will be mentioned below, one of the foremost anatomists of the age. Like Hooke, he was a Christ Church man, though his High Church Anglican and Royalist principles, not to mention his marriage to Mary, the daughter of the Revd Dr Samuel Fell, the ex-Rector of Freshwater and fervently anti-Puritan Dean of Christ Church, had led to his, and Dr Fell's, eviction from their University and College posts by the Parliamentary Visitation of 1647–8. Though his friendship with Wilkins had perhaps softened the official ire against him, Willis would not be publicly and honourably reinstated into both Christ Church and the University's Sedleian Professorship of Natural Philosophy until the Restoration of the Monarchy in 1660. During the period that he enjoyed Hooke's assistance around 1656 or 1658, therefore, Willis was living privately at his house, Beam Hall, in Merton Street, Oxford (which also served as Oxford's unofficial Anglican Church during the Cromwellian period, where 'he entertained Religion then a Fugitive', and services were sung and the Eucharist celebrated by Anglican clergymen who had been evicted from their benefices by the Puritans), and earning his living by practising medicine.[14] It may strike us today as incongruous that the room in Willis's Oxford house which had been set aside for these religious services is only likely to have been a few yards away from the private laboratory in which he probably pursued his chemical researches and also perhaps dissected human cadavers. Indeed, the seeming irony was also probably being alluded to by Willis's friend, translator and obituarist, Samuel Pordage, when he remarked that 'the Parish Church where he [Willis] dwelt' seemed to be closely associated with 'those excellent Tracts by which he first became known to the World, *viz.* of *Fermentation*, of *Feavers* and of *Urines*'.[15] Yet as was the case with all of these men, Thomas Willis's science and his Christian faith were intimately and self-evidently related. For once the immortal soul had taken flight to God and left its inert fleshly carcass behind, then why should not a pious anatomist dissect that same carcass to learn yet more about how wonderfully the Divine Designer had put our bodies together? It was not for nothing that Willis dedicated his monumental *Cerebri Anatome* (1664), which was the foundation stone upon which subsequent scientific studies of the brain were based, and in which he coined the term 'neurology', to his friend Gilbert Sheldon, Archbishop of Canterbury. Willis saw research into the body and brain of man — 'the living breathing Chapel of the Deity'[16] — as a sacred study, and as a confutation of atheism, while his *De Anima Brutorum* (1672) discussed the life-principle of animals and the immortal human soul's relationship to the mortal brain.[17]

For Thomas Willis, whom Anthony Wood (his next-door neighbour in Merton Street, Oxford) generously styled 'the most famous physician of his time', was a clinician, chemist and experimentalist of genius. While his causal explanations for medical conditions and chemical reactions are very different from those accepted by scientists today, anyone who reads his published treatises, surviving 'Oxford Lectures', and medical 'Casebooks' cannot help being impressed by his uncanny insight into how living creatures worked, not to mention his technical brilliance as an experimentalist.[18]

It was also at this time, when Willis was conducting researches into chemistry as well as anatomy, that Hooke was mentioned specifically as being Willis's chemical assistant. In particular, Willis believed, two centuries before the discovery of microbic action, that the chemical processes which took place during fermentation were a natural phenomenon worth serious investigation, the results of which he published in *De Fermentatione* (1659, 1681). Among other things, fermentation seemed to generate heat without the fire, in contradiction to Aristotle's ideas of the four elements. Willis also suspected that the increased heat of patients in a state of high fever (an all too common condition encountered by physicians in the seventeenth century) was caused by the fermentation of the blood in their veins. But what Hooke did in any detail when in Willis's employ is not clear, though it is likely that it was with him that Hooke got his first taste of practical chemistry, and possibly of medicine, animal vivisection (which he always found distasteful) and human anatomy, for Willis's wider published researches covered all these topics.

Then in 1658 or 1659 Hooke passed into the employ of Robert Boyle, the wealthy, ascetic and deeply pious Anglo-Irish bachelor who had rooms at Deep Hall in the premises of the Oxford apothecary, entrepreneur and hotelier John Crosse on the south side of the High Street, Oxford.[19] Deep Hall was demolished soon after 1800, and the site absorbed into the subsequent extension of University College, but the location of their laboratory was marked in the twentieth century by an engraved stone plaque let into the wall of University College.

While Boyle undertook a variety of researches at Deep Hall and possibly elsewhere in Oxford during the 14 or so years between 1656 and 1668 that he maintained a residence in the city, those which he did into the physical, chemical and physiological properties of air between *c.* 1658 and *c.* 1662, and in which he was assisted by Hooke, were undoubtedly the most significant.[20]

For over two thousand years, from classical antiquity up to the seventeenth century, air had been regarded as a stable, simple, vital force of nature, and one of the four elements. According to Aristotle, it must invade all spaces not occupied by one of the other three elements, for the universe was full and intact, with no unoccupied interstices. Nature, therefore, abhorred the concept of the vacuum, for in a balanced universe actual vacuums could not exist. It was seriously disconcerting for the loyal adherents of Aristotle's physics after 1643, therefore, when Galileo's pupil Evangelista Torricelli found an empty space in the sealed glass tube above the mercury in his newly-invented barometer. Philosophers across Europe tried to devise ways of establishing the properties of the 'Torricellian vacuum', and the French mathematician Giles Persone de Roberval had the idea of inserting a deflated and sealed carp's bladder into the barometer tube before it was filled with mercury and inverted. The apparently deflated bladder suddenly inflated in the vacuum, as the residual air inside it expanded to fill the vacant space.[21]

To investigate this new and conspicuously instrument-generated phenomenon of nature, however, it was necessary to make a vacuum that was physically larger and more accessible than that inside a barometer tube. Otto von Guericke in Germany pumped air out of a sealed barrel, invented the famous 'Magdeburg Hemispheres' which needed teams of horses to separate when evacuated, and showed that air pressure bore down intensely from all directions upon an evacuated space.

But it was when Robert Boyle began his own researches into air and vacuum around 1658 that Robert Hooke made his début on the stage of published scientific research, for it was with Boyle that his creative scientific genius first received formal and published acknowledgement. Originally, Boyle had gone to Ralph Greatorex for the construction of an airpump which was to be more versatile than von Guericke's, but it had been a failure. How Hooke, an entirely academically-trained individual, without any formally-imparted craft skills behind him that we know of, had designed and supervised the construction of a successful pump is hard to explain, for Greatorex was the leading pumping engineer in England, and a man who had made a considerable amount of money in draining the Fens.[22] But it is possible that Hooke may have had some instruction from Christopher Brookes, the Manciple, or 'domestic manager', of Wadham, who was also an instrument-maker, whom Wilkins had taken on in 1651 because of his technical skills. We do know, however, that Hooke was sent to London 'to get the Barrel and other parts for the Engine which could not be made at Oxford'[23] (and which were most likely the work of a London gunsmith with experience in making strong, pressure-tested, evenly-bored tubes), so it is more likely that Hooke was the deviser rather than the actual constructor of Boyle's airpump. But while Boyle gave Hooke acknowledgement in his published researches, it is clear that even with Hooke's machine, vacuum seals were difficult to maintain, and pistons, cylinders and valves were treated with 'Sallad Oyl' to make them more airtight. It took about three minutes of hard pumping to get a good vacuum, and some additional action was probably necessary if it was to be kept for several minutes, as the air squeaked and whistled through the imperfect seals. Hooke does not seem to have done the manual work during the experiments, however, for Boyle refers to 'our pumper' as a third party.[24]

Though the original airpump has long since disappeared, several detailed working drawings of it were made and published as engravings, and from these two full-size working replicas were built for the Science Museum, London, and for the Museum of the History of Science, Oxford, in the 1920s. In November 2002, I attempted to test the Science Museum, London, machine with the skilled help of Dr Michael Wright, and though the perished 80-year-old leather seals on the piston did not allow us to get a vacuum, Dr Wright and I quickly mastered the technique whereby, when one of us operated the piston ratchet and the other opened and shut the valves, it would have been feasible to obtain a vacuum relatively quickly.[25]

Apart from any new features that were special to the pumps or valves, Hooke's machine contained three design features that were of the greatest significance. The first of these was a large glass vessel some 15 inches in diameter, called the 'Receiver', which contained the space to be evacuated. Secondly, a brass stopper some four inches in diameter set and cemented into the top of the glass Receiver made it possible to gain easy access to the experimental area, and seal everything up before the pumper set to work. Thirdly, an ingenious secondary brass stopper with conical sides passed through the large stopper, so that when liberally coated with salad oil, it could be turned around without breaking the air seal. This rotating stopper could be used to pull a thread to actuate some experiment *in vacuo*. With Hooke's machine, therefore, the experimenter had easy physical access to a fairly large experimental site that was entirely visible through the thick walls of the glass Receiver. It was to be used to conduct a series of experiments which needed clear vision and the ability to ignite and move things.[26]

Candles, glowing coals and slow-match all went out in the evacuated Receiver, though the coals could be revived to glow again spontaneously if air was re-admitted in time. Boyle and Hooke were especially interested in the behaviour of smoke that rose from the extinguished candle wicks. Did some all-pervading ether that was much more subtle than air still bear the smoke upwards before it touched the inner walls of the Receiver and descended? Was the Aristotelian doctrine true in so far that all fire-products rose upwards—even *in vacuo*? But it was impossible to be certain about the true cause of the smoke's behaviour, for both Boyle and Hooke realized that they did not have perfect vacuums, and a very small quantity of residual air was always likely to be present.[27]

What the airpump experiments did demonstrate beyond all doubt, however, is that the element air was greatly elastic and capable of much rarefaction and compression. Hooke found, for instance, that when a burning candle was placed inside a sealed vessel of a particular internal volume, the flame went out in about three minutes. But when the air was compressed by pumping more of it into the same vessel, the candle burned for 15 minutes, thereby showing that the active principle in air which sustained a flame in any given volume could be altered mechanically. When Boyle repeated Roberval's carp bladder experiment with the much larger bladder of a lamb in the capacious Receiver of the Hooke airpump, he concluded that air could expand by a factor of 152 to fill the vessel.[28]

In many ways, however, Boyle's interests in air differed from those of Hooke, for while Boyle was primarily concerned with the physical properties of aerial elasticity, and how it related to his ideas on atomic structures, his assistant's interests tended to be more chemical. In particular, they related to combustion and what Hooke conceived as a corrosive property in air.

Almost certainly, Hooke's interest in air and its role in combustion stemmed from Boyle's original researches into saltpetre, or nitre, as the seventeenth-century chemists called potassium nitrate. Boyle had conducted these experiments around 1655, several years before the airpump was built, as part of an investigation into the nature of saltpetre's 'inflammable principle' and its relation with air. A crucible of saltpetre, or nitre, had been heated until it melted. Pieces of charcoal were then dropped into the fused mass, and each piece immediately combusted into flames and smoke. After a certain amount of charcoal had been dropped in, however, no more combustion took place, and the washed residue was found to be inert. But when this residue was mixed with nitric acid, Boyle observed that the easily identified crystals of saltpetre began to 'shoot' in it once again, and when prepared and dried, it could be heated so as to consume more charcoal.[29]

It seemed, indeed, that the so-called 'inflammable principle' could be driven off by burning, and then restored by a 'nitrous' chemical substance. Was burning, therefore, not an innate force of nature, but a chemical process where ingredients or properties were exchanged between the air and chemical substances, or else were dissolved or corroded by an aerial solvent? If this was the case, it seriously challenged Aristotle's doctrine of combustion as caused by the element fire, and further implied that Aristotelian air was not a pure element, but possessed a component or aspect that had especially chemically reactive properties. Hooke and Boyle called this component a *menstruum*, or solvent, from a medieval alchemical term used to describe dissolution, corrosion and breaking down into solution, and popularized in the seventeenth century by the German chemist, Van Helmont. *Menstruums* were believed to attack most avidly the 'sulphureous' or combustive parts of bodies: parts believed, indeed, to be associated with sulphur from the characteristic smell that they left behind.

The challenge to Aristotelian ideas was strengthened when Thomas Willis, John Ward, and other members of the Oxford experimental club began to experiment with a relatively new preparation named *aurum fulminans*, or exploding gold (gold fulminate), which could go off without fire. All that one needed to do was to place a little *aurum fulminans* (which, from contemporary accounts of how to prepare it was probably a form of gold tetrachloride) on to a spoon, and cover the spoon with a heavy coin. If one gently tapped the spoon on a table top, the chemical exploded violently and blew the coin up to the ceiling.[30] Willis even developed an *aurum-fulminans*-related model for muscle action, based on the idea that a form of explosion without fire must take place when muscles work and to create body heat.[31]

If *aurum fulminans* could create fire by means of mechanical percussion without a spark, it was also shown during the winter of 1659–60 when Hooke was working with Boyle that gunpowder, which is rich in saltpetre, could sometimes be made to smoulder or burn more slowly in the evacuated Receiver of the airpump without the need for air. During the first experiments, the low altitude, weak winter sunlight and the irregularities in the glass made it difficult to ignite the gunpowder by means of a burning-glass focusing the sun's rays, and the results were inconclusive, but Hooke devised a frame that could be secured inside the Receiver, upon which was fastened a cocked pocket pistol with a pinch of gunpowder in its flashpan. By turning the airpump's secondary brass stopper, with its well-oiled airtight seal, a piece of string could be used to pull the trigger of the pistol. The ensuing sparks and localized ignition of the priming powder *in vacuo* seemed to indicate that atmospheric air was not necessary for the discharge of gunpowder. For while *in vacuo* the gunpowder (when it could be fired at all) only fizzled as opposed to exploding as it did in ambient air, what appeared to be present was some fire-related, or 'nitrous', agent within the powder itself.[32]

By the early 1660s Robert Hooke had come to articulate a theory of combustion in which the elastic medium of air possessed two quite separate properties: an idea probably inspired initially by prior experiments with Boyle, where it had been noticed that when charcoal was heated in a limited volume of air within a tightly-sealed vessel the charcoal was not consumed.[33] Air appeared to contain a part 'that is like, if not the very same, with that which is fixt in Salt-peter', which was capable of reacting with 'sulphureous' substances to produce combustion or explosion, along with a non-reactive or inert part. He outlined the theory and the experiments by which he tried to substantiate it in 'Observation XVI; Of Charcoal' in *Micrographia* (1665).[34] Hooke therefore envisaged air as containing a powerful dissolving agent, or a *menstruum*. For whenever 'sulphureous', or potentially inflammable, substances like wood were heated to a particular point, their atoms were furiously descended upon by the *menstruum* of 'nitrous air' or 'aerial nitre'. The potentially volatile 'sulphurs' within the substance were thereby consumed, and an inert ash remained. But commonly combustive materials like wood and candle wicks could not burn *in vacuo* because even when locally heated by a burning-glass no aerial nitre was present to dissolve them. Gunpowder seemed to ignite *in vacuo*, however, because it contained its own inbuilt supply of 'fixed' nitre, in its saltpetre, or what modern chemists call potassium nitrate (KNO_3). More than a decade and a half later, early in 1679, Hooke was still experimenting with these concepts of combustion and aerial *menstruums*, when he informed the Royal Society, which had built upon his own and Boyle's earlier work, that if a piece of charcoal was placed 'in an Iron Case with a Screw-stopper' (and hence in a limited air volume) it could not be made to burn even if heated to red heat. Indeed, when a piece of charcoal which weighed 128 grains, apothecaries' weight, was heated for two hours in an air-sealed vessel, it was subsequently found to have lost only 1½ grains.[35] It would seem, therefore, that what caused things to burn was not some elemental fire principle so much as a 'sulphureous' body's exposure to the corrosive *menstruum* in air, for when the air was excluded, no amount of external heat applied to the piece of wood or charcoal would make it burn.

In his Cutlerian Lecture *Lampas*, delivered to the Royal Society in 1677, Hooke developed his ideas on combustion when he analysed the parts of a lamp or candle flame, though these experiments on flames went back at least to February and March 1672, when he described their examination by means of magnifying mirrors and lenses. Hooke noticed that the actual point of combustion appeared to be at the bottom part of the conical flame, where the oil rising up the wick became excited by the heat above it. At a critical point, it was devoured by the aerial nitre, and produced the tulip-shaped inner flame, where the rising 'sulphureous' particles or atoms made contact with the aerial nitre to produce a glowing combustive envelope, or interface. He also realized that the interior part of the flame did not emit light, but only the tulip-shaped combustive envelope around it. The interior of this envelope consisted of heated but non-luminous sooty particles that had failed to go off, as it were, and simply rose as greasy smoke. It was within this dark, sooty interior that the non-light-emitting part of the wick lay, and Hooke noticed that when this spent wick fell over, and broke through the combustive envelope, it glowed red, as it entered the aerial nitre that surrounded the flame.

As he later recorded in *Lampas*, Hooke obtained this information by inserting thin plates of glass and 'Muscovy glass' (mica) into the flame, both from above and bisecting the flame sideways, to reveal the light and dark zones of the interior. He also used powerful sunlight to project an image of a candle flame on to a whitewashed wall, whereby he could discern the dark interior and heat zones revealed in the convection currents within the resulting shadow.[36]

Hooke's researches into combustion and the airpump naturally led him to the physiology of respiration. In the wake of William Harvey's discovery of blood circulation in 1628, physiologists were investigating the relationship between blood circulation and respiration, and it came to be seen that the death of birds and small animals that were placed in the receiver of the airpump demonstrated the importance of 'vital' air to life in a way that we take for granted today. The circulatory physiology of William Harvey recognized that the lighter colour of arterial, as opposed to venous, blood could have something to do with its having had contact with air in the lungs, although no one could be sure what occurred in precise chemical terms, considering the evidence available in the 1660s. Did the air cool, nourish, cleanse or invigorate the blood, and was the colour change between dark venous and lighter-coloured arterial blood due to nutrition or ablution? Hooke's anatomical experiments on the hearts, lungs and respiratory tracts of living dogs had convinced him by 1667 that fresh air was fundamental to synchronized heart–lung action and to life itself (as will be further discussed in chapter 6), although he 'could not perceive any thing distinctly [during these experiments], whether the Air did unite and mix with the Blood' in the lungs.[37] Even so, Hooke was clearly pondering the possibility of invigorating atmospheric air combining with the blood during the cardio-respiratory cycle, and thereby giving arterial blood its florid colour, though he had no firm experimental data to indicate which parts of the air were involved.

Then, using his second airpump of *c.* 1666 (to which the 'ingenious Mr Hooke' had suggested new design features),[38] Robert Boyle placed a small pot of warm lamb's blood, fresh from the slaughter house, into the Receiver of the airpump. On reducing air pressure, however, '...the Blood was so Volatile, and the expansion so vehement, that it boyled over the containing glass'.[39] Whether the experiment was performed on fresh arterial or venous blood is not specified, but it was immediately obvious that some airy substance was present in the blood, and that in reduced atmospheric pressure the bond was broken and the blood and the air separated.

But what is clear is that the airpump which Hooke had devised for Boyle around 1658, and the improved one after 1666, opened up a new world of experimental possibilities to scientists who were investigating the nature of breathing, burning, blood-colour changes interpreted in the light of Harvey's circulatory physiology, and the connections between them. During the 1660s and 1670s in particular, the airpump became a potent symbol of the new instrument-based mode of scientific research in the life and chemical sciences, in very much the same way as Galileo's telescope had done for astronomy half a century before.[40] The asphyxiating properties of the newly-discovered vacuum were so impressive, moreover, that when a visiting dignitary from Denmark was entertained by the Royal Society around 1663, an anonymous wit commemorated the event in the following verse:

'To the Danish Agent late was showne,
That where noe Ayre is, theres noe breathe;
A glass this secret did make knowne
Where[in] a Catt was put to death.

Out of the glass the Ayre being screwed,
Pusse died, and ne're so much as mewed.'[41]

To study the effects of low atmospheric pressure upon a human being, Hooke suggested the building of a 'rarefying engine', or airpump which, instead of having the usual glass 'Receiver', had a tightly-corked wooden chamber within which a man could be accommodated. On 27 June 1667, Hooke costed the construction of such a machine at £5, and while it was speedily built, the only immediate reports made about it to the Royal Society concerned the difficulty in obtaining a tight air seal.[42] It was not until March 1671 that any serious results were announced, when Hooke described to the Society how the pumper had reduced the air pressure inside the 'engine' to one-quarter of normal atmospheric pressure, as 'estimated by a gage', and he, inside the apparatus, began to experience deafness and pain in his ears. Fascinated as he was with the relationship between breathing and burning, Hooke noted that, during the experiment, he shared the inside of the engine's sealed chamber with a lighted candle which 'went out long before any of that inconvenience in his ears'.[43]

Within a decade, the Aristotelian explanations for burning and breathing and the scientific plausibility of two of the four classical elements had been fundamentally challenged. Then in the early 1670s, John Mayow, a medical doctor of Wadham College, began to take Boyle's and Hooke's researches into breathing and burning further, publishing the results in his *Tractatus Quinque* of 1674. Mayow's ingeniously-devised and meticulously-executed experiments on mice and candles enclosed in glass vessels over water seals led him to realize that any given volume of air seemed to consist of two separately reacting parts: one part sustained both life and combustion, while the other, greater in volume, supported neither. Mayow had also realized that this 'nitrous' or nitre-related combustive life-sustaining part of the air further caused the bright redness of arterial blood, which seemed to give substantiation to Hooke's 1667 surmise that arterial blood took on its floridness in the lungs. But it would be historically incorrect to see Hooke in the wake of Boyle with his aerial nitre as a proto-discoverer of oxygen, for seventeenth-century researchers had no real concept of chemically specific gases, and still couched their ideas in terms of 'vital principles' such as inflammability or the corrosiveness of *menstruums*, and what we might call allotropic or variant states of air. Yet the very fact that air might have allotropic states, that it was vastly elastic, and that it might somehow be able to 'fix' its vital parts such as its corrosiveness in stable chemical substances such as saltpetre, signalled a fundamental shift in ideas about the natural world. All these new ideas had been gained not by speculation, but by Baconian acts of 'putting nature to the torture' in carefully-planned inquiries that hinged upon a newly-invented piece of apparatus—experiments, and experimental conclusions, indeed, that would have been unimaginable in 1640!

Undoubtedly, it had been the invention of the airpump which made these new researches and discoveries in combustion and respiration possible, both by creating new experimental environments and by encouraging Boyle, Hooke, Mayow and their friends to look at breathing and burning from a new perspective. Even so, by the late 1670s, they seem, after numerous variations and permutations, to have taken the experimental capabilities of airpump research about a far as the contemporary apparatus would allow, and in that series of demonstrations and discussions about combustion, explosion and the nature of saltpetre placed on record by the Royal Society between January and March 1679 one encounters little that is new, for Hooke re-asserts that 'air is a menstruum, that dissolves all sulphureous bodies by burning',[44] that saltpetre has a fixed 'alcalizate' and a volatile, aerial part,[45] and he repeats familiar experiments on charcoal,[46] while the explosive properties of gunpowder and *aurum fulminans*[47] are once again tested. Indeed, much of this had already been published in *Micrographia* or in Boyle's researches, and while Hooke was undoubtedly influenced by the 'nitrous' airs and experiments in Mayow's *Tractatus Quinque*, he would not fundamentally change his mind about the chemical character of breathing and burning for the rest of his life.

By the time that he had reached the age of 25 in 1660, however, it was quite evident that Robert Hooke was a man of remarkable versatility and promise: as scholar, artist, experimentalist and one who could design and perhaps personally construct complicated pieces of scientific apparatus. Moreover, Hooke's skills were now likely to be wasted if he continued in Oxford, especially as most of the scientists began to meet in London again, following the restoration of King Charles II in 1660.

Though Robert Hooke left no direct written responses to the events of May 1660, beyond a remark made in 1676, in which he tried to date an event by saying that it took place 'presently after His Majesty's happy Restauration',[48] the restoring of the monarchy and the crowning of King Charles II on 29 May 1660 was one of the pivotal events of British history. Following Oliver Cromwell's death in September 1658, the Parliamentary régime had lacked a strong leader, and while Richard Cromwell succeeded his father to the Lord Protectorship, his lack of decisiveness meant that some of his father's old generals and other influential figures of the Protectorate now seemed poised to plunge the country into another cycle of civil wars as they contended for mastery.

But this potential débâcle was averted when Edward Hyde, Edward Montague (soon to be ennobled as Lords Clarendon and Sandwich respectively), and a group of moderate Parliamentarians made the bold step of inviting the 32-year-old Charles Stuart, the eldest son of the executed King Charles I, who was then living in exile in Holland, to return to England and restore the monarchy. It was this same Edward Hyde who, as Chancellor in 1663, would confer Robert Hooke's Oxford MA upon him; while Samuel Pepys the diarist, the astute right-hand man of Edward Montague and a future friend of Hooke, was one of the English delegation which sailed out to Holland to accompany the King home. This Restoration of the Monarchy, far from meeting any resistance, had the effect of galvanizing most of the nation into an enthusiastic support for the young King. The once-reforming Parliament had failed to give stability and justice to the people, and had become deeply corrupt, faction-ridden and detested, while the end of Puritan religious influence was met with general rejoicing.

The Restoration of the English monarchy in the spring of 1660 was also to become one of the turning points of Robert Hooke's life, as it was to be for so many of his friends and patrons in the Oxford Philosophical Club. The King's re-establishment of the Church of England, moreover, meant that Hooke's astronomical encourager, Seth Ward, left Oxford for Exeter to take up the first of his two bishoprics. A few years later, in spite of his marriage connection with the Cromwell family, John Wilkins would follow to the see of Chester. Robert Boyle who, in addition to being a scientist of international reputation, was so personally devout and theologically learned that he was referred to as a 'lay bishop' graciously turned down offers of preferment, while Christopher Wren — the son of a former Royal Chaplain and the nephew of the Bishop of Ely — was soon to become Surveyor of the King's Buildings. Thomas Willis was restored to his dignities in Christ Church before proceeding on the invitation of Gilbert Sheldon, Archbishop of Canterbury, to a distinguished London medical practice in 1666,[49] and Willis's brother-in-law Dr John Fell became Dean of Christ Church and later Bishop of Oxford.

With friends like these around him, one can perhaps come to understand how the curate's son received his MA degree without examination, and entered upon a London-based scientific career that would make his name familiar to the *savants* of France, Holland and Italy by the time that he was 30.

It was, however, thanks to his initial reputation as an intelligent and experienced 'operator' to Willis and Boyle, and no doubt to the assistance of Wilkins, not to mention the fact that he was a man of acknowledged talent as a practical exponent of the 'new philosophy', that Robert Hooke became the obvious choice for Sir Robert Moray's proposed office of 'Curator of Experiments' to the newly-chartered Royal Society of London in 1662. Yet it was with some 'Reluctancy' and modesty, so Hooke tells us, that he embarked upon such a high-profile scientific career, not because of other professional plans, but 'because I was to follow in the footsteps of so eminent a Person as Dr Wren', who was Wilkins's first scientific protégé, and the genius of whose 'Mechanical Hand' and 'Philosophical Mind' made him our English Archimedes, not to mention the connections with Royalty that seemed to be involved in natural philosophy which also, as Hooke said, 'did affright me'.[50] While it is easy to be cynical regarding the hyperbole and high-flown language with which Hooke described his friends and encouragers, one must remember not only the immense sense of debt and gratitude which he felt towards them, but also his very great and genuine respect, for 'Dr Wilkins...is indeed a man born for the good of mankind, and for the *honour* of his Country.... So may I thank God, that Dr. Wilkins was born an Englishman, for wherever he had lived, there had been the chief Seat of *generous Knowledge* and *True Philosophy*'.[51] Likewise, Robert Boyle deserved 'all honour, not only as my particular Patron, but as the *Patron* of *Philosophy* itself; which he every day *increases* by his *Labours*, and *adorns* by his *Example*'.[52] It is all the more surprising in the midst of these 1665 encomiums, therefore, that Hooke says nothing about Thomas Willis. Willis on his part gives no acknowledgement to Hooke—as he does to Richard Lower, Thomas Millington, Edmund King and Christopher Wren—for assistance in those anatomical researches which culminated in his great neurological masterpieces, *Cerebri Anatome* (1664) and *De Anima Brutorum* (1672), though one might argue that this omission stemmed from Hooke's not being known as an anatomist at that date. On the other hand, Willis's chemical study, *De Fermentatione* (1659), with which it is almost certain that Hooke assisted him in Oxford, carries no acknowledgement either. One is left wondering whether Hooke and Willis somehow failed to 'hit it off' and become close friends in the way that Hooke formed firm lifelong friendships with Wilkins, Boyle and Wren.

On the other hand, we must not forget that at the start of his London career around 1662, Hooke did not sit with the Fellows—Boyle, Wren, Wilkins and others—as an equal, but as an employee (or 'servant' in seventeenth-century language) with a salary yet to be paid. Unlike them at this early stage in his career, he was not a '*Gentleman*, free, and unconfin'd'[53] possessing independent means, as the original terms of his employment required him to produce demonstrations at Royal Society meetings, as well as receive 'orders' for the undertaking of particular research investigations.

On the other hand, to assume, as some writers have, that Hooke was regarded by the Fellows as some sort of menial or hired labourer is entirely wrong, and derives more from the deliberate down-playing of his significance in the post-Newtonian era than it does from the social realities of the mid-seventeenth century. While he was not born to independent wealth, Robert Hooke, we must never forget, was, as the son of a clergyman and a youth of good family on both sides, undoubtedly born a gentleman, albeit of modest means and 'poor', relatively speaking. Indeed, considering the more conventional career prospects that an astute ex-Westminster boy and Christ Church MA might reasonably have had before him by the early 1660s, one suspects that the founding Fellows of the Royal Society must have realized what a treasure they had in the young Hooke and, in spite of the apparent burdens they placed upon him, were keen to see him established so as to guarantee his continuing services.

This recognition of Hooke's usefulness and promise, moreover, was not confined solely to the Virtuosi of the Royal Society, but must have extended to the wider maritime world as well, for as he tells us in *Helio-scopes* (1676), in a narrative essay that was published posthumously in 1726, and elsewhere, Hooke had in 1662 become closely involved in the land, and perhaps the sea, trials of Huygens-type pendulum clocks, in an attempt to find the longitude at sea. He also worked closely with Lord Kincardine, who was promoting the marine chronometer in which the isochronous actions of a pair of short pendulum clocks, gimbal-mounted so as to counter the ship's motion. Quite simply, by 1662 Robert Hooke was becoming a national asset, and a man to be encouraged.[54]

Then soon after the publication of *Micrographia*, in 1665, Hooke's academic standing became completely regularized when he was appointed Professor of Geometry at Gresham College, a post which inevitably brought him into the social orbit of the City Livery Companies, and which he held in conjunction with his Curatorship, with rooms and accommodation in the same building in which the Royal Society held its meetings. By the time of his Gresham professorial appointment, as he plainly states on the title page of *Micrographia*, he was already a full Fellow of the Royal Society anyway, and proudly writing 'FRS' after his name. By 1665, therefore, Hooke had acquired all of those physical circumstances and early badges of prestige that would see him through the remaining 38 years of his life; and while, within a decade, his architectural and other work would begin to earn him the 'great Estate' to truly become a '*Gentleman*, free and unconfin'd' if he so wished, his continuing office of Curator of Experiments also made him the first salaried research scientist in Britain.

CHAPTER 3

THE CURATOR OF EXPERIMENTS

The Royal Society, to which Hooke was unanimously elected Curator of Experiments upon the original proposal of Sir Robert Moray on 12 November 1662,[1] was not only to become one of the truly enduring ornaments of King Charles II's reign, but was also to embody and perpetuate a new and radical approach in understanding the natural world. What is more, King Charles II took a genuine if sometimes bemused interest in the Society's aims and researches, for he was one of the most intellectually gifted people to sit on the British throne. It is true that Charles tended to be lazy, sometimes duplicitous, and given to womanizing, but he possessed a first-class critical intellect, a genuine and informed interest in the arts and the sciences, and greatly enjoyed the company and conversation of clever men, be they wits, bishops or scientists. Dr William Harvey, the discoverer of the circulation of the blood, for instance, had once acted as his informal tutor during the Civil Wars, and the King so enjoyed the company of the outrageous philosopher Thomas Hobbes that he granted him open access to the Palace of Whitehall.[2]

This new Society, which met in the King's name, drew in some of the most gifted men of the age, its avowed objective being to break free from the limitations of traditional learning, and discover the world afresh through the 'Experimental Philosophy'. Inspired as it was by the writings of Francis Bacon, the Society aimed to explode superstitions, discredit magic, and cast light on all that was obscure. God, they argued, had created an ordered, logical world, and it was the duty of the 'Virtuosi' or scientists both to demonstrate the beauty and usefulness of God's creation, and to teach it to all men, for it must never be forgotten that their motivation owed as much to religious inspiration as it did to secular curiosity. Science had a twofold purpose, as Bacon had argued in his *Novum Organon*: to produce 'Experimenta lucifera', or 'illuminatory experiments', which showed the wonderful way in which God had made the world; and 'Experimenta fructifera', or 'fruit-bearing experiments'.[3] One might think of them as *pure* and *applied* science which, when carefully cultivated and employed, could improve the human condition. This objective was to be achieved, Hooke stressed in 1665, in the wake of Bacon's own precepts, by 'avoiding *Dogmatizing* and the espousal of any Hypothesis not sufficiently grounded and confirm'd by Experiments'.[4]

Intellectual freedom was also central to this ethos, and the ability of the researcher to follow the investigation or line or argument wheresoever it led, without any prior ideological commitment. Indeed, this principle of independence was encapsulated in the Royal Society's very motto, *Nullius in Verba*, which was a contraction of the line *nullius addictus iurare in verba magistri* from the *Epistles* of Horace — a Roman author whose works would have had a schooldays familiarity to all educated men in the seventeenth century — and which means 'not bound to swear allegiance to any master' or to any school of philosophy.[5]

Amongst the early Fellows of the Royal Society one finds clergymen, physicians, academics, men of affairs, private gentlemen, noblemen (and even close working relationships with certain craftsmen such as Thomas Tompion), and the 1671 Fellows List includes well over 200 names.[6] The common bond between these men was a passion for 'Experiment', and a desire to break the traditional bounds of knowledge and, in particular, to liberate learning from the 'wordy' syllogisms of Aristotle. Many of the scientific men who had been in the universities, and especially in Oxford during the 1650s, came together in London to form the Society following the Restoration of the monarchy: Dr John Wilkins, Sir William Petty, Sir Christopher Wren, Robert Boyle, Thomas Willis, followed by Dr Richard Lower, John Aubrey and many others. These men, moreover, joined ranks with their London-based experimental friends who had continued to meet in Gresham College following the main migration to Oxford after 1648 for, as they were aware, the very first English experimental gatherings had taken place at Gresham College after 1644 or 1645. But from its very inception, on 28 November 1660, the new Royal Society attracted a large and vibrant membership, which now met weekly in Gresham College, Bishopsgate Street, in the heart of the City of London. Quite often, indeed, the Society was simply referred to in everyday London parlance as 'Gresham College', and its Fellows as 'Gresham Witts', not from any tendency to comedy within the Fellowship, but in keeping with the older usage of the Anglo-Saxon word *wit*, to mean one who possessed acute senses. Its deliberations soon became one of the talking points of City life, as the early Royal Society enjoyed the patronage of King Charles himself, and a roll-call of already famous men.[7]

All that the new Society lacked was money, for while the hard-up King was able to bestow honour, three Royal Charters in 1662 and 1663 that confirmed its high status, and a few ceremonial artefacts to the Society, he could give it no funds. Indeed, so pinched were its finances that the Fellows made a weekly subscription to cover necessities such as candles and other incidental costs, though as Hooke recorded, it was the personal munificence of several noblemen and 'merchants' within the early Fellowship that paid for several original experimental investigations.[8] Yet this relative poverty turned out to be a blessing in disguise, for while the Society enjoyed a position of the greatest social eminence, its lack of Royal or Government funding gave it an intellectual independence and freedom from official control which contrasted sharply with bodies such as the Florentine Accademia del Cimento and, later in the 1660s, Louis XIV's French Académie des Sciences. No rich patron could close the Society down by suddenly withdrawing support, as occurred in Florence in 1667; nor, as happened in Paris, could a powerful minister of state lean on the well-paid Académiciens and encourage them to undertake lines of research in which the government had a vested interest. While King Charles, as we have seen, had a genuine and active interest in science, and enjoyed the company of learned men, the self-financing Royal Society was free to pursue whatever researches its Fellows chose. Nor should one forget that as a *Society*, with a body of equal, mutually-electing *Fellows*, its constitutional basis was closer to the self-regulating model of an Oxford or Cambridge College than to that of a centralized, hierarchically-organized continental academy. This Fellowship mode of organization proved unrivalled not only in establishing the intellectual freedom of the Royal Society itself, but also in creating a precedent for subsequent learned societies in Britain, even when the Sovereign was the formal patron and allowed the prefix 'Royal' to be placed before its name.

So when Hooke took up his duties, they seem to have been something of a labour of love, for though he was ceremoniously inducted into the Society on 12 November 1662, there was no immediate proposal of a salary, and he still received his income, and probably his accommodation, from Robert Boyle. In July 1663, for instance, Hooke was addressing his letters to Boyle, when Boyle was in Oxford, the country, or otherwise out of London, from 'Pall Mall', which would have been the residence of Lady Katherine Ranalaugh, Boyle's sister, at whose house the great chemist and his assistant resided.[9] The Society, however, 'ordered that Mr Hooke should come and sit among them, and both bring in every day three or four of his own experiments, and take care of such others as should be recommended to him by the Society', but stipulated that he should be 'expecting no recompense till the Society get a stock enabling them to give it'.[10]

Undaunted by the parsimony of the Society, Hooke threw himself into a programme of research which would run at white heat for nearly 20 years, and was to continue somewhat abated until his death in 1703. It was not until June 1664, however, that there was any proper discussion of money for his Curatorship, and then it came from one of the Fellows in his private capacity, when Sir John Cutler decided to endow an annual

'Mechanick Lecture', to be researched and given by Hooke with a £50 stipend.[11] The sum proposed by Cutler was the same as that paid to the Gresham College Professors, and it seems that Cutler wished to establish a lectureship that was similar in character to the seven Gresham Professorships: a lecture, moreover, which would demonstrate the significance and usefulness of mechanics in very much the same way as the existing Gresham lectures demonstrated the importance of astronomy, geometry, grammar and the other classical sciences. Then in November 1664, the Council of the Society got around to discussing Hooke's situation, and in January 1665 seemed to have 'got stock' enough for it to confirm his position as Curator at the salary of £30 per annum.[12] The salary offered by the Society was originally intended to be larger, but the Council decided to take advantage of Sir John Cutler's unexpected generosity, and gave Hooke a smaller amount, although the Curatorship and Cutlerian Lectureship would mean twice as much work.

Even this parsimony would have been bearable, had not both the Royal Society and Sir John Cutler been better at talking about money than at actually parting with it. For it soon became apparent that Cutler and the Royal Society were talking at cross purposes when it came to what they expected Hooke to do for his respective payments, as a result of which the Cutlerian stipend was not properly paid, in spite of the fact that Hooke had been researching and delivering his Cutlerian Lectures since the autumn of 1664.[13] By November 1666, moreover, the Royal Society had still only paid him £50 for four years' work as Curator. But at least one presumes that monies amounting to around £925 had been paid by Cutler over the years, for when Hooke won his Chancery lawsuit against Cutler's executors in 1696, the sum which he was claiming was the outstanding £550.[14] In consequence, the Cutlerian Lectures must have brought a sum in the region of £1500 to Robert Hooke over a 32-year period, though as he admitted, the anxiety concerning the lawsuit had done much to wear him down.[15]

In the light of this treatment, one might sympathize with Hooke's sometimes sharp tongue, choleric temper and reluctance to squander money once he began to make it. While it is true that Hooke's background had been modest rather than poor, the Freshwater Hookes had not lived in the same style as did those men who became the original Fellows of the Royal Society, and one can fully understand Robert's feeling financially disadvantaged and watchful of his money when in their company, because not until 1665, following his appointment to the Gresham College Geometry Professorship, did he start to receive a regular salary at all, beyond what Boyle paid him. Being controlled by different trustees, comprising the Mercers' Company and the 'City Side' trustees nominated by the City of London, with an established capital endowment behind it, the College professorship gave him a modest but reliable income, a rather grand set of rooms (more correctly, perhaps, a house, with several spacious rooms, double cellars, and an attic), and the basic requisites for Hooke's donnish lifestyle. He does not seem to have been the Electors' first choice, for the original decision seems to have been in favour of Dr Arthur Dacres; but it is possible that Sir John Cutler and the 'City Side' (Cutler being a member of the Grocers' and not the Mercers' Company) were influential in securing it for Hooke.[16]

Yet even prior to his formal *ex officio* residence as a Professor, it is clear that Robert Hooke had been living at Gresham College since the beginning of September 1664, for as he told Boyle in a letter of 6 October 1664, 'I have also, since my settling at Gresham College, which has been now full five weeks, constantly observed with the baroscope [barometer]'.[17] Though we do not know for certain, it is likely that he had moved into the temporarily vacant rooms of Dr Walter Pope, Gresham Professor of Astronomy and half-brother of Dr Wilkins, who had been granted permission to travel abroad to visit Italy for two years from the late spring of 1663. First Isaac Barrow and then Robert Hooke read Pope's astronomy lectures during his absence.[18]

After his professorial appointment, however, Hooke officially moved into the Geometry Professor's rooms, and spent the rest of his days in Gresham College, with the possible exception of the period immediately following the Great Fire of September 1666, when the Lord Mayor and part of the London City government moved into temporary accommodation in the College, during which time the Professors lodged elsewhere.[19] There, during the last decade of his life, 'being us'd to a collegiate or rather monastic life, which might be some reason for his continuing to live so like a hermit or cynic',[20] Hooke became increasingly ill and withdrawn, before drawing his last breath in his College rooms on 3 March 1703. In 1665 his professorial duties in themselves seem to have been relatively light, and since they were closely related to his much more onerous Royal Society work, he seems to have had no problem in continuing as Curator of Experiments.

Financial wranglings apart, if one really wishes to get a sense of the genuine esteem in which Hooke was held by the time that he was 30, in August 1665, one has to look not only at the company he kept, but also at the company within which he clearly moved on equal terms. For in the summer of 1665, as the Great Plague raged in London, Hooke had left the city, like most Fellows of the Royal Society, and happened to be staying at Durdans, the county seat of Lord Berkeley, near Epsom, Surrey, where he had joined Dr Wilkins's party. Here, on 7 August 1665, John Evelyn (whom Hooke acknowledged as his 'friend' in *Micrographia*) joined a company of plague-exiled Virtuosi, where, he recorded, 'I call'd at Durdans, where I found Dr Wilkins, Sir Wm. Petty, and Mr Hooke, contriving chariots, new rigging for ships, a wheele for one to run races in and other mechanical inventions; perhaps three such persons together were not to be found anywhere in Europe, for parts and ingenuity'.[21] Here is Professor Hooke being put on a par with two of the most illustrious scientists of the older generation by a gentleman who himself was a model of both intellectual and social discernment. Indeed, this Hooke is a far cry from the mere clever mechanic of post-Newtonian mythology. Durdans, the mansion, seems to have been a general haunt of delight, especially for intellectual and artistic people, and on 26 July 1663 Samuel Pepys had recalled it as a place 'where I have seen so much mirth in my time', the first of which occasions dated back to his childhood.[22]

Before examining Hooke's scientific ideas and researches conducted at Gresham College, however, it is important to look at their context: for what had gone wrong with the immemorial explanations for natural phenomena as taught in the universities of Europe, and why did Baconian experimentation and Galileo's and Harvey's discoveries take on such an explanatory cogency as to bring about the creation of the Royal Society? In this respect it is important to remember that Hooke's contributions to physical science came at the end of a period of a century and a half during which the once-coherent structures of classical science had received one blow after another. What is more, the explanations for the natural world which medieval European scholars had inherited from classical Greece had been essentially static in character, and based upon a series of apparently self-evident principles. All changes of earthly matter, for instance, could be explained by the classical writers, and by Aristotle in particular, by the interaction of the four elements, Earth, Water, Air and Fire, as the principles of solidity, wetness, volatility and heat endlessly mixed and separated, for these four elements also lay at the foundation of all living things. The hearts of all living creatures, for example, in this pre-Harveian circulatory physiology, were thought to generate a spontaneous, or innate, heat, that was radiated throughout the body by the blood; while the life-principle of air, or *pneuma*, was believed both to intermingle with the blood in respiration, and also to help cool down the heart. For in this system heat rose, cold congealed, 'grass became flesh', in accordance with the Biblical adage, and flesh decayed once its life principle had departed and went on to replenish the earth.[23]

While this flux of elements prevailed upon the earth, the heavens by contrast were made from a perfect, stable fifth element, or 'quintessence'. Because they were made of one single and changeless substance, the stars and planets moved with a geometrical precision that nothing on earth could ever emulate, thus exemplifying that deep dichotomy between terrestrial and celestial substance that lay at the heart of all classical science.

The ancient Greeks, and most significantly Aristotle, had devised a complete taxonomy of nature based upon these principles by 330 BC, and for the next 1900 years it proved capable of answering most of the questions that could be addressed to it. Aristotle's science was a magnificent intellectual achievement, though it embodied a conservative approach to knowledge, and like librarianship or museum curation saw its first duty as absorbing, classifying and preserving the known rather than exploring pastures new.

After 1492, the assaults upon its all-encompassing explanatory credibility began to increase. The discovery of America fundamentally discredited ancient geography, by physically demonstrating the existence of a continent of which the ancients had been ignorant; and one can understand why Bacon and Wilkins in particular so often invoked the names of Columbus, Magellan, Drake and other navigators as leading exponents of fact-driven physical inquiry, for what could be more physical or experimental than sailing a ship to discover hitherto undreamed-of continents? Tycho Brahe's supernova of 1572 and Galileo's telescopic discoveries after 1609 similarly shook classical astronomy by showing that the composition and structure of the heavens were *not* what the ancients said they should be. Rapid developments in optics and mechanics, moreover, such as Galileo's ground-breaking research into falling and oscillating terrestrial bodies, seemed to indicate that whole classes of phenomena could be studied amidst the four chaotic elements of the earth that were just as mathematically or mechanically exact as those observed in the heavens, while in 1628 William Harvey discovered that the blood circulated around the bodies of living creatures under the action of the heart, in contradiction to classical physiological teachings. All of these discoveries flew in the face of the classical writings, and showed that the 'moderns' might well know more than the 'ancients'; and very importantly, none of these discoveries were the fruits of speculative philosophies, but rather of meticulously-conducted physical investigations. Passive observation could classify, but experiment could break into realms of new knowledge. In the words of Sir Francis Bacon, who more than anyone else championed the cause of experiment, and whose writings directly inspired Robert Hooke and the early Fellows of the Royal Society, argued that nature must be 'put to the torture', and made to yield its secrets to the astute investigator. It was not for nothing, moreover, that Bacon's distinguished legal career took place during one of the most sanguinary periods of English constitutional history. Just as the cross-examining lawyer in a Jacobean treason trial, with the assistance of torture, could force access to realms of conspiratorial knowledge otherwise closed to him, so the divulged names of fellow-conspirators could set further chains of inquiry in motion until this physical forensic process finally exposed the full 'truth' of the case.

As the judicial inquisitor needed his special tools of assault and persuasion to make his victim speak, so the scientific experimentalist needed his, for the laboratory — which included the newly-invented telescope, microscope, airpump, thermometer and many other instruments which became Hooke's 'artificial organs'[24] that refined the perceptions — was the torture chamber wherein long-secretive nature would be cross-examined, revealing one truth after another until the whole 'truth' of nature was brought into the light.

As indicated above, the radical reappraisal of how nature worked that was taking place in the early seventeenth century was also rich in perceived religious implications. Far from being persecuted by the Church, indeed, the Scientific Revolution was in fact seen as fulfilling Old Testament prophecies. Hooke, for example, saw the newly-invented instruments as re-strengthening those human organs of perception that had been blunted by ignorance, folly and improper usage, and expressed the prophetic character of the New Science very succinctly in the Preface to *Micrographia* in 1665, when he wrote:

> 'And as at first, mankind fell by tasting of the forbidden Tree of Knowledge, so we, their Posterity, may be in part restor'd by the same way, not only by beholding and contemplating, but by tasting too those fruits of Natural Knowledge, that were never yet forbidden.'[25]

In the spirit of Bacon's and the Royal Society's motto, *Nullius in Verba*, mankind must be allowed to use its own faculty of honest inquiry to reach a more profound understanding of the Divine Creation, and even to use its free will and ability to experiment. While this self-help route out of Original Sin might have been dubious theologically, it nonetheless gave science a powerful social agenda; and when Thomas Sprat FRS, a future Bishop of Rochester, was to write his *History of the Royal Society* (1667), he pointed out that the Bible itself contained several prefigurements for the New Science, and even implied the future use of scientific instruments: for had not God taught the Sons of Adam 'to build Cities, to play on the Harp and Organs, and to work in Brass and Iron'? [Sprat, *Epistle Dedicatory*], and was not Tubal-cain in the Book of *Genesis* 'an instructor of every artifice in brass and iron'?[26]

Yet this new mastery of nature to which the age was laying claim had a darker (or more prophetic or revelatory) dimension, depending upon one's perspective. After the Fall of Mankind which came about as a result of excessive curiosity, when Adam and Eve ate the fruit from the prohibited Tree of Knowledge, as recounted in Genesis, humanity had been bounded within a fixed scheme of knowledge, though ancient prophecies had indicated that new enlightenment would come to man shortly before the end of the world. Many seventeenth-century scholars believed that they had computed from the prophetic books of the Bible that Armageddon was now at hand, and no prophecy fitted the age better than that from the Book of Daniel, XII.4, translated in the Authorized Version of 1611 as

'Many shall run to and fro, and Knowledge shall be increased.'

Francis Bacon had also had the same passage, in its earlier Latin form 'multi pertransibunt & augebitur scientia', engraved on the frontispiece of his *Novum Organum* (1620), where the text was accompanied by a picture of a modern galleon—the type of versatile three-masted sailing ship which had been used to make the post-Columban voyages of discovery—sailing out beyond those Pillars of Hercules which had marked the traditional western boundary of classical geographical knowledge.

The geographical discoveries, the religious wars of the Reformation, numerous new inventions, supernovae, Jupiter's moons, the execution of King Charles I, the discovery of the microscopic realm and of the vacuum, and the refutation of the truths of Aristotle's science: all of these could be construed as fulfilments of Daniel's and many similar prophecies. The search for religious meaning lay at the heart of seventeenth-century intellectual culture, and to dismiss it from our understanding of their science produces a picture as lopsided as that which would result if a historian in 350 years' time wrote about the twentieth-first century in a way that ignored the significance of economics.

Yet one of the truly confusing baffling issues facing scientific researchers at this time lay in deciding which *modern* method of investigating nature was the correct one, for in addition to Baconianism, mid-seventeenth-century Europe was a veritable market-place of competing philosophies in the wake of the confusion that followed the eclipse of Aristotelianism. Though historians of science generally speak of the rise of the 'mechanical philosophy' at this period, one should remember that this is a portmanteau designation for several quite distinct 'systems' that shared the speculative premise that the thing which we now define as 'energy' was transmitted by particulate collision. One of the most uncompromising of the mechanists was Thomas Hobbes (better known today as a political philosopher), who argued that matter and the laws of motion could be made to explain everything, from celestial mechanics to the appearance of ghosts. The Frenchman René Descartes saw all physical, but not spiritual, phenomena as occasioned by an endlessly agitated ether, the vortices and swirls of which carried along the particles that produced physical motion. Pierre Gassendi, the French disciple of Galileo, revived the so-called Godless doctrine of atomism, and conceived of matter in terms of the geometrical arrangement of fundamental particles guided by the hand of God. Especially popular in England were the ideas of Francis Bacon, which were concerned less with the innermost structures of matter in motion and more with developing the correct experimental method and arranging the results into taxonomic schemes. Not only was Robert Hooke familiar with all of these systems, but according to Antony à Wood, he was taught Cartesianism by Boyle, while Aubrey tells us that Hooke 'loved' Thomas Hobbes, although he only seems to have met him once.[27]

Robert Boyle, in his chemical investigations, was drawn to a Christianized version of the atomic theory of Democritus, where the geometric arrangement of the atoms somehow defined the chemical characteristics of the substance. As a Baconian, Boyle devised meticulous courses of experiments by which he hoped to test these ideas. As Boyle's assistant, one can expect Robert Hooke to have been influenced by his master's ideas, though there are important points of divergence.

While it goes without saying that Hooke was an experimentalist in the Baconian tradition, it is obvious to anyone who reads Hooke's writings that while he stressed the importance of logical sequence, record and repeatability in his researches, he was no methodological purist. As every modern scientist now knows, no original investigator can be the rigid adherent of a pre-determined method, for creativity in science is more than recipe-following. Robert Boyle and Robert Hooke, however, were probably among the first scientists to encounter this fact of life, for while they were by no means the first men to perform experiments, they were some of the first to undertake whole complex courses of connected experiments and, in Hooke's case, conduct them in disciplines as diverse as astronomy, chemistry, physics and physiology, and strive to find coherent connections both within and between them. *Micrographia*, which published the results of a series of observations and experiments conducted mainly between 1661 and 1664, should be required reading for every science undergraduate, for it amply demonstrates how brilliantly eclectic, yet how tightly controlled, a series of physical investigations can be. It showed how the microscopical examination of ice crystals could lead to a discussion of atomic structures; how the first recognition of the cellular structure of wood initiated research into the role of air in combustion; and how the anatomical description of a fly developed into an experimental essay in aerodynamics, acoustics and wave-patterns.

In published researches covering nearly 40 years, Hooke was constantly casting around for a consistent, underlying principle that could be shown to bind the whole of nature together: a 'Grand Unified Theory', as it were. That nature did contain common lucid principles would have been taken as axiomatic by Hooke, for as the entire universe was the product of one divine intelligence, then, as Galileo argued in 1615, it was inconceivable that God could be inconsistent in His Grand Design.[28]

As human intelligence was congruent with that of God — mankind being in God's image — it stood to reason that the key should be within man's reach, for as Kepler had said, science was thinking God's thoughts after Him.

Although Robert Hooke never came up with a consistent theory that could be made to stand experimentally in all cases, one can extract a series of principles which run as a thread through his thought. One of these was a version of the atomic theory of matter, though he was careful not to push it too far, for lack of clear experimental evidence. Yet Hooke's atomism is more dynamic and rooted in motion than that of his master. Boyle, as we have seen, held to a broadly geometrical concept of atomic arrangement, whereas Hooke's in many ways was more kinematic and based on 'force', or pressure, constantly exciting a medium so that the atoms became the *efficient* — or the immediate — causes of all material things.[29]

When one came to the medium, or ether, in which the atoms were suspended and through which they received their powers of impulse, Hooke seems to have held different ideas at different times, depending upon the results of particular researches. Was the ether itself a 'stagnant', passive agent through which atomic collisions took place (in the way that railway lines are passive agents down which colliding wagons move), or was it the ether that originated the motion, as was implicit in Descartes? Hooke considered that the primary forces of nature, such as light, magnetism and gravity, might act through ethers, or parts of the ether, that were peculiar to themselves,[30] and one of the things which separated Hooke's from Newton's thought was Hooke's assumption that these forces operated through what one might call 'local' foci, such as the moon's gravity or Jupiter's gravity, the sun's light or the earth's magnetic field, though being aspects of one universal agent. While Hooke's universe aspired to being coherent in so far as it was amenable to experimental investigation in all places, it was nonetheless suffused with examples of the particular and the peculiar. Yet one should not forget that the nature of 'force', and especially gravity, occupied his thoughts for many decades, and by the early 1680s Hooke was moving towards a more universally-acting concept of natural forces.

As a mechanist, therefore, Hooke needed a medium or media of some sort if a cause was to produce an effect, for without a physical connecting agent, no matter how tenuous, one was no better off than the magicians who explained cause and effect by means of occult sympathies. One fundamental way in which Hooke differs from modern (or post-Newtonian) scientists is in his concern with active principles and connecting mediums, for like most other seventeenth-century researchers, he was still a 'philosopher' who was interested in the *causes* of things, such as light and gravity. Though he, like Boyle, Descartes and Gassendi, had abandoned the Greek qualitative approach to nature, in favour of a mechanical, quantitative one, his thought processes were still haunted by the sources of cause and effect, albeit re-dressed in mechanical garb. It was not until the eighteenth century that scientists eventually bequeathed causes to the metaphysicians, and concentrated upon expressing the nature of effects in precise mathematical terms, and for this, in many ways, we have Newton and the Newtonians to thank.

If there was one single mechanical principle which Hooke saw as present in most parts of physical nature, it was vibration. In all branches of physical research, he saw vibration, or the activation of wave patterns, as the thing which moved from an active and particular source, through its appropriate ether, to produce a measurable effect. We will return to Hooke and vibration when looking at his work on spring and the elasticity of bodies, along with light and gravity, in chapters 9–12.

Though Hooke might have been happy enough to entertain the presence and differing characteristics of atoms and ethers when speculating about an ultimate metaphysic for science, he had a clear understanding of what had made the scientific discoveries of the age possible: an enhanced ability to perceive and quantify nature by means of instruments. It is in the long Preface to *Micrographia* in 1665 that he sets out most clearly that scientific manifesto which was to run through all his subsequent writings, speaking of instruments as devices which lent new investigative power to relatively imprecise human sense-perception: for,

'The next care to be taken, in respect of the Senses, is a supplying of their infirmities with Instruments, and, as it were, the adding of artificial Organs to the natural.'[31]

Passages like this enable us to understand what Aubrey meant when he said that Hooke 'loved' Hobbes, for one can almost think of these words as an amplification of *Leviathan* (1651), Book 1, 'Of Sense', where Hobbes had tried to define the process of human perception as operating through trains of mechanical contacts, a process now defined by Hooke as operating through and being strengthened by an instrumental technology.

In his stress upon the primacy of the senses in all perceptions of nature, from everyday experience to sophisticated research, Robert Hooke became one of the promoters of the British empirical philosophical tradition, and an influence upon figures like his Westminster and Christ Church contemporary John Locke. Hooke, we must not forget, had a large corpus of original published researches to his name long before Locke published his *Essay Concerning Human Understanding* (1690). Robert Hooke, moreover, did not merely talk about sense-knowledge, but made it the very king-pin of his experimental technique, realizing that, if one were going to investigate the cellular structure of plants, the surface of the moon, or the vacuum, then the senses would need artificial enhancement by means of instruments. The invention and use of instruments, indeed, runs through Hooke's entire career, from flying experiments with Wilkins around 1656 to his devising of an airpump for Boyle around 1659 and on to his last recorded scientific utterance in December 1702, when he tried to design an improved instrument to measure the horizontal solar diameter, 'but discovers not the way'.[32] It is true that Aristotle had placed an emphasis upon the senses when examining natural phenomena, but to Aristotle the reality of a thing was defined in the totality of its parts as perceived by the gross or unaided human senses, for Aristotle had no lenses, micrometers or barometers to aid him. But what the new, instrument-based, experimentalists introduced into sensory perception were new parameters whereby a scientific reality could be defined precisely. Was, for instance, the correct definition of a horse simply a large quadruped, or was it the mechanics of its skeleton and muscles, its heart-rate related to body weight, or particular characteristics of cells and blood as seen under the microscope?

Indeed, this analytical, mechanical and experimental approach to nature inspired a whole generation of scientists and philosophers across Europe. For everything under the sun caught the interest of these men, who were coming to look at the natural world with a new sense of wonder. They were the 'moderns', men inspired by what they believed to be the intellectual adventure of the ancient Greeks, and generally contemptuous of things from the 'Middle Age', which they believed separated the Greeks from their own time. The medieval period, especially for the new Protestants of northern Europe, came to be envisaged as a time of darkness, superstition and dogma. They had little time for the great minds and achievements of the High Middle Ages, writing off most of their intellectual culture as mere word-splitting or concept-spinning without reference to experimental reality. Medieval philosophy was seen as intellectually bankrupt, being concerned, it was believed, only with formal logic and rarefied speculation, and not with 'putting nature to the torture', taking her secrets from the source, and using them for the transformation of the human condition. In his views about the Middle Ages, Hooke was entirely a man of his time, singling out Pope Sylvester II in AD 999, who devised such 'Curious works in mechanicks, as waterworks hydraulick organs clepsydras and clocks', and Friar Roger Bacon as the beacons who shone through 'the Darknesse of those times'.[33]

It was not without irony, however, that in spite of Renaissance Protestant intellectuals' zeal to differentiate themselves from the world of medieval scholastics the three inventions which Bacon and Hooke saw as fundamentally transforming the world—gunpowder, the magnetic compass and printing—were all in regular use by 1470.

Fortunately, modern scholars are less historiographically naïve when assessing the scientific revolution's debt to medieval culture, and few would now regard the 'Middle Ages' as a synonym for backwardness. But a lot of scholarly water has flowed under the bridge of historical interpretation since the days of Bacon and Hooke, especially in the late twentieth century.

The application of scientific solutions to practical, everyday problems fascinated the early Royal Society just as much as the pure science itself, and this interest was clearly manifested in John Evelyn's *Fumifugium* (1661), which was the first scientific study of the problems caused by smoke pollution in London. In particular, Evelyn pointed out the dangerous effects of burning sea-coal, the smoke of which damaged or destroyed local-grown fruit and vegetation, rotted tapestries, cloth and buildings, and did untold mischief to the human lungs, killing old people and even putting the health of the Sovereign and his government at risk. The real culprits, argued Evelyn, were not the thousands of domestic fires in London so much as the brewers, lime-burners, salt and soap manufacturers, and other industrial entrepreneurs whose business consumed vast quantities of coal. Evelyn was also England's leading authority on the propagation of trees, and in his book *Sylva, or, a Discourse on Forest Trees*, first published in 1664, he dealt with the problem of England's diminishing timber reserves, and what could be done about it. In John Evelyn, indeed, with his acute perception of the relationship between nature and the demands which man puts upon it, the Royal Society unknowingly gave voice to the first environmental scientist.

Desiring to make all useful knowledge public, Hooke and the Royal Society projected an encyclopaedia or register dealing with a *History of Trades* ('History', of course, being used in the sense of the Latin word *historia*, or an organized account), in which the skills and techniques of all the manual arts would be rendered accessible. In spite of the obvious good intentions of this project as far as the status of public knowledge was concerned, its extent was simply too vast for the limited resources of the Society. Indeed, Robert Hooke's own fascination with useful skills was in many ways symptomatic of the Society as a whole, for a significant number of the Fellows saw the 'Mechanick arts', when scientifically applied, as the agency through which their vision of 'relieving man's estate' would be brought about, while a similar vision had also lain at the heart of Sir John Cutler's wish to establish a 'Mechanick' lecture for Hooke. For from his earliest years and throughout his life Hooke's fertile imagination expressed itself through the devising of gadgets, recipes and craft skills, linked into the broader philosophical premiss that the improvement of the human condition lay in the cooperation of the scientific intellectual and the enlightened tradesman. The man who would come most obviously to epitomize this ideal type of enlightened tradesman between 1675 and 1710 was the horologist Thomas Tompion, whose clocks and watches were seen as definitive specimens of advanced precision engineering, owing their success to the practical application of recent 'philosophical' discoveries in the pure physics of pendulums and springs.

It is unfortunate, however, that the very word 'Mechanick', which Hooke regarded as a creative appellation (deriving as it does from the Greek *mekhane*, 'contrivance'), and in the spirit of which Sir John Cutler's intended endowment of his Lectureship for Hooke, came to be increasingly seen in the post-Newtonian world of the eighteenth century as denoting something inferior. Something more akin, indeed, to the inferior *banausoi*, or artisans and handicraft workers, deemed by Aristotle in his *Politics* to rank well below those natural governors, the magnanimous intellectuals, in a well-balanced state.[34] Yet to Hooke, with his classical education imbibed at Westminster and at Oxford, who knew his Aristotle better than most men, and who held a Professorship in Geometry, the word *Mechanick*, and the experimental procedures that went along with it, signified not intellectual inferiority, but rather intellectual *adventure*[35] — for had not the investigation of the world and the universe by instrumental procedures and by *mechanical* analogies advanced knowledge further in one hundred years than abstract philosophy had in two thousand? *Mechanick*, indeed, was an intellectually empowering concept in the 1660s, and in no way to be confused with simply sawing timber or beating iron. I would suggest, however, that one of the ways in which Hooke's reputation was to suffer after *c.* 1700, when the much more cerebral science of Newtonian gravitation constituted the dominant *leitmotif* in the west, was in the association of his name not with adventurous scientific research, but with the mere *mechanic* activities of a clever artisan.

One of the Royal Society's most intellectually far-reaching ventures that bore permanent and profoundly important fruit, and with which Hooke came to be intimately associated, was the creation of a regularly-published scientific journal. The *Philosophical Transactions* of the Royal Society of London first appeared in 1665, and publication has continued down to the present day, making the journal not only a benchmark in the early history of scientific communication, but also, along with the Bible and *Book of Common Prayer*, a long-term record in the history of publication. During part of his career in the 1680s, Hooke edited the *Transactions* (which between 1681 and 1682 were temporarily known as the *Philosophical Collections*) and which aimed to collect together all the latest research which the Society could lay its hand on, and publish it. No branch of 'curious inquiry' lay outside the Society's grasp, so long as it was amenable to rational, experimental study. Foreign correspondents such as Christiaan Huygens, Giovanni Domenico Cassini and Antoni van Leeuwenhoek communicated original discoveries in planetary astronomy and microscopy; English provincial doctors reported the appearance of 'monstrous births' such as those conjoined unfortunates that the Victorians would later call 'Siamese twins', or commented upon clinical curiosities; country gentlemen described interesting pieces of local natural phenomena, such as the foul airs encountered in deep pits and mines;[36] strange inflammable 'fumes' that came out of the ground near Wigan, on the Lancashire coalfield;[37] the toxicity of American rattlesnakes gave rise to curious experiments in Virginia;[38] while many of the company of Virtuosi sent in reports and experiments on all manner of phenomena. Needless to say, Hooke was a regular contributor, communicating over 20 papers after 1665 and, as Curator of Experiments, it was his duty to initiate lines of research, and encourage others to do likewise. One must also note that Hooke not only contributed to the intellectual content of the early numbers of the *Transactions*, but even provided some of its artwork. As a skilled draughtsman, he sometimes drew the plates that were to illustrate the various experiments, and when Richard Towneley published the first description of an astronomical micrometer in *Philosophical Transactions* in 1667, the Society expressed its thanks: 'For the draught of the *Figures*, representing the *New Instrument* it self, and the Description of the same, we are obliged to the ingenuity of Mr *Hook*'.[39] When Zacharias Conrad von Uffenbach was in London in 1710, he mentions that one could buy the published set of *Philosophical Transactions* for £12; however, the most original and interesting volumes were those issued over the first six years of the journal.

In his *History of the Royal Society*, 1667, the Revd Dr Thomas Sprat, another Wadham College man, listed the Society's projected lines of research over its early years, and these included:

> 'A Method of making a History of the Weather. By Mr Hook', 'A Proposal for making Wine. By Dr Goddard' and 'The History of making Gunpowder.' For 'The Society had reduc'd its principal observations into one common stock: and laid them up in publique registers, to be nakedly transmitted to the next Generation of Men: and so from them to their Successors.'[40]

Once again, one encounters that persistent theme in the early Royal Society: that the truth should not merely be revealed, but that it should be both *registered* and *made public*.

There were also schemes to improve the design of carriages, and to accurately measure the distance between places on the earth's surface, whilst Dr Wilkins devised a geared 'way-wiser' which had been commented upon by John Evelyn in Oxford in 1654, to measure the distance travelled by a coach—the precursor of the odometer of the modern motor car. Sir William Petty, moreover, laboured to perfect his double-bottomed ship—an early form of catamaran—which would hopefully make navigation not only faster but also safer, and which was tried with some success in the stormy waters of the Irish Sea, making two voyages between Holyhead and Dublin before being lost.[41] Of more enduring importance, however, were Petty's contributions to the then infant sciences of economics and statistics, where he pioneered the scientific study of human population groups, of the effects of taxation, and of the importance of labour in establishing the prices of commodities.

In its early days, moreover, the Society attempted to build up a 'Musaeum', or collection of 'rarities', and a research laboratory, which were probably inspired by those set up in Wilkins's Lodgings in Wadham during the 1650s. While we today can immediately recognize the reason why a scientific institution needed a research laboratory, we might not, at first sight, see the need for a 'Musaeum'. Three and a half centuries ago, however, scientists were often uncertain as to what was 'normal' and 'abnormal' in nature, and a Musaeum, especially when organized along Baconian lines, could serve a valuable purpose in defining the boundaries that separated different types of phenomena. The Royal Society's 'Musaeum' was only one of several in Holland, France and Italy by the 1660s, although the collections of natural and man-made artefacts originally assembled by John Tradescant and donated to Oxford University by Elias Ashmole, becoming the Ashmolean Museum, Oxford, in 1683, was one of the few to survive, institutionally, down to the present day, even if many of its original exhibits were subsequently lost or destroyed.[42] When Uffenbach visited England in 1710, he was shown around both the Royal Society and Oxford Ashmolean Museums, and recorded a poor impression of both in his Diary, due primarily to their sloppy state of curation. Even so, Uffenbach found the Ashmolean Museum to be in a better condition than that of the Royal Society, though the Ashmolean's 'Custos...Mr Parry, cannot show strangers over the Museum for guzzling and toping'. Instead, that was left to the *Sub-Custos*, who for sixpence would let in anybody '...even women... [who] run here and there, grabbing at everything and taking no rebuff from the *Sub-Custos*'.[43]

Indeed, the very business of collecting, organizing and analysing natural phenomena (not to mention the Ashmolean Museum's unofficial 'science for all' policy) lay at the heart of seventeenth-century experimental science, for precisely documented astronomical observations, laboratory experiments, and anatomical dissections were just as much a part of the accumulated 'treasures' of a body like the Royal Society as were its collections of crystals, chemicals, stones, Roman coins, and instruments. What the Royal Society stood for was a conception of human knowledge based upon what was believed to be ostensible physical fact as opposed to theory and abstraction, and this in turn presupposed an accumulative, taxonomic approach to scientific data. Perhaps nowhere does one find this approach to the new knowledge more chillingly displayed than in the case of Samuel Pepys's bladder stone, for after Pepys's stone had been removed during an unanaesthetized operation in March 1658, he had the tennis-ball-sized encrustation mounted in an exquisite cabinet, and displayed to his friends as a part of his own private collection of books and philosophical rarities.[44]

The Museum, the Laboratory and the Observatory, therefore, became symbolic of the new scientific culture. During his years as Curator of Experiments, of course, it was Hooke's job to look after the Royal Society Museum in Gresham College, and receive and label specimens donated by the Fellows. Some of these specimens could excite Hooke's curiosity, and launch him upon a new series of original researches, as did the piece of 'petryfied' wood which John Evelyn showed to him in the early 1660s, and which seems to have initiated Hooke's interest in the organic origin of fossils, of which more will be said in chapter 8.[45]

It was, however, the exploratory, accumulative, and instrument- and specimen-based experimental approach of Hooke—whom one might think of as England's Leonardo in his sheer diversity—which most differentiated him, and his Royal Society colleagues, from students of the natural world in previous centuries. Indeed, one can never avoid being struck by the parallel between Hooke and Leonardo da Vinci, who had lived and worked in Italy nearly two centuries before, for both men shared interests in the mechanics of flight, physiology, architecture, draughtsmanship and the contrivance of ingenious devices. Where, perhaps, they differed most was in the respect that Leonardo lived before the concept of scientific instrumentation, as an avenue to progressive discovery, had really developed, for Leonardo had no 'artificial organs'—telescopes, microscopes, airpumps and such—with which to pry ever deeper into the hidden structures of nature, and one can only guess what the additional wonders he might have come up with had he possessed them.

It is one of the ironies of history, however, that Hooke would have known nothing beyond the generalities about his Italian predecessor's scientific researches, because during Hooke's lifetime Leonardo would have been known primarily as an Italian painter in oils—a new technique which Hooke admired because of the brilliance with which it could depict nature—either from the biography by Giorgio Vasari or from the 1651 edition of da Vinci's *Trattato dell Pittura*. Indeed, Hooke knew both works well, for in April 1673 he recorded bringing over the three volumes of Vasari's *Lives of the Painters* from Arundel House to

Gresham College,[46] while in October of the same year he mentioned having 'bought Leonardi da Vinci for 15 sh[illings]', which must have been the *Trattato*, for this was the only one of Leonardo's works then in print. But in Hooke's lifetime, Leonardo's scientific notebooks remained as manuscripts in private collections in Italy, and would not begin to appear in print for another two centuries.

Yet while the experimental researches of Leonardo da Vinci had been conducted in a context of secrecy, those of Robert Hooke unfolded as ongoing public knowledge — indeed, almost as public spectacle — as the Royal Society became fashionable, and everybody who was anybody felt impelled to visit the rooms of the Society, or, as in the case of King Charles II himself, to be kept closely informed of what was going on. 'Gresham College' soon became a favourite showpiece of the City, and no visiting provincial or foreigner, such as the 'Danish Agent' who witnessed Boyle's and Hooke's cat in the airpump experiments around 1662 or 1663, could miss seeing the Virtuosi in their natural habitat. On 30 May 1667, for example, Margaret, Duchess of Newcastle, whose extravagant behaviour had won her the nickname 'Mad Madge', visited the Society. Hooke and the other Virtuosi showed her a wide range of experiments, including the properties of magnets, microscopic observations, and 'one that did, while she was there, turn a piece of roasted mutton into pure blood, which was very rare'. Samuel Pepys, diarist and Fellow of the Society, was present at Lady Margaret's visit, and wrote a detailed account of what went on. The independently-minded Duchess no doubt felt in good company amongst the Fellows of the Royal Society. It is unfortunate that Hooke did not begin keeping his own Diary until 1672, for it would have been interesting to read a record of Lady Margaret's visit under Robert's own hand.[47]

One indication of how well known the Society had become was the increasing frequency with which it was lampooned by the satirists, for hack poets, popular playwrights, and critics who depend on the public response for their livelihood deal with the topical, and have no time for the obscure. About 1663, an anonymous writer published *The Ballad of Gresham College*, which reviewed the Society's activities to date, and was clearly written by one who knew the Virtuosi well.[48] Then around 1665 Samuel Butler also poked fun at the Society in his *Satire on the Royal Society*, and Robert Hooke is almost certainly the butt of the verse which runs:

> 'A Learned man, who once a week
> A hundred virtuosos seek,
> And like an oracle apply to
> T'ask questions, and admire, and lie to.'[49]

Butler, like most of the satirists, clearly made himself familiar with the functioning of the Society, in order that his shots might hit closer to the mark. The characters in his poems seem to be based on specific Fellows or projects of the Society, along with its published or recorded experiments:

86 'To measure wind, and weigh the air   [wind gauge and barometer]
   And turn a circle into a square   [Wallis]
   To make a powder burn the sun,   [aurum fulminans?]
89 By which all doctors should b'undone.'[50] [= used as medicine?]

The weighing and measuring of the air were most certainly projects close to Hooke's heart, involving as they did his barometric inventions, and the devising of his and Wren's self-recording 'weather-clock', while the 'squaring of the circle' was a geometrical problem which had been the subject of a lively controversy between Thomas Hobbes and John Wallis. It is possible that the powder burning the sun was a reference to the touch-sensitive *aurum fulminans*, or exploding gold, that was being experimented with from the late 1650s onwards, and variants of which, such as *aurum potabile* or drinkable gold, were also being tried as a medicine.

In another Royal Society satire, *The Elephant and the Moon*, Butler almost certainly found inspiration in Sir Paul Neile's telescopic researches, and probably in Dr John Wilkins's book *A Discourse concerning a New World and Another Planet*. Although Wilkins's book had been published back in 1640, it was still highly topical, for it was in this book that Wilkins had opened a discussion in English about the possible existence of intelligent creatures living on the moon. But such a discussion provided perfect satirical ammunition:

'A Learn'd Society of late
The glory of a Foreign State.
Agreed, upon a summer's night
To search the Moon by her own light.'

The satirized Virtuosi in the poem went on to find what they believed to be a great beast on the lunar surface, and reckoned that they had made a wonderful discovery, and began to speculate at length upon the consequences of their findings. But presently, after the Virtuosi had adjourned for the night, it was the servants who, in the best traditions of classical and Shakespearian comedy, got to the bottom of things when (lines 324ff.):

'The footboys, for diversion too
As having nothing else to do,
Seeing the telescope at leisure,
Turn'd virtuosos for their pleasure;
Began to gaze upon the Moon,
As those they waited on had done.'

Like their masters, the servants saw the great creature through the telescope, but through a combination of plain common sense and shrewd observation, they quickly realized that the 'Elephant' was no more than a mouse which had crept into the instrument. Finally, the servants confirmed their discovery by dismantling the lower end of the tube, and (501ff.)

> ' ...unscrew'd the glass,
> To find out where th' impostor was,
> And saw the Mouse, that, by mishap,
> Had made the telescope a trap.'[51]

Could it be that all the Virtuosi, in 'presuming to know more than their forefathers', as John Aubrey put it, were going to trip up and make themselves look foolish?

Ridiculing the 'Gresham College Witts' came to be a sure way of raising a laugh for hack writers, and even 50 years later Jonathan Swift was successfully employing the formula in *Gulliver's Travels* (1726). In his account of the voyage to Laputa, Gulliver visits the Grand Academy of Lagardo and the College of Projectors or Project Devisors. Once again, this section of Swift's satire contains a thinly-veiled jibe at the Royal Society, especially when Gulliver meets distracted and self-absorbed scientists engaged in seemingly ludicrous experiments—for one man is trying to bottle sunbeams, so that they may be released on cold winter's days, another is trying to make food out of excrement, while yet another hopes to teach blind men how to mix colours. A further Projector is hoping to transform architecture by building houses starting from the roof and working downwards. Again, most of the projects presented by Swift are parodies of suggested or attempted Royal Society experiments or observations, some of them published in the *Philosophical Transactions*. Stephen Hales's experiments which first investigated the roles of air, fertilizers and light in plant growth quite likely provided some of the ideas to which Swift's bizarre imagination gave ludicrous new twists,[52] while the architectural jibes had an obvious butt in Hooke and Wren, especially in Wren's eventual solution to the problem of how to crown St Paul's Cathedral with a dome which seemed to hang above the London skyline, and which was only completed in 1708.[53]

But one of the most pointed satires of all was Thomas Shadwell's scurrilous play, *The Virtuoso*, which began to attract large audiences in 1676. The 'hero' was one Sir Nicholas Gimcrack, a Virtuoso, and it was from the clear allusions to Robert Hooke in the play that the Curator of Experiments was occasioned considerable embarrassment, for some of the best laughs were inspired by his own scientific master-piece, the *Micrographia*.

CHAPTER 4

MICROSCOPES AND METEOROLOGY

In his Diary for 25 May 1676, Hooke wrote 'Mr Hill gave Sir J. Hoskins, Aubrey and I an account of Virtuoso play', and followed it on 2 June by: 'with Godfrey and Tompion at play. Met Oliver there. Damned Doggs. Vindica me Deus. People almost pointed.'[1]

In 1676, Thomas Shadwell's *The Virtuoso* was enjoying a great popular success, and Hooke was persuaded to go and see it, much, it would seem, to his subsequent embarrassment, for it must have seemed natural for the crowd of London theatre-goers to associate Hooke, who was firmly established as a figure of note by 1676, with the antics of Sir Nicholas Gimcrack, Virtuoso and microscopist.[2]

The microscope had been invented some 30 years before Robert Hooke was born. The Yorkshire scientist Henry Power had already published microscopical observations before Hooke[3] but, as Hooke reminded his readers, *without* providing any illustrations to amplify his descriptions,[4] and in his *De pulmonibus* (1661) Marcello Malphigi of Bologna had already demonstrated the use of the instrument to provide clinching evidence of Harvey's theory of blood circulation when he discovered the capillary vessels which linked the arterial and venous systems of the body in the lungs of a frog. Yet for over half a century after its invention, the microscope had been a poor relation to the telescope in terms of its ability to produce fundamental scientific discoveries. Not until Robert Hooke published his own microscopical researches, in 1665, stemming as they did from those original microscopical observations which, as Curator, he was obliged to devise and show to the Royal Society at its meetings between March 1663 and November 1664,[5] and no doubt inspired by the microscopical drawings which Sir Christopher Wren had prepared for the King, was it made manifest to the scientific world that the microscope revealed an organized realm of nature that was as diverse in its structures and as vast in its scale as was the telescopic universe. For centuries, indeed, and long before the invention of the telescope, philosophers had speculated about the vastness of space, but no one had thought seriously about the existence of living creatures smaller than cheese-mites or inanimate objects smaller than dust particles. It is true that the atomists had posited the existence of minuscule particles that composed matter, but these had been subjects of a philosophical discussion, and held no hope of physical detection.

These assumptions were fundamentally challenged and overturned when Hooke directed his creative scientific imagination upon the microscopic realm for, in reality, *Micrographia, or Some Physiological Descriptions of Minute Bodies made by Magnifying Glasses* was destined to become one of the most original books of the seventeenth century. As a scientific clarion call and opener-up of new worlds, it was no less significant than Galileo's *Siderius Nuncius* had been in 1610, though as microscopy was not associated with any philosophical agenda in the way that Galileo had allied the telescope with Copernicanism, *Micrographia* became more of a book of wonders than a manifesto for controversy. Published at the beginning of the plague year of 1665, it was an instant success, and without doubt it stood as the most dramatic proof to date of the power of the new Experimental Philosophy produced in England; and like Galileo's *Siderius Nuncius*, but unlike most original scientific books, it captivated the imagination of the layman.

When, therefore, Hooke's *Micrographia* first appeared in the bookshops in January 1665 at an expensive 30 shillings per copy, it had a quite remarkable impact. It bowled Samuel Pepys right over when on 21 January he saw it in his bookseller's, 'so very pretty that I presently bespoke it',[6] from whence he proceeded to sit up until two o'clock in the morning 'reading of Mr Hooke's Microscopical Observations, the most ingenious book that ever I read in my life.' Pepys was already an avid scientific dilettante who on 13 August 1664 had bought a microscope for the 'great price' of £5–10–00.[7] The instrument was purchased from Richard Reeves (whose lenses were especially praised by Hooke),[8] who along with Mr Spong, another craftsman, succeeded in selling Pepys a number of expensive scientific instruments, as Pepys recorded in his *Diary*.[9] More than anything else, *Micrographia* whetted Pepys's appetite for the New Science. He subsequently joined the Royal Society in February 1665, and in 1684 became its President. *Micrographia*, therefore, became one of the formative books of the modern world, and like all influential pieces of writing, was capable of triggering responses on many different levels of understanding.

Within the scientific community, *Micrographia* constituted one of the most articulate and beautifully-presented justifications for experimental science ever devised, and in its 'Preface' Hooke left the reader in no doubt as to how the Experimental Method was able to yield more answers to natural questions than any previous philosophy. Mere observation, after all, could take one no further than Aristotle had gone in his descriptions of animals or natural forces, but when observation was refined by means of specially-designed instruments, and used to investigate specific questions within the wider scheme of nature, then remarkable discoveries could be made. *Micrographia* not only provided a wealth of new data for science to consider, but showed how whole courses of experimental investigations could be built upon it. A seemingly simple observation of a piece of charcoal under the microscope, for instance, could lead to a recognition of the presence of cells in the wood, to an investigation into burning, and to Hooke's work on the dissolving properties of aerial nitre. In spite of their visual beauty and elegance of description, none of the Observations in *Micrographia* are simple; instead, almost all of them are detailed starting-points for further and often far-reaching physical investigations in one way or another: in biology, in the nature of light and colour, in atmospheric refraction, and even in astronomy. Hooke also used the book to demonstrate that sense knowledge could be reliable when used within the correct disciplinary restraints, and what the human body could physically perceive via its 'artificial Organs' or instruments left little doubt that the experimental method actually worked.[10]

In the 'Preface' to *Micrographia*, Hooke stressed that it was only through the senses that true knowledge of the physical world could be gained, and not by speculation. But, as everyone knew, the human senses were imperfect, and could be led astray, so that it was necessary to devise disciplined and openly verifiable techniques, used in conjunction with reliable instruments, to successfully investigate the 'Secret Workings of Nature', for

> 'By means of Telescopes, there is nothing so far distant but may be represented to our view; and by the help of microscopes, there is nothing so small as to escape our inquiry; hence, there is a new visible World discovered to the Understanding.'[11]

This world is ordered, logical and capable of being comprehended in terms of observation and measurement by refined tools. As much as Hooke admired the great discoveries made so far, the experimental philosophy was still in its infancy, and 'What may not therefore be expected from it if thoroughly prosecuted?' Man could perhaps recapture his ancient dignity, for while Adam and Eve had sinned by eating of the Tree of Knowledge, yet we, by acting not from wilful disobedience, but from a motive of reverential contemplation, might win a degree of redemption. It was the advancement of the 'Experimental Philosophy', epitomized in England by the Royal Society and led by John Wilkins, Robert Boyle, Sir Christopher Wren and other men, whom Hooke most profoundly admired, that was to be the agent whereby the state of mankind would be improved.[12]

Even if one did not feel inclined to read Hooke's clear and unmistakable prose, one could browse through the 38 'schemes', or full-page plates, that accompanied the text. Here, for the first time, the layman could see the cell structure of a piece of cork, and see as much detail in 'A Flea, a Mite, [or] a Gnat' as if it were 'a Horse, an Elephant, or a Lyon'.[13] It is books such as *Micrographia* which open up new realms of wonder to all levels of the human imagination, and Hooke should not have felt too offended when one of these turned out to be the popular stage, as in the case of Shadwell's *Virtuoso* play.

No other work captures so well Hooke's excitement and awe in the face of a new science as does *Micrographia*. In its pages one finds fresh inspiration, a contempt for bookish philosophy, an enthusiasm to look at the world through new eyes, combined with a sense of piety and humility when brought face to face with its Creator. Throughout his career, Hooke always had a talent for combining a vivid and even boisterous curiosity with a genuine humility, which is seen at its best in *Micrographia*. It was published at the beginning of Hooke's thirtieth year, and here one finds his creative powers rising to their peak, while his personality was still unscarred by the feuds and disappointments of his later life.

If *Micrographia* was so important within the scientific community, it must be remembered that its influence on members of the cultured laity, like Pepys, was equally profound. The book possessed a dazzling, immediate quality, being written in an easy style that would have been accessible to any innumerate who could read Shakespeare or the Bible, for Hooke, like John Wilkins, could write vivid and powerful prose. It was, moreover, the first proper *picture-book* of science to come off the presses, for its 60 Observations were accompanied by 38 beautiful engravings of the objects seen with the new instruments. These 60 Observations included a total of 57 microscopical, two astronomical, and one set of atmospheric observations. Hooke's artistic gifts, moreover, had been essential to *Micrographia*, for only a man who could faithfully interpret and delineate the awkward images that were produced by the compound microscopes of the 1660s could envisage such a book in the first place. Modern science is replete with visual images, and in our own time the televisual image is the most powerful medium through which its ideas are now communicated to the lay public. So we must not forget that this tradition of visual communication in science largely begins with Hooke's *Micrographia*.

All of Hooke's microscopical observations were made with a compound instrument using two lenses. One of these was a small-diameter short-focus object lens that was placed as near to the specimen as its focal length permitted. This was matched with a field or eye lens of larger diameter placed several inches away and close to the observer's eye. The magnification of the overall optical system would depend on the respective focal lengths of the two lenses and the way in which they were matched together. Each lens would be mounted in a turned wooden or metal cell and secured at their correct operational distances by a cardboard tube.[14] While Hooke knew of the use of the single-lens microscope, and in subsequent papers noted the discoveries made with such instruments by the Dutchman Antoni van Leeuwenhoek, he never used such a single-lens microscope for his *Micrographia* observations.[15]

Hooke-type compound microscopes, however, must have been very tricky to use. They had no fine-focus components, and getting the instrument correctly adjusted and focused upon an often three-dimensional specimen — for Hooke did not use glass slides to control his specimens — must have demanded enormous patience. This I can confirm from personal experience, having built a Hooke-type microscope of a pattern shown in *Micrographia*. But one operational trick quickly became apparent in my own reconstructions, however: it was easier to leave the microscope tube fixed in one position, and obtain one's focus and correct viewing angle of the specimen by moving the specimen itself and by changing the direction of the light source. Even so, the prospect of performing all 57 of the major observations of *Micrographia* with such a difficult and optically-aberrated instrument only increased my respect for Hooke's patience and manipulative skill.

Hooke began his book by discussing the microscope as an instrument, and the problem of obtaining adequate lenses that both admitted sufficient light to allow a high magnification, and yet cut out those optical distortions which blurred or gave false colours to the image. With good lenses, he was able to obtain images magnified several hundred diameters, and he mentions working on specimens that were a mere one-hundredth of an inch from the microscope's object glass,[16] as would have been necessary to observe the inner structure of plant cells and the details of insect anatomy. More commonly, however, he probably worked at around one hundred times magnification. His lenses 'were of our *English* make, but though very good for the Kind' did not give images of the best quality because they were only figured with spherical and not elliptical curves. This required the stopping down of their apertures, making them 'so very small, that very few Rays are admitted', with the consequence that the images were not as bright and clear as Hooke desired.[17] Even so, microscopes were expensive, and on 2 January 1676 he bargained for two of the instruments, at £5 and £15 respectively, from an individual referred to as Cox, but who was no doubt the optician Christopher Coxe, or Cocks, who was Reeve's successor as London's foremost lens maker.[18]

To make an observation, Hooke would, as he tells us in the text, carefully examine the object many times under varying conditions of light until fully certain of its form, for only then could worthwhile conclusions be drawn. When daytime observations were being made, the specimens would be examined in sunlight, otherwise a brine-filled glass ball was used to focus the light of a lamp on to the specimen; but he did find that providing adequate illumination for a specimen when working at very high magnifications (and thus, with the microscope lens only a fraction of an inch from the specimen) was a real difficulty, and described a special condenser he had devised to direct the necessary quantity of light.[19]

Yet what clearly delighted Hooke, and challenged him both as a scientist and as a draughtsman, was the enormous diversity of ways in which a single object could appear, depending upon the strength and direction of its illumination:

> 'For the Eyes of a fly in one kind of light appear almost like a
> Lattice, drill'd through with abundance of small holes; which may
> be the Reason why the Ingenious Dr Power seems to suppose them
> such. [Yet] In the Sunshine they look like a Surface cover'd with
> Pyramids; in another with Cones; and in other postures of quite
> other shapes; but that which exhibits the best, is the Light collected
> on the Object, by those means I have already described'[20]

by which, no doubt, he means the special close-focus illuminator mentioned above.

With this discussion about technique and method complete, Hooke then proceeded in the main text to examine the microscopic natural world. Much of the fascination of *Micrographia*, we must not forget, lay in the arresting new perspective that it cast upon common and familiar objects: a fine needle point looked like a rough carrot,[21] delicate silk looked like basket work,[22] and extinguished sparks resembled lumps of coal.[23] Indeed, what the microscope revealed was interpreted by contemporaries as a corrective to human vanity, for under the microscope mankind's most delicate creations, such as silks, razor edges, or steel points, looked crude, whereas nature's lowliest and most contemned creations, such as insects, suddenly became objects of exquisite refinement and ingenuity of design when seen under the lens. It was the microscopical observations of living things that caused the greatest sensation. Meals must have been spoiled on the sight of the 'eels' swimming in vinegar,[24] while the depiction of a flea with the anatomical precision of an elephant was quite shocking as well as beautiful; and one wonders how many nightmares were occasioned by *Micrographia* in that unbathed age, when even persons of the highest social standing often complained of the discomfort occasioned by personal vermin. One might also suggest that part of the fascination of *Micrographia* lay in the tacit horror of some of the natural secrets which it revealed. In the twenty-first century we have become blasé about the impact of scientific discovery, generally communicated by means of visual images, and it is hard for us to imagine the fascination value of a book like *Micrographia*, which opened up a hitherto invisible universe to the reading public.

What *Micrographia*, just like Galileo's *Siderius Nuncius* (1610) before it, did was challenge traditional assumptions about perception and the natural 'scale' of things. Just as Galileo's telescope had altered our perceptions about the physical size and significance of the earth within the vastness of the astronomical universe, so Hooke's microscope showed that fleas were veritable monsters compared with the eels in vinegar, and that the cellular structure of plant matter suggested the presence of a hidden universe, as perfect in its infinitesimal geometry as the greater universe viewed through the telescope. So did the 'scale' of the natural world have any meaning beyond the conventional parameters imposed by traditional naked-eye, common-sense perception? Indeed, the nature and scale of sense knowledge stimulated one of the philosophical questions that would be explored in the decades after 1665 by people like Hooke's Christ Church and Royal Society contemporary, John Locke.

In one respect, however, *Micrographia* differed radically from Galileo's *Siderius Nuncius* and *Letter on Sunspots* (1612), for whereas Galileo's astronomical books, through their explicitly controversial stance, had spurred on other astronomers to corroborate or reassess his announced discoveries, Hooke's *Micrographia* was very much of a one-off. Not only was Hooke's book wonderful rather than controversial, but microscopists and astronomers were very differently placed when it came to the accurate depiction of what they could see through their respective instruments. Having the geometry of the 360° circle and its degree, minute and second subdivisions, along with telescopic eyepiece reticules, and (after Richard Townley and Robert Hooke in 1667)[25] the filar micrometer with which to record celestial detail, the astronomer required no significant artistic skill in order to communicate his findings.

The microscopist, conversely, was entirely dependent upon his personal skills as a draughtsman in capturing the proportions, details, and light and shade of the object under his lens if he hoped to communicate his findings with verisimilitude. If the microscopist wished to depict the multi-faceted structure of the eye of a fly, or the minute anatomy of a louse, then he needed artistic skills of the very highest order, not to mention infinite patience and great manipulative dexterity. Robert Hooke possessed all of these talents in abundance, though their combination in one person is very rare indeed. This is perhaps why *Micrographia*, instead of becoming a clarion call to further researches in published microscopy, became instead something of an icon. While contemporary microscopists such as the Dutchmen Antoni van Leeuwenhoek, Jan Swammerdam and others all went on to make fundamental new discoveries in microscopy after the 1660s that built upon and went beyond Hooke's work, and while Hooke was aware that Malphigi's use of the microscope to discover the cardiovascular capillaries had preceded his own, none of these scientists possessed his artistic skills. Their findings — of the pulmonary capillaries, the 'animalcules' of bacteria, living spermatozoa, or the globules found in blood — were communicated to the world either in words or else in very simple drawings, and not until the development of photography in the mid nineteenth century was the microscopic realm visually communicated with the same breathtaking clarity as Robert Hooke had achieved.[26]

Hooke's observations in *Micrographia* are arranged into six broad classes, starting with the realm of artificial, or man-made, objects, such as the needle, fabric and razor edges mentioned above. These man-made objects account for the first seven observations, and are followed by minerals. There are eight mineral observations, including highly-magnified studies of sand grains, the shape of spark cinders produced by hitting flint on steel, and a study of the optical and mineralogical properties of glass. One curious observation, reflecting on the medical preoccupations of the period, is No. 12, 'Of gravel in Urine'. When one notices how many men of that time suffered from 'the Stone', including Samuel Pepys, who was successfully operated on in 1658, and John Wilkins who supposedly died of it in 1672,[27] one can appreciate the particular relevance of this observation. Hooke examined these stones from the human urinary tract under his microscope, and observed the layered or stratified structure which existed within them.

Observations 16 to 31 are taken from the vegetable kingdom, beginning with a study of charcoal. Yet this is much more than a simple study of charcoal grains, for Hooke proceeds, in the six closely-printed pages devoted to it, to discuss the whole business of combustion. Starting, no doubt, from what he had learned when working with Boyle in Oxford around 1659, Hooke discounts the generally held theory that fire is a substance contained in bodies, and argues that combustion rather consists of the consumption of 'sulpherous bodies' (wood, coal, and such) by some dissolving power present in the air.

Sulphurous bodies, moreover, only begin to 'burn' after reaching a certain temperature, and Hooke recognized the difference between friction heat and burning proper, though as he believed that mechanical motion lay at the heart of all things, he probably considered that they were connected. In the seventeenth century 'air' was generally thought to be a homogeneous medium, though in Observation 16 Hooke drew some interesting new conclusions, some of which have been touched upon in earlier chapters.[28] For one thing, he argued that his experiments undermined the ancient doctrine that fire was an *element* that drew flame and heat upwards to itself. Instead, he proposed that combustion was an act of chemical dissolution, for air was a *menstruum* or solvent, capable of breaking down the 'sulpherous' structures of bodies which, once excited by heat, combined with the *menstruum* and glowed to form flames. Yet the dissolving power of any given volume or 'parcel' or air was limited and only capable of releasing a fixed quantity of sulphurous matter, unless more air, and by definition more *menstruum*, was supplied by means of bellows or other blasts which drove more air to the flame.

On the other hand, when the white crystals of saltpetre (or potassium nitrate) were heated until they melted, a *menstruum* was given off 'that abounds more with the Dissolving particles, [than common air] and therefore as a small quantity of it will dissolve a great sulphureous body, so will the dissolution be very quick and violent'.[29] From the standpoint of modern chemistry, we know that when potassium nitrate is roasted it gives off abundant oxygen, the compound containing a triple oxygen bond. We also now know that the reason why burning wood is rapidly consumed to ash, whereas charcoal slowly toasts and retains its cellular structure, is because the first burns in the open air — where an unlimited supply of the *menstruum* can get at and dissolve it — while the second is in a confined, sealed space where a limited air supply restricts 'dissolution'.

As was indicated in chapter 2, however, it would be incorrect to see Hooke as a proto-discoverer of oxygen, for neither he nor Boyle had any coherent concept of a chemically-specific gas, yet what was clearly there by 1665 was the realization that air was no more to be considered as a simple Aristotelian element than was fire. The air was 'elastic' in so far as it could be rarefied or compressed in any given volume, while it also seemed to contain a 'nitrous' aspect capable of acting as a *menstruum* with sulphurous bodies, and yet which seemed weaker in its dissolving properties than the 'air' given off by toasted saltpetre.

Hooke's work on 'airs' formed part of a broad range of interests in the early Royal Society, as provincial gentlemen wrote to the *Philosophical Transactions* to report notable local phenomena, such as the inflammable airs that came out of the ground around the Wigan coalfield in Lancashire and the suffocating 'Damps' found in pits near Leeds,[30] and Richard Townley sent in his observations upon the behaviour of the barometer on Pendle Hill, Lancashire.[31] These reports were to stimulate other researchers to the devising of new experiments, such as those undertaken by Dr John Mayow, and the Revd Stephen Hales who, in the early eighteenth century, began to collect and classify the properties of a range of natural and laboratory-generated 'airs', the results of which he published in his *Vegetable Staticks* (1727).

In his other vegetable observations, Hooke looked at the cellular structure of cork, and the microscopic structures of mould, moss, sponges and leaves. Cutting off a thin sliver of cork with a razor-sharp pen knife, and placing it under his microscope, Hooke noticed that cork was made up of 'cells' or 'pores' — his term *cell* deriving from their enclosed, yet individually connected, honeycomb-like appearance. Cork cells, containing what he supposed was air, were suggested by Hooke to lie at the heart of cork's buoyancy in water, though he also realized that a similar cellular design was to be found in wood (as noted in his charcoal observations), plant leaves and other botanical structures. Though Hooke was both the discoverer of and coiner of the term *cells*, we must not forget that he had no concept of their true biological function, nor had he any idea of cell replication and its relationship to growth, for these ideas would only develop in the wake of the researches of Rudolf Virchow after 1858.[32]

Instead, Hooke saw the cells in terms of Harvey's circulatory physiology, as 'channels or pipes through which the *Succus nutritious*, or natural juices of Vegetables are convey'd, and seem to correspond to the veins, arteries and other Vessels in sensible creatures' or higher animals.[33] What did excite his sense of wonder, however, is how these nutrition-bearing cells nonetheless sustained the 'vastest body in the World', such as great trees. Being by instinct a physical rather than a natural history scientist, Hooke must have been one of the first researchers to use an optical instrument to quantify the inner geometry of living structures, for after counting some sixty cork cells against a 1/18th inch division on a measuring scale, he calculated that there must be 'above twelve hundred Millions' of separate cells in a single cubic inch of cork.[34] Yet these cork cells were large when compared with the 'exceeding small' cells found in wood charcoal, of which he counted 150 against his 1/18th of an inch measuring scale, and of which he computed that there were 5,725,350 in a single *circle* of wood one inch in diameter.[35] Indeed, Hooke's fascination with the geometry of cell structure in living things has a clear parallel with his interest in the natural geometry of crystals, and the way in which contiguous arrangements of circles and spheres can produce the flat planes and exact angles of mineral structures, as described in *Micrographia* Observation 13.

Observations 32 to 36 are of hairs and feathers highly magnified, considering their various functions and providing beautiful illustrations to amplify the text. Never before, indeed, had commonplace objects been described and portrayed in this manner, revealing a whole new dimension of the physical world, nor, moreover, had such far-reaching conclusions been drawn from their study. Then came the largest single category of Observations in *Micrographia*, running from 37 to 57, which consisted of insect studies. These constituted some of the most sensational observations in the book, especially his illustrations of the fly, ant, louse and other creatures.

Hooke discussed insect anatomy and function in great detail, and in view of his old interest in the 'Art of Flying', was particularly concerned with discovering how these creatures remained aloft. In Observation 38,[36] he described how he attached a fly to the rounded quill end of a feather with a spot of glue so that it could remain secured in one place, thus enabling him to watch with a powerful magnifying glass how its wings moved as it tried to fly away, while at the same time utilizing his sensitive musical ear in an attempt to tune a string so that its pitch was perfectly in unison with the fly's hum. Hooke's microscopic examination of a living fly that was attempting to escape the feather to which it was glued also led him to discover the 'curious *Mechanism*' or '*pendulums* or extended drops' that hung down from under each wing. He wondered whether their rapid vibrating motions acted like cocks, or valves, that admitted air to the fly's respiratory system, and whether they had something to do with sustaining the muscles. Indeed, when we bear in mind his long-standing interest in elasticity and in the devising of 'artificial muscle' when working with Dr Wilkins in Oxford during the 1650s, we see how his microscopical studies of insect musculature formed part of a long-term fascination with the mechanics of winged flight.

A comparison between the flight patterns and wing structure of different insect species then led to a set of ingenious and highly significant conclusions: insects which have shiny membrane-structured wings, like the fly or bee, beat them much faster than insects like the moth or butterfly, the surface of whose wings are 'feathered', or covered with furry, pollen-like particles. Hooke considered that the 'feathers' or wing pollen trapped air and hence imparted a natural buoyancy to the creature, making it less necessary to expend energy to keep aloft. Insects like bees and flies, however, with hard, glassy, shiny, membraned wings, depended entirely on a rapid wing motion which would compress the air beneath them and hence force them upwards. This mode of flying required the expenditure of greater mechanical energy from the muscles, producing a higher vibration rate, or buzz, enabling the creature to fly much more rapidly than moths and butterflies.

Yet while Hooke's observations of flying insects and their structures were the most advanced and carefully-thought-out that had ever been made, we must not forget that there were other people marvelling at them as well. As the optician John Spong informed Samuel Pepys on 7 August 1664 — at a time when *Micrographia* was going through the press — 'by his Microscope of his own making he [Spong] doth discover that the wings of a moth is made just as the feathers of the wing of a bird, and that most plainly and certainly'.[37] Very clearly, others were studying insects through microscopes in the early 1660s, though with his characteristic ingenuity, Hooke was going beyond passive observation into a realm of experimentation which brought up questions about air pressure, elasticity and the behaviour of mechanical structures under variable tension.

The final category of observations was not made with the microscope, but with the telescope and other instruments. Observation 58, for instance, consists of a set of studies in the behaviour of light passing through air and other media, while No. 59 is 'Of the fix't stars', where Hooke studies the Pleiades cluster and describes stars visible in the Orion Nebula as observed with a 12-foot focal length refracting telescope. The last Observation in the book, No. 60,[38] is devoted to the moon, where Hooke not only describes its surface features with great detail, using a telescope with either a 30-foot or a 36-foot focal length object-glass, but discusses the formation of its craters and mountains, as will be shown in more detail in the next chapter.

In the 'Preface' to *Micrographia*, in Observation No. 58, and elsewhere in that book, Hooke had made clear his interest in meteorology, and in the devising of instruments, such as the barometer and various gauges, whereby it might be quantified: interests, indeed, which he shared with his friend Sir Christopher Wren. What is more, the Royal Society was encouraging the Virtuosi to compile accurate weather records, in the hope of eliciting the physical causes of climate, and Hooke had already written a 'History of the Weather' which had been published as part of Thomas Sprat's *History of the Royal Society* in 1667, and offered guidance and technical advice to Virtuosi who aspired to keep a regular 'Account' of the weather.[39]

In his 'History of the Weather', Hooke had suggested a form of wind gauge and various other instruments, such as the hygrometer, thermometer and barometer, and had proposed a formal plan or method whereby the Virtuosi could use these instruments, in conjunction with general observations of cloud, damp and vegetation changes, to record meteorological phenomena systematically. Some of his meteorological interests, moreover, were tied in with microscopy, as in *Micrographia*'s Observation 27, which was devoted to a study of the way in which the 'beard' hair of oat husks responded to moisture changes in the atmosphere, following upon an examination of the fibrous structure of oat beard hair. However, as with several other lines of investigation in *Micrographia*, Hooke's studies in meteorology ran along independent lines of their own and were connected to his wider interests in spring, vibration and the nature of 'force'. One of the most long-standing of these wider studies, to which Hooke returned time and again between the 1660s and 1680s, was the physical and optical behaviour of the air itself. It was this collection of researches, among other things, that forced Hooke to wrestle with an aspect of the mathematics of varying densities, as he tried to compute the changing optical behaviour of the air as it became progressively more dense under increasing pressure between space and the earth's surface, which in turn led him to coherent (if not always accurate) explanations of why the stars twinkle, and how colours are formed in air. Hooke's interest in colour will be looked at in more detail later, but it was air pressure, and how it could be measured with a barometer, that most fascinated him with regard to meteorology.

Hooke recorded in a note subsequently found in his posthumous papers and published by William Derham that the Torricellian tube's, or barometer's, meteorological sensitivity was first recognized accidentally by Robert Boyle when pursuing a line of research at the suggestion of Sir Christopher Wren around 1659.[40] Boyle had been trying to see if, in accordance with Descartes' theories, the mercury level in the barometer changed with the lunar cycle, only to find that it seemed to be the weather rather than the moon that produced the most notable changes in the mercury. Though Hooke did not mention his own involvement in these experiments, the fact that they were performed at Boyle's lodgings in Oxford — probably at Deep Hall on the High Street — at exactly the same time that Hooke was working as Boyle's Assistant in the airpump experiments, makes it highly likely that Hooke must have been involved with barometric meteorology from the very start. In *Micrographia* Observation 8,[41] for instance, Hooke records having performed experiments with mercury in long glass tubes since at least August 1661, while some time later, in London, he took a barometer up the 200-foot-high tower of Old St Paul's Cathedral, to find that the mercury level fell by 1/50th or 1/60th of its ground level height, or about ½ inch. Some years later, after the Great Fire had destroyed Old St Paul's, Hooke performed similar experiments on top of his own and Wren's 'Piller' or Fire Monument in Fish Hill Street, which was 202 feet high, only to observe a similar ½ inch drop in the mercury when compared with its level at the bottom.[42] But perhaps more than any other scientist of the day, Hooke devoted attention to the barometer as a convenient scientific instrument, describing his 'Wheel Barometer' in *Micrographia*,[43] and communicating his subsequent refinements of the instrument to *Philosophical Transactions*.[44]

It was these observations of the increasing density of the atmosphere as one approached the earth's surface, moreover, which led Hooke into a series of laboratory experiments involving the mixing of dense and less dense liquids — such as alcohol and water — in an attempt to replicate the formation of zones of different densities within the same transparent medium. This was to lie at the heart of Hooke's theory of atmospheric *Inflection*, or 'multiplicate refractions',[45] whereby a ray of light from an astronomical body became progressively bent into a curve as it approached the earth's surface. Hooke was to use this model of *Inflection* to explain the redness of sunrise and sunset, the distorted shape of the sun on the horizon, and the twinkling of the stars, and it will be dealt with more fully when looking at Hooke's theory of light and colours. It is important to remember that for Hooke, and for all scientists of his time, colour and light were by definition *meteorological* phenomena, standing as they did upon the prior studies of aerial phenomena extending back through Roger Bacon and Alhazen to Aristotle's *De Meteorologica*: not only was the sun the only powerful light source available, but it came to us through a deepening atmosphere, with its suspended water vapour, to produce a bewildering array of natural wonders, such as sunsets, rainbows, mock suns, refractions, and reflections. Thus meteorology was not just the science whereby one might understand the weather, but was also an attempt to make comprehensible the whole gamut of the phenomena of pressure, light, heat, transparency, vaporization and optical sensation. It seemed eminently logical, therefore, to a seventeenth-century scientist to include such studies in a book devoted to the microscope, and hence *Micrographia* contains observations made with barometers, telescopes and other instruments, all of which tried, in their respective ways, to quantify nature and interpret her operations with greater exactitude, because these instruments were 'artificial organs' which made our natural senses more sensitive and more precise.

At a very early stage, at the beginning of the 1660s, Hooke and Wren recognized that a scientific study of meteorology in the specific sense of weather studies hinged upon the taking of regular and exact observations of such variables as barometric pressure, rainfall and temperature. The Royal Society's ambitiously-planned 'History' of the weather was an aspect of such intended instrumentalization, and one of the first daily weather records to be undertaken was by Richard Townley, at his mansion, Townley Hall, near Burnley, Lancashire, in conjunction with his air observations on and around the adjacent Pendle Hill. Hooke and Wren realized somewhere around 1662 or 1663 that weather changes could be so many and varied in the course of a day that a noon or other single reading of the instruments could not yield sufficient data upon which to establish a true science of the weather. This led Hooke and Wren to design a 'weather clock', which, in its conception at least, was to become the first self-recording meteorological instrument, and perhaps the ancestor of all self-recording scientific instruments. Various references to the weather clock occur over the years, and in December 1678 Hooke was optimistically telling the Royal Society that this self-recording instrument — capable, he was claiming, of automatically recording no less than *eight* different aspects of the weather — would be ready in four to six weeks' time.[46] In the final form of the design, as posthumously communicated by William Derham from descriptions in Hooke's papers, an eight-day pendulum clock mechanism advanced a paper-covered cylinder and a recording plate, upon which punch marks traced the various pressures and temperatures at 15 minute intervals.[47] The instrument was not a practical success, but its design indicated enormous foresight, and recognized not only the need for taking weather records at very short intervals if its complex changes were ever to be scientifically understood, but also that tedious, routine observations were done better by a machine than by a man.

In his meteorological interests, as in all his other scientific investigations, Hooke tried to isolate by means of experiment the 'mechanical' principles which he believed underlie nature. If these principles could be grouped, and instruments devised to measure and monitor their changes, then Hooke hoped that the apparent capriciousness of the weather, no less than that bending of light in air which seemed to produce colours, could be reduced to laws no less regular and predictable than those of astronomy. It was an extraordinary vision of the future possibility of scientific understanding.

Hooke's Diary records various forms of meteorological phenomena, including a freak hailstorm on 18 May 1680. The 'Hail fell from the size of pistol-bullets to the bigness of Pullet eggs', and when cut open, the stones were found to have a hard core, like a pea, covered with ice in stratified layers, which he suggested were made up of water picked up as they fell.[48] Yet there again, in *Micrographia* Observation 14, he had devoted several pages and plates to the examination of snowflakes and frozen water, in an attempt to understand the geometrical structure of frozen water crystals. Here, indeed, one sees the imaginative connection that must have existed between Hooke's experimental researches and the Gresham Geometry Professorship which he held, for in the study of crystals one finds terrestrial nature's own geometry no less exquisite in its perfection than the cosmological geometry of the stars in their courses. While not directly discussed in *Micrographia*, Hooke had struggled over the years to understand that meteorological phenomenon later to be named 'dewpoint', or those circumstances of temperature and barometric pressure which make the air deposit its ambient moisture as dew or mist. The problem was brought clearly home to Hooke in the peculiar circumstances reported by the young Edmond Halley on the island of St Helena in 1677, and related in detail by Hooke to the Royal Society the following year. It seemed that when Halley ascended the mountain on St Helena, with the intention of making astronomical observations from a high elevation (and becoming the first astronomer to do so), he found that even under a clear sky the air contained such an 'abundance of mists and moisture [that the damp] unglued the Tubes' of the wood and vellum telescopes that he was using. Thin films of water also formed on the lenses which ruined optical definition.[49]

In consequence of all of the above studies and lines of experimental investigation, *Micrographia* was much more than a book of microscopical observations. It was, rather, a fount of inspiration that would send natural inquiry off along many different paths of research into over half a dozen sciences. *Micrographia* encapsulated Hooke's entire philosophy of science, and with it the rationale behind much of what the Royal Society was trying to achieve, in its stress on the mechanical principles which underlie nature, and how they could be measured, comprehended and extended to benefit mankind by instruments and controlled experiments. Indeed, it is hard to imagine what would have happened to the Royal Society's visionary schemes for science at this period without Robert Hooke's driving energy, practical ingenuity and sheer genius to give them form and substance.

Yet one thing which is conspicuous by its absence in *Micrographia* is any evidence of a particular interest in natural history itself on Hooke's part. In spite of his numerous observations and careful drawings of things from the vegetable and insect kingdoms as seen under the microscope, Robert Hooke appears to have had relatively little interest in living things *per se*. The habits, habitats and diversity of living fauna never seem to have stirred his imagination, as they did that of many of his scientific contemporaries: only the physical structures of things, be they the highly-magnified cells in plants or the vivisected respiratory tracts of dogs, seem to have appealed to him. Neither his Diary nor his numerous other writings, moreover, suggest that Hooke was especially entranced by the sheer beauty of a world teeming with life. He does not appear to have been particularly drawn to animals, nor (as was common amongst many bachelor academics) to have kept pets, though it does seem that by 1693 there was a feline around his Gresham rooms, and that its disappearance warranted the Diary entry 'Lost cat'.[50] Yet while Hooke may not have been an 'animal man', the revulsion which he felt when faced with the scientific necessity of having to vivisect a second living dog in 1667, about which he complained to the Royal Society, clearly indicates that he abominated unnecessary cruelty.[51] It was nature the great machine, rather than the realms of beasts and plants, which most decisively triggered Robert Hooke's own sense of delight, for at heart he was a physicist.

Just as one of the hallmarks of an outstanding scientific discovery in our own time—from black holes to DNA—is its influence upon the popular mind and culture, so Hooke's *Micrographia* captured the creative imagination of the Restoration, and nowhere more embarrassingly for him than, as we saw at the beginning of this chapter, in Thomas Shadwell's box-office success, *The Virtuoso*, for Shadwell's play used the recently-published discoveries and activities of the Royal Society to provide part of the plot and a main ingredient of the comedy in this farce of duplicity, seduction and experimental philosophy. The butt of most of the jokes was Sir Nicholas Gimcrack, a foolish amateur scientist, or *Virtuoso*, who wasted his energies and fortune on seemingly absurd enterprises: Sir Nicholas 'spent two thousand pounds on microscopes to find out the nature of eels in vinegar, mites in cheese, and the blue of plums which he has subtly found out to be living creatures'.[52] One might indeed be so bold as to suggest that the unintended combination of Hooke's genius and Shadwell's wit had created the first specimen of an enduringly popular fictional character: the deranged scientist.

Micrographia, therefore, was far more than a collection of careful observations made, as the title said, 'by the aid of magnifying glasses'. It was one of the first fruits of the new science to strike deep into the non-scientific imagination, and show how cardboard tubes containing paired lenses, especially when conjoined with mercury-filled glass tubes and other pieces of apparatus, could provide a key to a vast new realm of knowledge. When this realm was communicated through the medium of clear English and beautiful engravings, it could keep senior civil servants like Pepys from their beds till two o'clock in the morning, and supply material for popular plays. Second only to Newton's *Principia*, which was published 22 years later and was a very different type of book, *Micrographia* was the most formative work of the age, and assured the 30-year-old Robert Hooke's reputation as a scientist of genius.

CHAPTER 5

HOOKE AND THE ASTRONOMERS

Of all the individual branches of science to which Robert Hooke made significant contributions, astronomy was the most extensive. Astronomical matters concerned him, in one way or another, for well over 40 years, as he dealt with the subject theoretically, observationally, and from the viewpoint of instrumentation. Of the 21 papers that he contributed to *Philosophical Transactions*, over a dozen deal with astronomy. Three separate sets of astronomical observations appear in *Micrographia*, and during his frenetic decade of architectural activity, the 1670s, he produced his most significant astronomical publications, while his last recorded scientific investigation, in 1702, was an attempt to measure the solar diameter more accurately.

Much of this concern stemmed from the fact that, in the seventeenth century, astronomy was by far the most advanced of the sciences. Not only did it stand upon observational and theoretical premises that ran deep into classical antiquity, and had been further perfected by both Christian and Islamic scholars in the Middle Ages, but the debate about the motion of the earth around the sun, initiated by Nicholas Copernicus in 1543, gave cosmology a new urgency during the European Renaissance. Yet astronomy's status as the most advanced of the classical sciences derived from other considerations as well.[1]

Grounded as it was upon the premise that the planets moved within a 'closed' circle of 360°, astronomy always possessed exact coordinates against which to make measurements and test theories. It was intimately linked thereby to the intellectual purity of Greek geometry, and in its use of bronze rings or quadrants that were accurately divided into 360° or 90° respectively with which to measure the changing nightly positions of the planets against the stars, astronomy was the oldest, and by 1600 the most instrumentally sophisticated, of all the sciences. By contrast, as we shall see in chapter 6, the life sciences such as medicine, natural history, and botany, lacking as they did any fundamental standard of measurement or natural constants, seemed to be stuck in cycles of endless subjective comparison. Yet somehow, the colours, habitats, virulences and empirical rules by which one might hope to classify a species or a disease type always displayed the infuriating tendency of never being *exactly* the same in two consecutive instances: no two individual cats, oak trees or cases of fever were exactly the same, whereas the motions of Jupiter or the Sun, by contrast, could be accurately calculated for decades into the future.

If astronomy was already the most mature of the sciences in 1600, its developments, both technically and conceptually, were to race along at an incredible rate during the seventeenth century, and one can fully understand why Hooke, and physically-minded scientists like him, paid so much attention to it. Astronomy was the model *par excellence* for the new sciences that were developing, in so far as it dealt with the philosophical problems of infinity and mankind's place in Creation on the one hand, while on the other showing that ingenious new devices, such as improved telescope technology, held the key to an unlimited store of natural wonders yet awaiting discovery. Indeed, the key to many of these seventeenth-century discoveries was the telescope, which Galileo had used to great effect in his *Siderius Nuncius* and subsequent works after 1610, and which Dr John Wilkins had done much to popularize in England.

While Galileo's telescopic discoveries — such as Jupiter's moons, the phases of Venus, sunspots, and the dense star fields of the Milky Way[2] — could not in themselves provide a solid proof for the earth's rotation around the sun, they certainly undermined the credibility of the Ptolemaic geocentric cosmology, for the new telescopic discoveries seemed to be at odds with classical geocentricism. Pre-telescopic astronomy, for instance, had had no real awareness that the planets might be spherical worlds that were in some respects similar to the earth. Nor was it aware of detailed topographical features on those worlds, such as the complex surface features of the moon or Mars, the belts and spots of Jupiter, the rings of Saturn, or solar-system satellites that did not rotate around the earth. And when figures like Wilkins were even willing to consider flight to those worlds or, as Christiaan Huygens would later suggest, contemplate the technological sophistication of their inhabitants,[3] one sees the way in which new astronomical discoveries, made with telescopes, unleashed numerous imaginative possibilities for philosophers, theologians and ingenious lateral thinkers.

Perhaps one of the most thought-provoking possibilities suggested by the telescope concerned the very nature of the universe itself, for it was hard to retain any belief in the classical universe of concentric spheres, in which all of the stars rotated at the same distance around the earth on the 'Eighth Sphere', when increasingly powerful telescopes revealed ever more shoals of dim stars, invisible even through less powerful instruments, let alone to the naked eye, which suggested that the universe extended in all directions. Now this idea of an infinite universe was not new to western thought, for around 1340 Thomas

Bradwardine, Archbishop of Canterbury and Oxford philosopher, had suggested that there was no reason why an infinitely powerful God could not, as a theoretical possibility, create an infinite universe if He chose; but what one had by 1640 was physical evidence, as well as theological corroboration, for a seemingly infinite universe.[4]

While it is true that Hooke does not discuss the philosophical problems inherent within the Copernican, heliocentric universe, nor does he discuss infinity (indeed, he seems to take the facts of both geocentricism and cosmological vastness as read), he must have been intimately familiar with them as current ideas. While Dr Wilkins had been cautious in his *Discovery of a New World in the Moon* (1638) about speculating upon the characteristics of the 'Selenites' or moon-men (of whose very existence he knew he had not a shred of physical evidence), he was nonetheless fascinated by the spiritual state of any notional beings living on other worlds. Did they also share the human race's fallen state of Original Sin, and hence, like us, needed Redemption, or were they in an uncontaminated state of Grace?[5]

By the time that Robert Hooke was doing creative astronomical research, in the 1660s, the instrumental capacity of the science had increased greatly even since the days of Galileo half a century before. One of the most significant innovations which had made this possible was the rapid development of practical optics—mainly in Holland and England, and in Italy after *c.* 1650—which had facilitated the production of telescope object glasses of greater aperture, clarity and power, the significance of which, as we shall see, was quickly grasped by Hooke.[6] These lenses, of three, six or seven inches in diameter and anything from 30 to over a 100 feet in focal length, enabled Christiaan and Constantijn Huygens, Johannes Hevelius, Giovanni Domenico Cassini and Robert Hooke to make the second great wave of telescopic discoveries after those of Galileo: to determine the axial rotation periods of Mars and Jupiter from their newly-mapped surface features, produce the first high-definition maps of the moon, discover the rings and satellites of Saturn, and to make the first studies of nebulae and star clusters and the first accurate delineations of cometary nuclei and tail structures.

The writings of the period also make it clear that, by 1670, astronomers were aware of living in a different conceptual universe from that of their intellectual grandfathers. Not only had the telescope opened up, and continued to reveal, new realms of observational knowledge, but the rapid improvement that had taken place in precision angle-measuring instruments since the time of Tycho Brahe (who died in 1601) meant that the celestial geometry upon which analyses of planetary motion were built had likewise improved by leaps and bounds. The three laws of planetary motion announced by Johannes Kepler between 1609 and 1618 had, furthermore, created a whole new language and intellectual

framework within which subsequent astronomers would come to understand relative speed and motion in space — not least Robert Hooke and Sir
Isaac Newton in their respective and bitterly-contested contributions
towards gravitation theory.[7]

And when one looks at the writings of Hooke and his astronomical
contemporaries, such as John Flamsteed, John Wallis, Sir Jonas Moore,
Sir Isaac Newton, Thomas Streete and others, it is clear that they were
aware of a particularly English astronomical tradition that, in addition
to popularizing the ideas of Tycho, Galileo and Kepler as John Wilkins
had done, had first absorbed the great continental discoveries and then
gone on to conduct new original researches which had further amplified
them. In particular, they were aware of a group of private astronomical
researchers that had been active in south Lancashire and Yorkshire
between 1635 and 1650 and had included in their number Jeremiah
Horrocks, William Crabtree, William Gascoigne and Jeremy Shakerley.[8]
Between them, these men discovered that the lunar and Venusian orbits
(in addition to that of Mars) were ellipses after the manner of Kepler,
observed several new classes of astronomical phenomena, and
invented — in the case of Gascoigne — two new instruments, the telescopic
sight and filar or eyepiece micrometer, which vastly enhanced the astronomers' capacity to measure very small angles and thereby apply exacting
observational tests to theoretical predictions. If one wants confirmation of
the true international significance of these men's contributions, especially
those of the three north-country astronomers Horrocks, Crabtree and
Gascoigne, then one has only to remember that Johannes Hevelius of
Dantzig published part of the deceased Horrocks's observations in
1662, that John Wallis, almost certainly assisted by Hooke, edited
Horrocks's correspondence with Crabtree for publication by the Royal
Society in 1672, that an account of the deceased Gascoigne's micrometer
was published by Richard Townley (or Towneley) with a detailed
drawing by Hooke in *Philosophical Transactions* in 1667, and that Flamsteed
openly admitted that the work of Horrocks and his north-country colleagues had laid much of the foundation upon which his own work as
Astronomer Royal would stand.

Robert Hooke, therefore, approached astronomy not only with a
clear intellectual agenda behind him as far as the Tychonic and Galilean
European inheritance was concerned, but also with a sense of his
English inheritance, as he firmly reminded Cassini in the 1690s when
the naturalized Frenchman, who had been head-hunted from Bologna
for Louis XIV's Académie, allegedly failed to give sufficient credit to
the English in his recently-published history of astronomy.[9] While
Hooke had been too young to have known Horrocks, Crabtree, Gascoigne
and Shakerley personally, he included amongst his Royal Society friends
and colleagues men who had. These included John Wallis, who had been

an undergraduate contemporary of Horrocks at Emmanuel College, Cambridge, and who was to hold the Oxford Savilian Professorship of Astronomy from 1648 to 1703.[10] Hooke also enjoyed the friendship of Sir Jonas Moore FRS, now Master of the King's Ordnance in the Tower of London who, as a young man in Lancashire in the 1640s, had rescued part of the recently-deceased Crabtree's and Horrocks's unpublished scientific papers from destruction.[11] Along with Flamsteed, Hooke knew Richard Townley, the Roman Catholic country gentleman of Burnley, Lancashire, whose father Charles had died alongside Gascoigne in the Civil War battle of Marston Moor in 1644, and whose uncle Christopher Townley, who had surived the battle, had rescued more papers and instruments of the northern group, including the above-mentioned micrometer described by Townley and illustrated by Hooke for the Royal Society in 1667. Consequently, Hooke had an awareness both of being within the 'new astronomy' of the European Renaissance, and of continuing a vigorous and original English astronomical tradition. This new astronomy was unequivocally heliocentric, recognizing the universe to be immensely vast, and the sun to be only one of countless other shining bodies and, perhaps, centres of rotation. What is more, this new astronomy, while having a respectable ancestry extending back to the ancients, nonetheless stood on a series of very recent discoveries, made with the telescope and modern measuring instruments. This new astronomy, like the rest of the emerging sciences, saw primary observational data, obtained with optical and mechanical instruments, as the foundation of all of its aspirations and attempts to explain the natural world.

Within the context of these new attitudes, therefore, one can understand why Hooke was such an assiduous collector of data relating to the natural world, and why he took particular pleasure in using the telescope, as well as the microscope, to add to it. In particular, the surfaces of planetary bodies were of great interest to Hooke, especially during the 1660s, when he was using various long telescopes (including those of 12, 36 and 60 feet in focal length) to observe them. Robert Hooke, along with Hevelius, Cassini and Huygens, was among the first astronomers to carefully observe the surface of Jupiter at a relatively high magnification. In May 1664, for example,[12] Hooke reported a small round spot on the biggest Jovian belt, which he believed, unlike a satellite shadow, to be a permanent feature. It moved over two hours, and while Hooke later claimed to have used it to measure the planet's period of axial rotation, it was Cassini, in fact, who first published a value for Jupiter's rotation. In June 1666, however, Hooke reported another permanent spot, and differentiated its appearance from that of a satellite shadow.[13] What is immediately clear is the sheer speed with which astronomical discoveries were being made and verified across Europe, especially as the *Philosophical Transactions*, the Parisian *Journal de Scavans* and other periodicals were placing the discoveries into the international public arena as fast as printing presses and horses would allow. In January 1673, for instance, Hooke noted: 'New planet observed about ♄ [Saturn] Planet',[14] while two months later he reported to the Royal Society 'Cassini Discovery of two new comets about ♄ read and presented but not his letter'.[15] The 'Planet' and 'Comets' about Saturn were satellites, subsequently named Iapetus and Rhea, discovered by Cassini in 1671 and 1672 respectively.[16]

Saturn, indeed, had been the most puzzling planet in the solar system ever since Galileo's early telescopic discoveries revealed that it had what were initially termed *ansae*, 'handles', or what Christiaan Huygens in 1659 would announce to be a ring, along with his discovery of Saturn's first satellite, Titan. Sir Christopher Wren and others in the Oxford group had studied Saturn, and Hooke had made his own exceptionally fine observations of the ring with a 60-foot telescope, and what he believed could be its shadow falling upon the planet in 1666, while the scientific brothers William and Dr Peter Ball, using a 38-foot refractor near their house at Mainhead, near Exeter, in Devonshire, concluded that Saturn had a *pair* of somewhat asymmetric rings. But it would be Cassini, now working in Paris, who would finally resolve the nature of Saturn's ring in 1675, when his superior telescopes enabled him to discover that the planet had at least two concentric rings with an open division between them; and posterity would come to name that division after Cassini.[17]

In addition to what one might call sustained planetary researches, Hooke's Diary contains numerous records of astronomical observations, some of which would not have been of especially great importance scientifically, but which no doubt struck him as being singularly beautiful, such as when on 10 April 1673: 'Venus seen with the ☾ [Moon] at [St?] Paules'.[18] And on 19 October 1678, he 'Observed with [John] Aubery the Moon Eclipsed', which happened to be a fine total eclipse.[19]

Observation 60 in *Micrographia*[20] consists of an examination of a group of lunar craters made in October 1664 with a 30-foot (more likely 36-foot) focal length telescope probably built by Richard Reeves, which seemed capable of working with its full and unstopped aperture of $3\frac{1}{2}$ inches.[21] In addition to the very considerable amount of detail that Hooke includes in his survey of that part of the lunar surface which Riccoli had named Hipparchus, there follows a discourse on lunar geology, in connection with which Hooke conducted a series of experiments in which he firstly dropped lead bullets into a 'well temper'd mixture of Tobacco pipeclay and water', to produce crater-like impact holes, then secondly blew air bubbles through a pot full of molten alabaster, the bursting of which also resulted in crater-like depressions. The pistol-bullet impact craters in the pipe clay looked remarkably like the 'pits' or craters on the lunar surface, but in that age, which still knew nothing about the existence of meteoritic debris, Hooke was cautious of taking this explanation too far, 'for it would be difficult to imagine [from] where these bodies should come'. Instead, he was more inclined to consider a 'Moonquake' or geological explanation for the lunar craters, such as those which he could simulate by the bubbles bursting on the surface of boiling alabaster.

These experiments, I would argue, make Hooke the founding father of the science of lunar geology and, in his consideration of both meteoritic and volcanic sources as the causes of the moon's 'pits', he established two ways of looking at the moon that would crystallize into two separate scientific 'schools' in the twentieth century — and by the twentieth century, the discovery of abundant asteroidal and meteoritic detritus within the solar system would provide the projectiles which could have bombarded the moon, to answer Hooke's question 'where those bodies should come' from.[22]

Hooke's lunar modelling was also significant for another reason, insofar as it assumed that the moon had once been different from what it is now. No new craters had appeared on the lunar surface in the 20 years or so since Hevelius and Riccioli had made their lunar maps, nor had there been any obvious 'Moonquakes' or great volcanic eruptions seen by astronomers on its surface. While the modern lunar surface, with its craters, mountains, and 'seas', was clearly a solid rocky place, how far back must one go to find a lunar surface that was sufficiently viscous to respond like pipe clay or molten alabaster when hit by a projectile, or have sub-lunarian 'airs' released through it, in a form of volcanic activity? Hooke's moon, indeed, is a very ancient place, formed by gravity, heat and possible impacts, which suggests that even as early as 1664 he was at least toying with the idea of large, and possibly pre-Adamite, time-scales about which he would display such ambivalence in his subsequent terrestrial 'Earthquake Discourses' delivered to the Royal Society, and which will be discussed when looking at Hooke's geological ideas.

Hooke's astronomical observations contain many references to the characteristics of the telecopes with which he made them, as one might expect from a man who possessed such a thorough-going instrumental and sensory approach to research. But it is in his observations of the Pleiades star cluster, also recorded in *Micrographia*,[23] rather than in any of his explicitly astronomical writings, that he initiated discussion about what modern astronomers refer to as the 'resolving power' of the telescope. Though Hooke was aware of the higher magnifications obtained with object-glasses of longer focal length, and that with a telescope of 12-foot focal length he could see 78 stars in the Pleiades, whereas Galileo with his much feebler instrument had only been able to see 36, and that with a 36-foot focal length lens of correspondingly larger diameter than that of his 12-foot telescope, he could see yet more, he seems to have recognized the crucial principle that it was object-glass diameter that was of primary importance in seeing faint objects. He experimented with a series of object-glass stops, and noticed that he saw the maximum number of stars through an entirely unstopped lens. Surprisingly, however, in this particular discussion Hooke rarely gives his object-glass apertures in inches, though in another place he does commend the excellence of Richard Reeve's (or Reive's) object glasses, in that Reeves could impart excellent *spherical* curves upon three-inch-diameter lenses of what seem (rather ambiguously in Hooke's sentence) to have been of 30-foot focal length 'whereas there are very few thirty foot Glasses that will indure an [unstopped] Aperture of more than two inches'.[24] But the reason why astronomers stopped down their object glasses in the first place, by placing brass or cardboard rings over their full apertures, was to cut out the peripheral light from their often imperfect spherically-ground lenses, and improve the sharpness of the resulting images, though at the price of making the images dimmer. Hooke was all too aware of this, for as he also mentioned in *Micrographia*,[25] the insertion of lens stops deprived the astronomer of dim stars in very much the same way as did air turbulence.

One problem faced by the historian of astronomy when dealing with historical observations, such as those made by Hooke, is knowing in precise terms how good a particular telescope actually was, so that we can separate what it was optically possible for Hooke to see from what he *thought* he saw. Of course, this problem is also implicit within the interpretation of microscopical observations, for all the 'artificial Organs' of optical instrumentation inevitably introduce questions of *perception* and even *imagination* as well as objective *observation* into the business of putting Nature to the Torture. We can reconstruct the visual parameters of Hooke's microscopical work by comparing his drawings with modern photographs of flies' wings, cells and such; and likewise, his drawing of the lunar crater Hipparchus enables us to reconstruct at least one of his telescopic observations: the Hipparchus complex of craters, depicted in *Micrographia* Observation 60, subtends an angle of about 90 arc seconds to a terrestrial observer, and when we examine this area of the moon's surface with modern telescopes, or compare Hooke's drawing alongside a same-scale photograph of the same region, we cannot help but be amazed by the quality and detail of his work. His 36-foot object glass, for instance, was capable of resolving details that subtended between 5 and 10 arc seconds across on the starkly monochromatic lunar surface, which tells us much about Hooke's skill as an astronomical observer and draughtsman, as well as about the resolving power of Richard Reeves's 36-foot-focus lens.

Hooke's Diary also gives some insight into the costs of these three- or four-inch-diameter lenses of long focal length, for on 2 January 1676, Christopher Cocks the optician who had effectively inherited Reeves's mantle was demanding £40 for a 60-foot-focus glass.[26] What is not clear, however, is whether this sum was for the lens only, or for the telescope fittings as well. If it was just for the glass, it brings home to us what expensive pieces of capital equipment such telescopes were, for Hooke's annual salary as Gresham Professor was only £50. In spite of these high prices, Hooke was a warm defender of English opticians, both as philosophical designers and as bench craftsmen, especially when, in the 1690s, Cassini was claiming priority of innovation for French and Italian telescope makers. As Hooke snapped, 'The Improvment of Telescopes, both for Length and Goodness,... was first performed here by Sir Paul Neile, Sir Christopher Wren and Dr Goddard, who instructed and employed *Mr Reives* [sic] in the manual Operation; and by that Means it was carried to the Perfection of making Object-Glasses of 60 or 70 foot long, very good, before any Mention was made of such being made in France... [or] indeed, had been made in Italy by *Divini* or *Campani*.' Hooke also claimed to have had a telescope by Richard Reeves which, length for length, 'was full as good, if not better' than the Italian glasses.[27] And while his claims for English priority in virtually all long-telescope innovations sound over-sweeping, to say the least, one should not forget that Hooke had been in the thick of instrument innovation since his Oxford apprenticeship during the 1650s, and not only had direct experience of both English and continental instruments, but also a close personal acquaintance with at least the English innovators and craftsmen whom he mentions.

The precise ownership of some of the telescopes to which Hooke had ready access is not always clear. We know, however, that back into the mid 1650s, Sir Paul Neile FRS had been patronizing English glass grinders such as Reeves, and in addition to Neile, Wren and Goddard, Robert Hooke owned a telescope, while Wilkins had one of 80 feet focal length while at Wadham.[28] One also tends to find different instruments of similar focal lengths in the possession of different individuals: for instance, Wren, Neile and Pepys each owned 12-foot instruments, while Hooke in May 1664 was also observing Jupiter with an instrument of the same specification.[29] These all seem to have been Reeves instruments, moreover, and no doubt were the products of the same set of grinding and polishing tools which were made to impart identical curves, and hence identical focal lengths, to different pieces of glass.

Hooke was not just a connoisseur of telescopic object glasses as a user, moreover, but had, by the time that *Micrographia* was published in January 1665, a wealth of experience pertaining to their practical manufacture as well, Within the 'Preface' to *Micrographia* he had left a detailed description, accompanied by an engraved plate, of a novel lens-grinding machine of his own invention, and when Adrien Auzout in Paris made criticisms of the device in the pages of *Philosophical Transactions*, Hooke had been quick to defend himself, displaying a detailed knowledge of practical optics — a very early example of technological innovation becoming the subject of correspondence in a major international journal.[30] Indeed, Hooke's fascination with the practical business of producing good astronomical optics was one of the consistent themes of his life. The third entry in his Diary, made on 3 August 1672, contains a cryptic reference to seeing Christopher 'Cox [or Cocks] turning [an] Object Glass with flints' as an abrasive.[31] Even while we are not aware of any good-quality object glasses that Hooke made with his own hands, his Diary records many attempts to produce speculum mirrors for reflecting telescopes. (Hooke's personal familiarity with lens and mirror grinding will be discussed more fully in chapter 9.)

Yet why, one might wonder, were the astronomers and opticians of the mid-seventeenth century so concerned with producing telescopes of 12, 36, 60 or more feet focal length? The reason lies in the physical size of the 'prime-focus' image produced by the lens, for the longer the focal length, the larger the prime focus will be. When Hooke was observing Jupiter in June 1666 with a telescope of 60 feet focal length, he claimed that the image of Jupiter as seen in the eyepiece was four times the size of the full moon as seen with the naked eye, or about 2 degrees of arc. At this time, in June 1666, Jupiter subtended an angle of some 41.48 arc seconds to a terrestrial observer, so that at 60 feet its prime-focus image would have been some 3.67 mm across. When Hooke looked at that image through an eyepiece of 4.15 inches, he would have obtained a magnification of around 173 times.[32]

But if the prime-focus image of an astronomical body is 3.67 mm across at 60 feet, it will only be about 2 mm in a 36-foot lens, and correspondingly smaller in a 12-foot lens. It becomes clear, therefore, that the greater the focal length of the telescope's object lens, the greater the magnification it will give, and hence the more planetary detail one should be able to see through it. And as the surface details of planets and the discovery of further non-terrestrial satellites formed such a central agenda for seventeenth-century astronomers, one can understand why the production of increasingly long-focus telescopes was so essential to their whole programme of research.

During the winter of 1664–5, the skies of the northern hemisphere were dominated by a brilliant comet, which was the most conspicuous since that of 1618. Indeed, Robert Hooke said as much in his *second* lecture on the comet which he delivered to the Royal Society on 1 March 1665, in which he suggested, according to Samuel Pepys, 'that this is the very same Comet that appeared before in the year 1618, and that in such a time [i.e. in another 47 years] probably it will appear again, which is a very new opinion'.[33]

Hooke observed the comet of 1664–5 with his long telescope from London, and (along with Johannes Hevelius working independently in Dantzig) conducted one of the first detailed telescopic investigations into cometary nuclei and tails, though Hooke and Hevelius developed different ideas of what comets were, and how they generated light. It was not until a second brilliant comet appeared in 1677, however, that Hooke published his cometary researches. From detailed studies of the two comets, Hooke concluded that their nuclei were solid bodies. A combination of the comets' movement through the aether and internal agitations within caused the nuclei to be gradually eroded away, to form tails and streamers across space. He also concluded that comets did not merely reflect sunlight, but generated light of their own, for there was never any sign of shadow in a cometary nucleus, even in those parts that were not facing the sun.[34]

The causes of cometary attrition and light-generation puzzled Hooke, and while he saw the aether as playing a role in this process, it was not so straightforward as that played by the 'nitrous' part of the air in the burning of a candle flame. In a cometary nucleus, observed Hooke, the brightest part was at the very centre, whereas in a candle or lamp flame (and here he drew attention to his work in *Lampas*), the centre was always dark, with the greatest light around it.[35] Nonetheless, he came to see the glow of comets, just like all other light-emitting bodies, as produced by the dissolution of their solid nuclei by a *menstruum* (the aether), which produced internal motions, and hence light, within the nucleus itself. Even so, one senses that in spite of the elegance and supposed universality of his dissolution of agitated bodies theory of light, Hooke was always aware of problems in his explanation of how comets emitted light.

Being the ingenious experimentalist that he was, Hooke attempted to model the structures of comets in the laboratory. He suspended a wax ball, which he had first encrusted with fresh iron filings, into a tall glass cylinder containing sulphuric acid. As the 'menstruum' of the acid attacked the iron filings all around the wax ball, he was delighted to see that the ensuing bubble streams formed themselves into a *medulla*, or stem, a series of elegant streamers, and a tail as they rose upwards. Could this be replicating the way in which some kind of solar *levitating* or outwards pressure dissolved the earthy bodies of approaching comets?

When Hooke made his first observations of the comet of 1664, he lacked an effective micrometer whereby he could measure the angular diameter of the nucleus. His genius for improvization was brilliantly displayed, however, when he watched the comet low in the sky, and compared the diameter of its nucleus with the apparent width of an ornamental iron rod supporting a weather-cock on the roof of a distant building behind which the comet passed.[36] By later measuring the iron rod, and its distance from his telescope, he was able to calculate that the comet's nucleus was about 25 arc seconds and its coma $4\frac{1}{2}$ minutes in diameter. For while Hooke had spoken of the possibility of inserting a fine '*rete*' or calibrated network of lines in the eyepiece of a telescope for making angular measurements in *Micrographia*,[37] inspired perhaps by the experiments of Huygens, it does not seem to have been until the description of Richard Townley's micrometer in 1667, based on Gascoigne's original of 1640, that an exact and versatile method of making angular measurements through a telescope became generally available.

All of this observational work on solar system bodies had been made possible by the improvement of telescopes. It must not be forgotten, however, that contemporary innovations in horology also came to play a major role in astronomy, for if the earth rotated upon its axis in an exact and even period of time — as measured against two successive meridian passages of a celestial body — then an accurate clock could be used to measure Right Ascension, or East to West, angles of the sun, or a star by converting angles to periods of time. The sun, after all, goes around the 360° of the sky in 24 hours: therefore, in one hour it will go through 15°, in four minutes, through 1°, and so on, through arithmetic subdivisions, down to seconds of arc. Earlier astronomers had hoped to be able to make practical use of this equivalence of time and angle, including Tycho Brahe, who, when living in Prague around 1600, had commissioned Jobst Burgi to build him a complex astronomical clock. The ultimate drawback with all of these efforts, however, was the mechanical unreliability of contemporary clockwork, lacking as it then did any effective natural regulator that ensured that the clock was perfectly even in its running.[38]

By the 1660s, however, astronomers and physicists were actively accumulating observations and experiments performed with instruments of new design, and seeing how the results squared with theory, and some of these experiments were concerned with the swing of the pendulum. It was Galileo Galilei, the Italian physicist and astronomer, who had first noted in the late sixteenth century that the swings of a pendulum were equal, or isochronous, irrespective of whether a pendulum of given length was made to swing through a large or a small angle. It was also noted that the longer the pendulum, the slower was its swing, when left to vibrate freely. This was to prove invaluable on two counts. Firstly, it provided a demonstration of the natual relationship that existed between the attractive power of the earth and a moving or falling object such as a pendulum bob. Secondly, the unchanging rate of swing for a pendulum of given length made it the perfect governor, or regulator, for a clock mechanism; and in 1658 Christiaan Huygens had published a description of the way in which the pendulum could be successfully harnessed to a gear train, and the pendulum swings made more dynamically stable by the fitting of a specially-shaped pair of control-cheeks that were formed into 'cycloidial' curves.[39]

Much of Hooke's early work in applied mechanics related to the pendulum and isochronal actions, and during the 1660s he claimed to have invented a much-improved escapement, whereby the pendulum, which swung through a much smaller and hence more dynamically stable angle than in Huygens's 1658 arrangement, could be made to regulate a clock with very great precision indeed. This was the 'anchor escapement', named from the shape of the rocking mechanism which released the gear teeth with each swing of the pendulum and received a corresponding impulse, and was to lead to a dispute over the priority of the invention of the device with William Clement, a working clock-maker who made a clock incorporating the escapement in 1671. It is likely that Hooke had worked out the mechanical principles before Clement, and in his usual manner had forgotten to secure priority before passing on to fresh work, for Hooke was involved in various horological experiments in the early 1660s, around the time of his invention of the balance spring for watches, and claimed, according to Derham, to have demonstrated an anchor escapement to the Royal Society 'soon after the Fire of *London*' of 1666.[40] Clement on his part most probably invented the device quite independently and perhaps being more commercially-minded than Hooke, recognized it as a profitable artefact, and put it into practice in his clocks straight away. No matter who invented the anchor escapement, however, what cannot be denied is that a clock with an 'anchor' and a dynamically stable long pendulum had been installed in Wadham College, Oxford, by 1670. College tradition says that this clock was a gift from Sir Christopher Wren to his old College,[41] and while no maker can be positively identified, College documents show that from 1671 onwards Joseph Knibb, master clockmaker of Oxford and London, was paid £1 per annum for its upkeep, and as it was usual for a maker to be responsible for the maintenance of his own work, it appears likely that Knibb made this very early anchor escapement.[42] And considering the warm friendship which existed between Hooke and Wren, and both men's association with the College—Wren as a former undergraduate and Fellow and Hooke as a friend of Dr Wilkins, the ex-Warden—it seems likely that Hooke's anchor escapement had been given physical expression by Knibb as early as 1670. This 1670 clock, which ran for 200 years, is now preserved in honourable retirement in the Museum of the History of Science, Oxford.

Quite apart from the exact priority of invention, therefore, suffice to say that by 1670 it was possible to make clocks of hitherto unprecedented accuracy keeping time to less than 20 seconds of time per day.[43] These clocks incorporated Hooke's anchor escapement in conjunction with the 'Royal Pendulum', which was a pendulum 39.25 inches long, which beat exact seconds in London.[44] The 'Royal Pendulum' (so-called because of its dedication to King Charles II) was in its own right an interesting example of the interrelated developments of experimental physics and technological applications at this period, when it was realized that the swing of the pendulum was not only isochronous in itself, but that the length to period of oscillation ratio was an unalterable fact of nature for any given place on the earth's surface. This discovery had been made by chance in 1672, when Jean Richer found that a clock which had been adjusted to beat exact seconds in the Paris Observatory ran slow by two minutes and 28.5 seconds per day when set up on the South American island of Cayenne, and was subsequently confirmed by Edmond Halley on the island of St Helena some five years later.[45] Pendulums of a given length, therefore, could only vibrate at a given speed in a particular place, and 39.25 inches was the length at which they swung exact seconds in London.

The refined pendulum clock and long pendulum were destined to come together in an instrument of the highest importance, and made possible many of the rapid advances in late seventeenth-century astronomy. By its means astronomers could attempt to confirm their assumption that the earth rotated homogeneously on its axis, as did John Flamsteed at Greenwich in the late 1670s as a preliminary to his work on the longitude, as well as use it in conjunction with other instruments to make a wide range of observations.

Indeed, in spite of the philosophically far-reaching discoveries on planetary surfaces made with long telescopes, much seventeenth-century astronomy was about measurement, as astronomers tried to find physical evidence for the earth's motion around the sun. But before unequivocal proof could be provided to show that the earth was moving in space, it was seen to be necessary to re-map the sky to a new standard of accuracy, so that star positions could be definitively established, and planetary movements monitored against them. This required the angular separations between all the stars to be accurately re-measured, using quadrants, sextants and other instruments equipped with improved graduated scales, in conjunction with the new pendulum clock. Traditionally these large instruments, of five or six feet radius (such as those used by Tycho Brahe), were sighted on to a star or planet using a pair of plain, naked-eye 'open sights', not dissimilar to the front and back sights of a rifle; but for some years inventors in Italy, France, Holland and England had, with varying degrees of success, been attempting to replace these open sights with telescopes. Though the optical problems of the telescopic sight had been largely solved by the Yorkshireman William

Gascoigne as early as 1640, when he discovered that by placing a pair of cross-hairs in the focal plane of a Keplerian telescope he had an exact optical point which could be aligned on a celestial object, his work had remained little known until after the founding of the Royal Society.[46] Hooke had certainly addressed himself to the problem before 1665, for in *Micrographia*,[47] as we saw above, he discussed the possible insertion of '*retes*', reticular markers, or nets of lines into the eyepieces as a way of making more accurate angular observations aimed at measuring planetary and other angular distances. Hooke had immediately grasped that the large, magnified image of a celestial body produced by the telescope would allow much more accurate measurements than those possible with naked-eye sights.

But very significantly, Hooke recognized that the human eye had a limited angular resolving power and, by means of experiments in which he measured the exact point at which a series of black and white squares painted on to a board seemed to merge into one, established that the normal eye cannot separate or resolve angles of less than one minute of arc.[48] (This corresponds to the angular diameter subtended by a ten-pence piece one inch in diameter at 286 feet.) Therefore, the only way in which astronomers could make measurements more accurate than one minute of arc was by replacing the open sights on their quadrants and sextants with telescopes, which greatly increased the eye's resolving power. These new telescopic sights would have cross-hair sights in their fields of view, thus allowing the astronomer to aim the instrument with greater precision at a star or planet. Hooke claimed that he had used such an instrument to make accurate measurements of the path of the comet of 1664, and he soon became Europe's chief advocate for the new telescopic sights.

Although Hooke and other astronomers in London, and Jean Picard, Adrien Auzout and a group of French astronomers in Paris were busy with the development of these sights, their reliability was strongly criticized by the elder statesman of European practical astronomy: the 62-year-old Johannes Hevelius of Dantzig, who enjoyed the honour of being a corresponding Fellow of the Royal Society. In his book *Machina Coelestis*, 1673, Hevelius pointed out that, because of certain optical problems, telescopic sights could never be wholly reliable. Hevelius argued that as the observer moved his eye to different parts of the stream of light emerging from the eyepiece lens when observing with a telescope, so the image appeared to move slightly, so that it was impossible to place any reliable point of measurement — such as cross-hairs — in the geometrical centre of the optical field. If one could intercept different parts of the exiting cone of light, how did one know when one had the exact centre? Hevelius backed up his criticisms with his own superlatively accurate astronomical observations, made with naked-eye sighted instruments in his Dantzig observatory.[49]

Though it was soon realized that Hevelius's criticisms were in fact based upon an incorrect though seriously formulated optical theory on his part, Robert Hooke was stung by the Danzig astronomer's critique of what he was coming to see as 'his' telescopic sights. Often capable of over-reacting when faced with criticism, Hooke responded to Hevelius with considerable annoyance. Indeed, in the international scientific controversy which blew up in consequence, one can see how quickly pride was hurt and tempers flared amongst the advocates of the 'Experimental Philosophy' when they felt their individual contributions being threatened.

Hooke sprang back to defend himself with a book entitled *Some Animadversions on the first part of Hevelius, his Machina Coelestis*, which was the published version of his Cutlerian Lecture for 1674, and drew on the 1668–69 research that preceded it. In many ways, *Animadversions* resembled *Micrographia* in so far as it was a book of brilliant inspirations and connections of ideas based upon physical observations. In it Hooke went far beyond defending himself against criticisms to produce a remarkable series of inventions in the technology of instrument design which he saw as crucial to the wider business of advancing science. Not only does he provide a full optical and mechanical justification for telescopic sights, but he also develops a new method of constructing astronomical quadrants using micrometers, mirror telescopes, spirit levels and other devices. Once more, Hooke supplied a starting-point from which many future inventions would spring.

One of the most original features in *Animadversions* was a proposed method for graduating the degree divisions on a quadrant. Quadrants were used by astronomers to measure vertical angles in the sky as a way of obtaining the meridional and other altitudes of bodies, and consisted of a metal quarter-circle, the edge, or limb, of which was engraved with 90° and their subdivisions, using a complex compass geometry based on the length of the quadrant radius. Hooke suggested, however, that much greater accuracy could be obtained if, instead of using normal engraved degrees that had been laid off with compasses, and were therefore susceptible to innumerable and accumulative errors of draughtsmanship, the actual *edge* of the quadrant was divided by a precision mechanism. In one such configuration which Hooke proposed a carefully-turned metal roller of a given diameter was attached by a radial arm to the quadrant's centre, so that when the arm was moved, the roller would be made to move upon the quadrant's limb, so that its turns could be made to lay off degrees and their parts. Hooke first thought of this idea some time before 1674,[50] though even in 1679 he records discussing the technique with Sir Leoline Jenkins.[51] But the mechanical method of division into which Hooke invested so much of his time and ingenuity did not use rollers to lay off the angles, but fine gear teeth and a precision micrometer screw. The teeth, which were to be incised into the metal edge of the quadrant, would then be engaged with an endless screw set tangentially against it, and fixed on to the quadrant's radial sighting arm (which, in turn, was equipped with a telescopic sight), so that by turning the screw one made the sight arm advance up and down the quadrant. By counting exactly the full and part turns necessary to sight a given star through the telescopic sight attached to the radial sighting arm, the astronomer could find the angle by simple calculation. If one knew in advance the exact number of teeth into which the quadrant edge was divided (1600 teeth being recommended for a quadrant of 3 feet radius), it was easy to work out from the graduated dial, which counted the full and part turns of the tangent screw, how many turns were needed to carry the screw through whatever angle one wished to read off on the quadrant. For instance, 17.88 full turns on the screw would advance it exactly one degree of arc upon the limb, and 0.296 of a turn, a single arc minute.[52] These small parts of a turn of the tangent screw, and their angular correspondences, were read off from a large brass circular dial plate mounted on the end of the screw. Hooke recommended any Fellows of the Royal Society who desired such a quadrant to 'imploy Mr Tompion, a watchmaker of Water Lane near Fleet Street; this person I recommend having imploy'd him to make that which I have, whereby he hath seen and experienced the difficulties that do occur therein'.[53]

This instrument, described in such detail in 1674, gives us a clear proof of the fact that Hooke really did make inventions, sit upon them for years, and then subsequently claim priority only when someone else had come up with the same thing. While the screw-edge or micrometer quadrant itself was never the subject of a priority controversy, on the first occasion upon which the Royal Society resumed its meetings after the Great Plague adjournment, 14 March 1666, a clear precursor of the 1674 instrument was demonstrated. For '*Mr Hooke* produc'd a very small Quadrant for observing accurately to Minutes and Seconds, it had an Arm moving on it by means of a Screw lying on the Limb of the Quadrant'.[54] Had Hevelius's criticisms of telescopic sights in 1673 suddenly spurred Hooke to take a passing ingenious device from seven or eight years before, and transform it into a new system whereby one could make accurate measurements through a telescope? The screw-quadrant, I believe, gives us an insight into the dynamics of Hooke's creative process.

Hooke's quadrant, both in its 1666 and 1674 incarnations, while original in theory, was in many respects impracticable for the craft techniques of his day, and when his novel methods of graduation were tried by astronomers, such as William Molyneux, who commissioned a Hooke screw quadrant from the craftsman Richard Whitehead in 1685 for £20, they were found to be unsatisfactory in practice.[55] One of the few men who used a Hooke instrument on a regular basis, however, was John Flamsteed, the first Astronomer Royal.

The Royal Observatory at Greenwich, of which Flamsteed was Director, had been founded in 1675 by King Charles II, and Hooke, along with Sir Christopher Wren, had been closely associated with its workings right from the start, for Wren, as King's Surveyor, and Hooke, as his friend and associate (and Surveyor to the City of London), had designed the buildings. In hindsight it seems fitting indeed that the Royal Observatory should have been designed by two architects who were also working astronomers, for Wren had been Savilian Professor of Astronomy at Oxford before his move to London, and had retained that astronomical post until 1673.

Flamsteed, the newly-appointed Astronomer Royal, was a man who appeared to share several personality traits with Hooke. Both men were highly strung, sensitive to criticism, genuinely sickly or else hypochondriacal, with a tendency to be quarrelsome. There was also a distinctly depressive trait in their mental architecture, as well as stubborn pride. But most of all what they had in common was an 'addiction' to the new science and a determination to advance it as far as they could. Both men, moreover, would develop a deep enmity with Sir Isaac Newton. And like Hooke in the early Royal Society, Flamsteed was another victim of shoestring financing, who found that when he took up his appointment as Astronomer Royal, not only had his Royal Warrant omitted to provide the necessary public money to furnish the empty Greenwich Observatory with astronomical instruments, but that his ostensibly handsome stipend of £100 per annum had in practice to be spent on things which no office-holder should ever have to pay for out of his own pocket, while the sum itself was not even punctually paid. Like Hooke, however, Flamsteed worked on regardless from a financial foundation that in his early Greenwich years remains obscure, and probably depended on assistance from his patron Sir Jonas Moore until he was appointed in 1684 to the Rectory of Burstow, Sussex and after the death of his father Stephen, a substantial Derby merchant, in 1688, came into his inheritance.[56] As he told the visiting German scholar Zachiarias Conrad von Uffenbach in 1710, he would have achieved little had he not been 'the son of a rich merchant'.[57]

One can see how, in some respects, Hooke saw Flamsteed's empty observatory as an excellent testing ground for his telescopic sights, screw-edged quadrants, and to a lesser extent his pendulum clocks. Sir Jonas Moore generously paid out of his own pocket to have a sextant and two long pendulum clocks by Thomas Tompion with ingenious escapements designed by Richard Townley made for the observatory, the sextant incorporating several of Hooke's novel inventions. Though Flamsteed never gave credit to Hooke for the design of the sextant, in the Astronomer Royal's posthumously published *Historia Coelestis Britannica* (1725), where he gave a detailed written account of his instruments, the Hooke design features are immediately obvious, while it is likely that the 3-foot-radius quadrant loaned to him from the Royal Society and ascribed to Hooke also had screw-edge graduations.[58]

The sextant of seven feet radius made for Flamsteed by Tompion in 1676 was fitted with a pair of telescopic sights, and its angles were read off not from conventional degrees but from a screw-turns arrangement identical to that originally described by Hooke in 1674. What is more, even the mounting of Flamsteed's instrument seems to derive from the Curator of Experiments, for in the *Animadversions* Hooke had included an illustration of a screw-edged, telescopic instrument mounted on a polar or equatorial axis, powered by a driving clock, so that it could track objects as they moved across the sky without requiring the astronomer to make the constant adjustments that were necessary in a non-clock-driven telescope to keep the object under observation within the field of view.[59] Flamsteed's sextant, likewise, was set in the equatorial plane, and as if to emphasize the Hooke association, was even made, like the two clocks, by Tompion the horologist, who was Hooke's preferred precision engineer whom he had recommended people to 'imploy'. Indeed, the only significant way in which Flamsteed's sextant differed from the equatorial quadrant depicted in *Animadversions* was in the design configuration of its telescopic sights, and in the absence of a clock drive mechanism: a mechanism for which Hooke was to claim — on very good grounds — priority of invention against a French claim later advanced by Cassini.[60]

Brilliant in design as it was, however, the screw graduation of the new sextant was not satisfactory: the teeth on the geared limb of the instrument wore rapidly, and it soon began to produce faulty readings. After little more than a year, the Astronomer Royal was forced to equip the instrument with conventional degree scales and use the Hooke screw just as a micrometer for reading degree fractions, though once this had been done it gave good service for the next 40 years.

Hooke also designed a large iron quadrant of ten feet radius for Flamsteed at the Royal Observatory, but it was a failure. It was fixed to a brick wall in the plane of the meridian, so that when Flamsteed looked through the sighting telescope, he could observe objects as they transited the meridian, and measure their vertical, or declination, angles. It was also hoped that he could measure the right ascension angles between successive celestial objects simply by timing their successive meridian transits with a pendulum clock regulated to sidereal, or star, time, and then converting their time differences to angles by means of a simple calculation. In practice, however, the frame of Hooke's ten-foot quadrant turned out to be too flimsy, and not only were its readings unreliable but it was also dangerous to the user. It 'almost deprived Cuthbert of his fingers',[61] complained Flamsteed, when a heavy part of the instrument slipped when being operated by the assistant, Cuthbert Denton. Fortunately, most of the early instruments for the Observatory, including the Hooke pieces, were included in the series of engraved views of the new Royal foundation, made by Robert Thacker and Francis Place between 1676 and 1679. They give us a very clear sense of what the Hooke instruments at Greenwich looked like, along with how they were mounted and used.[62]

One can easily understand how the relationship between Hooke and the new Astronomer Royal deteriorated between 1676 and 1678, as Flamsteed was forced to abandon Hooke's brilliantly conceived instruments, which failed simply because the materials and craftsmanship of the 1670s could not manufacture them to a sufficiently high standard of precision. Then Flamsteed began to complain of being 'much troubled with Mr. Hooke who, not being troubled with the use of any instrument, will need to force his ill-contrived devices on us'.[63] One senses here the not infrequent conflict of interest and frustration that can occur when an inventor's brilliant and inspired creations are found to contain practical defects once they are subjected to the heavy demands of routine research in the hands of a busy scientist.

Though Flamsteed was a convert to and remained a life-long advocate of telescopic sights, he soon abandoned Hooke's other inventions, so that on 26 September 1679 the slighted Curator of Experiments realized that it was time to dissociate himself from the Royal Observatory, especially as Flamsteed's patron, Sir Jonas Moore, had died a month before, and could no longer protect his protégé.[64]

Hooke, accompanied by his assistants Harry Hunt and Thomas Crawley, turned up at Greenwich, recorded Flamsteed, where 'Mr Hooke produced an order to remove the instruments of the Royal Society to Gresham College'. 'They took away: the small quadrant of 5 inches radius with the screwed limb, another quadrant with two telescopes on it, a dividing plate, an instrument of 3 rulers, Mr Hookes [sic] 3 foot quadrant...'[65] The removal of the instruments made 'Flamsteed mad', Hooke confided to his Diary with a certain sense of glee.[66] With the reclaiming of the loaned instruments, excluding, however, the Moore gifts (which were later established by law to be Flamsteed's personal property), Greenwich Observatory would have been left under strength: while none of those instruments which had been removed 'by the ill-nature of Mr Hooke'[67] were vital pieces, they would still have been useful. The government, we must not forget, never did get around to providing any equipment for the Astronomer Royal, who ended up always having to provide his own.

Irrespective of the personality clashes that may have taken place between Flamsteed and Hooke, it is immediately apparent that virtually all of the Royal Observatory's instruments – including the two reclaimed screw edge quadrants and double telescope quadrant, along with the sextant and quadrant which Moore had given to Flamsteed personally – incorporated working parts or principles devised by Hooke. Yet what is truly remarkable is that a new research Observatory like Greenwich, which had been formally commissioned by the Crown to map the heavens as a preliminary to finding the longitude at sea for the Royal Navy, should have been privately equipped with a collection of instruments which were all brand new in design and quite untried in practice. We should be less surprised, therefore, by the failure of a novel quadrant or the inconvenient stopping of a clock with an experimental escapement (as in the case of the 13-foot pendulums with their Townley escapements), than by the fact that most of Flamsteed's instruments turned out to be successful, reliable and capable of supplying rich shoals of fresh scientific data which would remain in use throughout much of the eighteenth century.

Another instrument which Flamsteed set up at Greenwich around 1676 was a Hooke-inspired contrivance: the 'Well Telescope', a type of device which had first occurred to Hooke back in 1665 when he had been conducting experiments down deep wells on Epsom and Banstead Downs. As Hooke would have known from Aristotle's *De Generatione Animalium*,[68] one can see the stars shining in the daytime from the bottom of a deep well with the naked eye, and by 1669 if not before he had seen beautifully sharp star images in daytime with his 36-foot telescope.[69] This well telescope was, as its name suggests, a telescope with an 87.5 feet focal length object glass made by the Frenchman Pierre Borrel that was mounted so as to focus its image vertically down a specially-cut dry 'well' shaft in the grounds of the Royal Observatory to reveal stars in the zenith. Its particular purpose was to measure the supposed seasonal parallax of the star γ [gamma] Draconis, which passes directly overhead in London. To use the instrument, Flamsteed descended the well by means of a staircase, lay on his back at the bottom looking upwards, and applied his eye to the suspended eyepiece. On successive nights he would measure the angle of γ Draconis in relation to the vertical and to adjacent stars, using a micrometer to see if over a six-monthly period it changed its position by a few arc seconds. Unfortunately, the Greenwich instrument was not a success and by 1679 had achieved nothing conclusive.[70]

Although there is no certain documentary connection of the Greenwich Well Telescope to Hooke, the circumstantial evidence is overwhelming. For in 1668–9, Robert Hooke had used a 36-foot-focal-length object glass (probably by Reeves) to set up a vertical or 'zenith sector' telescope at Gresham College, London, with which to attempt to measure the parallax of γ Draconis. What is more, Hooke had not only published a detailed account of his instrument, eyepiece micrometer and experimental technique in *An Attempt to Prove the Motion of the Earth* in 1674, but had set his attempt into context. One of the agreed physical tests for the earth's motion around the sun, and hence the proof of Copernicus's theory, was the detection of a stellar parallax—or the apparent six-monthly rock of a star's angular position—as the moving earth slightly changed the observer's viewing direction as it passed through its $365\frac{1}{4}$-day orbit from December to June and back to December. The reason for conducting these observations in the zenith was that the light from a zenith star would pass straight through the earth's atmosphere and hence straight through the telescope, without being bent, refracted or 'inflected' by an unquantified amount, as would be the case if the star's light entered the earth's atmosphere at an oblique angle. The astronomers of the 1660s already realized that the universe was so vast, and even the bright stars so remote, that the relatively small cosmological distance that separated the two extreme points of the earth, in its seasonal orbit around the sun, would produce an angle so tiny that the slightest instrumental inaccuracy or atmospheric distortion would invalidate the entire observation. Yet the 'parallax' angle which Hooke had extracted from his 1669 observations was 27 arc seconds, which struck most people as being substantially larger than expected, and was treated with suspicion.[71] Sadly, Hooke was not able to continue with his zenith observations of γ Draconis, because an accidental breakage of the 36-foot lens brought everything to a sudden and dramatic conclusion.

Hooke's 1669 Gresham College observations had not really been successful partly, he realized, because, as the parallax angle was so very small, the experiment needed to be repeated with a lens with a longer focal length, which would cast the prime-focus light rays over a greater radial distance, and make the parallax easier to detect. Hooke also suggested that the vibrations picked up by the zenith sector inside a building, as in his Gresham College attempts, introduced problems in their own right, so that more stable results might be obtained by using a deep, dry *well*. Indeed, had he possessed the right object glass and, no doubt, the leisure to leave London for an extended period, he knew a well of exactly the right specifications, '... such as I have seen at a Gentleman's house not far from *Banstead Downs* in *Surry*, which is dugg through a body of chalk, and is near three hundred and sixty foot deep, and yet dry almost to the very bottom'.[72] Hooke was suggesting by 1674 that such a well could accommodate a lens of much longer focal length than the 36-foot glass he had used at Gresham College in 1669, thereby producing a much greater apparent angular displacement that would make a stellar parallax easier to measure.

This was, of course, a well in the same part of England in which Hooke had performed a series of gravity-measuring observations in the plague year of 1665, when he, Dr Wilkins and other Fellows of the Royal Society had quit London for the safety of Durdans, the welcoming country mansion of Lord Berkeley, upon whose ingenious company and experiments the visiting John Evelyn had commented so favourably. These gravity experiments in the Banstead Well will be discussed in chapter 12.

In the zenith sector observations described in *An Attempt to Prove the Motion of the Earth*, we get an insight into the working of Hooke's scientific imagination which complements and expands that which we have already gained from *Micrographia* and his other published researches. An investigation often started out from a single observation of nature or from a simple concept, and then blossomed into a full-scale inquiry into the very nature of scientific knowledge itself. Hypotheses would be proposed, and new technologies devised or designed to test them; when all available techniques had been tried, he would go on to suggest how future researchers might further direct their investigations. Hooke had a clear concept of scientific method based on observation, experiment and hypothesis testing, as he outlined in several of his papers delivered to the Royal Society.[73] While Hooke's instruments, and those of Flamsteed, had been too imprecise to obtain a reliable parallax angle, his *Attempt to Prove the Motion of the Earth* certainly inspired others. In the 1720s, for instance, both the Revd Dr James Pound and his nephew the Revd Dr James Bradley commissioned zenith sectors that were greatly superior to Hooke's, and set them up at a private observatory at Wanstead, Essex. Then in 1728 Bradley realized that while he too was detecting a 20-arc-seconds angle in the six-monthly motion of the star γ Draconis (which was remarkably close to Hooke's value of 27 seconds), the internal geometry of his observation led him to conclude that it could *not* be the sought-after parallax. What Bradley soon realized he had discovered and quantified in these 20 arc seconds was the Aberration of Light—a phenomenon which in itself provides a clear proof of the earth's motion in space quite separate from that of a stellar parallax.[74]

Hooke also attempted to evolve a successful design whereby the long refracting telescopes of the period could be shortened, to make them more easy to handle, especially in windy weather. Seventeenth-century refractors, with their 60 or more feet focal length object glasses mounted in long tubes on the tops of poles and supported by pulleys and ropes, must have been immensely temperamental instruments with which to work, because they were not only sensitive to every breeze, but their timber and cordage support systems were inclined to absorb moisture and go slack. Hooke tried to combine the optical advantages of long focal length lenses with the easier handling qualities of short ones, by using trains of plane mirrors in broad, flat rectangular-section tubes, so that the light would zigzag down to the eyepiece, rather like what it does in modern prismatic binoculars. Hooke's initial proposal for such a folded telescope came from an attempt to reduce the intensity of the light for solar work to produce 'A Helioscope,[75] to look upon the body of the Sun, without any offence to the Observer's Eye', and thereby deriving the title *Helioscope* for such a device. He suggested a folded telescope configuration incorporating four dark 'Reflecting Glasses', each of such a low reflectivity as to absorb three-quarters of the incoming light, so that only one-sixteenth of the light entering the object glass eventually reached the observer's eye, to produce, so he hoped, a better image of the sun than what could be obtained either by coloured filters or by projection. For the four solar reflecting surfaces within the Helioscope, Hooke suggested that highly-polished black glass, black marble, and 'Glass of Antimony' worked best to diminish the sunlight.[76]

Indeed, he had demonstrated such a device to the Royal Society as early as 1668, when he had 'folded' a 60-foot-focus object glass into a 12-foot tube, though he admitted that it was difficult to get adequate fully reflective mirrors for non-solar work.[77] Contemporary speculum mirrors were of poor reflectivity, hard to polish, and tended to tarnish quickly, while mercury-backed glass mirrors of the type used for looking glasses, while giving a brighter image, inevitably gave double and ultimately multiple reflections, for in the seventeenth century it was still not possible to surface-silver glass.

Unfortunately, therefore, the folded telescopes proved unworkable in practice, so that yet another of Hooke's ideas was frustrated by the limitations of contemporary technology. Nonetheless, his 'folded' telescopic optical systems were brought to a sufficient fruition as to be described and illustrated in detail in his published Cutlerian Lecture *A Description of Helioscopes . . .* (1676), and additional notes upon them in his private papers were published by Waller.[78] It was also in illustrations of telescopes and measuring instruments depicted in *Animadversions on . . . Hevelius* (1674) and in *Helioscopes* that Hooke first made use of the universal joint as a mechanism for the smooth transmission of mechanical power around corners, where he employed the device to actuate a pointer moving upon a graduated dial.[79]

Astronomy was in many respects the ideal vehicle for Hooke's approach to science. As an active experimental promoter of the 'Mechanical Philosophy', who insisted that physical, as opposed to mysterious or occult, forces lay at the heart of nature, the regularity of the heavens provided the perfect model for Hooke. That this natural regularity could be comprehended by devices of glass and metal seemed further to substantiate his philosophical stance, which argued that controlled observation and experimentation were the crucial links which placed the wider mechanism of nature within the mental grasp of man.

    Hooke's real and enduring contributions to astronomy were
concerned with instrumentation, observation and the establishment of
reliable methods of framing questions to nature. Yet as we have seen
before, it is quite wrong to see him as some form of inspired mechanic
who left theory to others: an image of Hooke that grew up in the post-
Newtonian age of the eighteenth and nineteenth centuries. Hooke's
relentlessly inquisitive mind was always trying to seek out and give
mathematical expression to the inner dynamics of and connections
within nature; and nowhere was this more obvious than in his astronomical writings where, as a self-conscious inheritor of the mantle of Tycho,
Galileo, Kepler and 'Our Horrocks', he wrestled with those forces that
held the solar system together, made the moon go around the earth,
and caused those 'quakes' that produced the surface features of the
moon. To his mind, instruments and observations were not ends in
themselves, so much as the providers of those sound and reliable pieces
of measured data from which one could proceed to unravel the inner
truths of nature. Being a profoundly intellectual man, what Hooke was
ultimately interested in was nature's intellectual coherence; and while
an investigation might start with ingenious devices, the instruments
were only used as a springboard to go beyond brass and glass to frame
and test wider theories. It was his ingenious devices that led him to
believe that 'vibracions' or wave forms were primary to the whole of
nature,[80] that the heavens had no hard barriers but rather a constant
series of proportionate relationships between bodies that were fearfully
difficult to compute and, perhaps most important of all, that 'gravity'
worked by laws that were related to the squares of the distances
between bodies. While Hooke's earlier gravitational ideas were conceived
in somewhat localized terms, such as the 'gravity' of the earth, moon, sun
or Jupiter, he was thinking in much more universal terms by 1680, as he
moved inexorably from experiments, models, and tests to the perception
of those grand and universal proportions that ran throughout the length
and breadth of nature.

CHAPTER 6

MEDICINE AND PHYSIOLOGY

Throughout his intellectually active life, Robert Hooke had a strong interest in medical subjects, and at the age of 56, in December 1691, received a Lambeth Doctorate in Medicine from Dr John Tillotson, the Archbishop of Canterbury, in recognition of his achievement. But medicine in Hooke's day was profoundly different from what it is now, in so far as it was still seen in many quarters as an essentially conservative art that rested, just like astronomy or geometry, on the supposedly enduring foundations laid down by the ancients. Practical medicine was perceived as innately limited in its therapeutic scope, as an almost total ignorance of the nature of infectious disease, combined with a body of physiological principles which owed more to philosophical deduction than to clinical empiricism, made successful treatment more a matter of luck than of scientific application. Diagnostics, prognostics and therapy remained rooted in the 400 BC writings of the Greek medical taxonomist Hippocrates, while a humoral physiology for the explanation of disease, based on hot, cold, moist and dry, rising, falling, thickening and thinning, which had come to be enshrined in the biological writings of Aristotle before 322 BC and was further amplified by Galen in the second century AD, provided a philosophically elegant but practically dubious rationale for how living things worked.[1]

Yet just as classical geography and astronomy had been challenged and transformed by the practical discoveries of Columbus, Drake, Tycho Brahe and Galileo, so ancient medicine had received its modern shocks. These shocks had come from recent dissections of human and animal cadavers, performed by Andreas Vesalius and his successors after 1540, and from the terrifying new disease syphilis, against which the classically-approved herbal drugs were powerless, and yet which seemed to respond to the new treatment of mercury salivation.

The fundamental breakthrough without which modern medicine could never have come into being, however, took place seven years before Hooke's birth, in 1628, when Dr William Harvey announced the results of an ingeniously devised and meticulously executed series of experiments which demonstrated that the blood circulated around

97

living bodies.[2] Before Harvey, medical men held to a set of classical doctrines which saw the heart as an innately hot biological furnace which heated and purified the 'life-blood' that rose up from the liver and ebbed and flowed in a roughly tidal manner through the *veins* (not the arteries) to nourish all the organs of the body. Creatures had to eat to keep producing more blood, which was in turn consumed in the maintenance and generation of flesh and body heat.

Harvey, however, showed that the heart was a muscular pump, the systolic (or contracting) force of which drove blood into the *arteries*, through the capillaries, and on into the veins, whereby it returned to the heart in an endless cycle. En route, moreover, this 'aimiable juice' passed through the lungs, which seemed to play a part in re-vivifying the dark venous blood and restoring its light colour, so that it could once again circulate and somehow remove impurities and distribute 'nourishment' throughout the body.

Robert Hooke, like all the other scientific men of the age, was fascinated by this new blood circulatory physiology of Harvey. Not only did it suggest that the body was a *machine*, acting under the influence of pumps, valves and moving liquids, rather than being a vaguely 'sympathetic organism', but Harvey's circulatory physiology chimed in elegantly with equally new discoveries in astronomy and magnetism, both of which seemed to hinge on near-circular orbits and force fields. And as we saw in chapter 2, Boyle, Hooke, Mayow and others, with their experiments upon 'aerial nitre', blood colour changes and airpumps, all quite rightly saw themselves as pursuing a research path that only made sense in the context of the 'circulatory physiology'. Harvey's impact upon English and European medical science after 1628, therefore, was fundamental.

One thing which Harvey's work did bring about was a radically new way of thinking about organic processes. The very painstaking experimental procedures by which Harvey had come to his conclusions, moreover, clearly proclaimed (in spite of his great and genuine respect for Aristotle) that experiment, dissection and physical investigation, as opposed to speculations based upon ancient texts, *must* be the way forward in the medical and biological sciences. By the time that Hooke was in his thirties, two major sets of researches that built upon Harvey's discovery were well under way in European medicine. One of these was conducted by the Dutchman Thomas Bartholin, who in the 1650s investigated the lymphatic and glandular systems—which seemed to be somehow connected with the cardio-vascular system—while the other was the cerebro-spinal research of Hooke's old Oxford master, Dr Thomas Willis. In his investigations into the structure of the central nervous system and the brain, in which he could well have been assisted by the young Hooke, whom he employed at one stage as his chemical

Assistant, Dr Willis was bringing into existence a new domain of research within experimental, scientific medicine.

Willis was, indeed, a skilled vivisectionist and, while there is no direct proof, it was probably from him that Hooke first acquired those manipulative skills which were to be so invaluable to the future Curator of Experiments.[3] Willis claimed, however, to have first become interested in *neurology* (Greek *neuron, logos*), a term which he himself coined for the scientific study of the brain, after noticing how a man whose cadaver he was dissecting had his left carotid artery completely blocked by a long-standing blood clot. Now since the time of Galen, it had been believed that each carotid took blood from the heart to the left and right hemispheres of the brain respectively, from whence it was distributed into each hemisphere via a *rete mirabile*, or a 'wonderful network' of delicate blood vessels. Yet if a man's left carotid artery had atrophied because of a long-term blood clot or thrombosis, why had not half of his brain died from lack of nourishment? This chance discovery led Willis to experiment with tying ligatures around one of the carotids of a dog, only to find that the dog suffered no neurological impairment as far as its behaviour was concerned. Then Willis discovered that higher animals, instead of possessing a *rete mirabile*, had, rather, a large circular artery at the base of the brain into which *both* carotids pumped blood. It was from this subsequently named 'Circle of Willis' that both hemispheres of the brain received their blood, so that if one artery did become blocked, then the other could compensate, increase its size, and take over its function. Willis's discovery only really made sense in the light of Harvey's theory of circulation—bringing yet another circular and perhaps *mechanical* model into science—and also revealed for the first time that one biological function could be automatically taken over and compensated for, if one of a pair of living structures failed.[4]

Thomas Willis was to go on, before his death in London in November 1675, to become what he quaintly termed as 'addicted to the opening of heads'.[5] And it was Willis whom we have to thank for the first modern descriptions of the interior of the cerebral cortex, for some of its enduring zonal definitions, and for the first experimental models that would lead to the realization that particular regions of the human brain were connected with specific mental phenomena. For instance, he believed from a combination of live and dissection experience (presumably of the same individuals) that people whose cerebellum regions, at the back of the brain, were soft, were musical in life. In this, and in several other of his supposed demonstrated connections between brain structures and mental responses, Willis was subsequently shown to have been wrong, for the experimental techniques available in the 1660s were still very crude. Even so, Willis is still regarded as the first medical researcher to speak seriously about the localized functions of the human brain in a way that we can understand today.[6]

Robert Hooke must have known Thomas Willis well, first in Oxford and presumably also in London during the 1660s. Yet while we know that the young Hooke had acted as Willis's assistant in Oxford, and probably lived in his house, Beam Hall, in Merton Street, it is hard to be sure, in spite of their having common research interests, how closely the two men collaborated in London during the 1660s and early 1670s, when both were Fellows of the Royal Society. But judging from the occasional and very fleeting appearances of Willis's name in Hooke's Diary it is clear that they never worked together between 1672, when Hooke began his Diary, and Willis's death in 1675.

What is undoubtedly true, however, is that Robert Hooke himself was a skilled dissector and vivisector, and that Thomas Willis and perhaps Hooke's slightly older Westminster and Christ Church contemporary Dr Richard Lower were the most likely candidates to have been his original teachers in these subjects. Yet quite apart from what he might have done in Oxford, the early records of the Royal Society provide ample testimony to Hooke's ingenuity both in proposing medical experiments, especially into the physiology of respiration, and in performing them. On 9 November 1664, for instance, following upon a whole series of discussions and experiments in the Royal Society into the physiological relation of the heart and lungs in breathing, and on an order given to him at the previous meeting, Hooke reported and submitted in writing an account of a complex vivisection on a dog. He opened up the living dog, cut away its thorax and diaphragm, and by means of the 'blast' imparted by a pair of bellows inserted into the poor creature's windpipe, was able to keep it alive for over an hour while observing the behaviour of the completely visible heart in an attempt to demonstrate the role of air and mechanical action in respiration.[7] When air was blown into the lungs with bellows the heart rate became normal, but the heart became convulsive and the lungs 'grew flactid' when the bellows action stopped. It was clear that the crucial thing in respiration was the presence of a constant supply of fresh air to the lungs for the maintenance of the heart rate and lung dilation, and *not*, as some had thought, some sort of innate muscular or lung expansion action; though in this experiment Hooke could still not be sure 'whether the Air did unite and mix with the Blood'[8] at a specific point in the heart–lung cycle. The Royal Society, however, wanted further experiments performed to try to establish exactly what happened to the blood during its passage through the lungs; but Hooke, who clearly hated performing experiments on fully-conscious living animals, kept getting things put off. And when a further 'Insufflation' experiment was ordered in July 1667, Hooke begged to be excused,[9] for as he had written to Boyle in November 1664 no known opiate could relieve 'the torture of the creature'.[10] One presumes, however, that Hooke was not the only person revolted by the animal suffering that these and similar experiments occasioned. Only a few months earlier, on 16 May 1664, Samuel Pepys recorded being present at an experiment on a dog into which, after some difficulty, Dr Timothy Clerke had succeeded in intra-venously injecting an opiate. After injection, 'the dog did presently fall asleep, and so lay till [h]e cut him up', at which point one assumes that the poor creature regained full consciousness.[11] The crude tinctures of opium used in the seventeenth century, made as they were in ignorance of the narcotically-active ingredient in opium juice, morphine, could vary greatly in strength and be very unpredictable in their effects.

When the Royal Society began its next set of researches into blood and respiration in 1667, the experiments were performed by two medical men, the expert vivisector Edmund King, and Peter Ball, but they seem to have botched the experiment, as a result of which more pressure was applied upon the Curator of Experiments.[12] Finally, on 10 October 1667, assisted by Dr Richard Lower, Hooke established experimentally and by cutting off a piece of lung tissue from the living dog, 'that the Blood did freely circulate and pass throw the Lungs, not only when the Lungs were kept constantly extended, but also when they were suffer'd to subside and lie still'.[13] In spite of the repugnance he had to vivisection, it is impossible to deny, considering his success where two medically qualified men had failed, that Hooke was very adept with the scalpel.

That group of men who would constitute the Royal Society after 1660 also performed another series of experiments that stemmed directly from Harvey's circulatory physiology, both at Oxford and London, in the late 1650s and 1660s. These experiments were concerned with the intravenous administration of drugs into living creatures, and with the transfusion of blood between individual animals and between animals and humans.

Hooke's old Oxford employer and lifelong friend, Robert Boyle, and his friend and architectural colleague, Sir Christopher Wren, left detailed accounts of these experiments, which Wren was said to have first tried in Oxford around 1659,[14] and though Hooke is not spoken of as having played an innovative role in them, it is hard to believe, especially consid-ering his active medical interests and close association with all the men involved, that he did not play some part. Some of the first experiments, according to Boyle, were blood transfusions between animals, such as dog to dog, like that performed in Oxford by Richard Lower, and reported to the Royal Society by Boyle. On this occasion, the carotid artery of one dog, a mastiff, was connected by a quill to the jugular vein of a second small dog. When the second dog was cut and began to bleed, the mastiff's blood went into him, 'and left the *Mastive* dead upon the Table', but after the little dog was 'untyed, he ran away and shak'd himself, as if he had been only thrown into water'.[15]

But it was also being asked whether a drug injected directly into the bloodstream of a living creature would act differently than if the dose were administered orally. Sometime around 1663, probably in Oxford, a warm mixture of opium and sack wine was injected into a dog, and it was noticed how quickly 'he began to nod his Head, and faulter and reel in his Pace, and presently... appeared so stupified' that the beast was expected to die, but, being 'whipped up and down a neighbouring Garden' was revived and lived.[16] It is clear that these experimentalists realized that intravenous injections worked so rapidly because 'the *Opium* being soon circulated into the Brain, did within a short time stupefy, though not Kill the Dog'[17] — a result which accorded perfectly with Harvey's discovery of the blood's circulation around the body of living creatures, and with Willis's researches into the supply of blood and nutrients to the brain. Unfortunately, no record survives regarding Hooke's presence or participation in the Oxford experiments conducted between 1659 and 1663, though by 1663 Hooke would almost certainly *not* have been present, for by then he was resident in London, though as the newly-appointed Curator of Experiments at the Royal Society, he would soon have known all about it. In November 1666, these dog-to-dog transfusions were still being developed in London for, as Pepys recorded on 16 November 1666, 'This noon I met with Mr Hooke, and he tells me of the dog which was filled with another dog's blood, at the College the other day, is very well', while Hooke 'doubts not its being found of great use to men'.[18] Indeed, as Timothy Clerke and others realized, the great potential of the intravenous injection of drugs lay in the fact that drugs thus administered bypassed the stomach and acted directly.[19]

By the time that Pepys was recording this conversation with Hooke, however, such injection and transfusion experiments had long been attempted by Fellows of the Royal Society, as Pepys's 1664 observation regarding Dr Clerke's opium experiments testifies. By early December 1665, the intravenous injection of the toxic purgative *Crocus Metallorum* had been tried on both a dog and a criminally-inclined servant in the entourage of an unnamed Ambassador resident in London. The dog did 'vomit up Life and All', and as for the man, 'it wrought once downwards with him'. Yet by this time, such experiments 'hath been frequently practised in Oxford & London'.[20]

These animal experiments were, in many ways, the prelude to blood transfusion experiments on humans. Yet the transfusion of blood into men and women was inspired not so much by a desire to replace blood lost in accidents or surgery, so much as to bring about personality changes; for in accordance with ancient beliefs about animal species and the Seven Ages of Man, a creature's blood was believed to be its defining characteristic. Could the old be rejuvenated by the blood of the young? Could the raving mad be cured by the blood of passive lambs or, as Dr Croone suggested to Samuel Pepys, might it not 'be of mighty use to man's health, for the amending of bad blood by borrowing from a better body'? Or, as Samuel Pepys himself humorously speculated on 14 November 1666, could 'the blood of a Quaker to be let into an Archbishop' and produce a sudden switch of stance towards Church government?[21]

But it was the Frenchman Professor Jean Denis of Montpellier who, in full acknowledgement of the prior published work of the English experimentalists, made the first successful transfusion from a sheep to a 15-year-old youth on 15 June 1667 in Paris.[22] Indeed, the *Philosophical Transactions* of the Royal Society gave detailed coverage of the French researches and of Denis's experiments.[23] Then on Saturday 23 November 1667, at a private experimental session for which preparation had been made for six months, and after 'some considerations of a Moral nature' had been deliberated upon, Dr Richard Lower, for a fee of one guinea (21 shillings), let around half a pint of sheep's blood into the basilic vein of the 32-year-old Arthur Coga, a debauched 'kind of Minister' of Cambridge, as Pepys had previously described him on 21 November, whom 'Dr Wilkins [said] that he hath read for him in his church'.[24] Yet Coga not only survived the transfusion, but a week later, on Saturday 30 November, gave a Latin address to the Royal Society, claiming 'that he finds himself much better since, and as a new man' and was apparently willing 'to have the same again tried upon him'.[25] Some three weeks after his first, Coga underwent a second transfusion, now before a packed meeting of the Royal Society — 'a strange crowd both of forrainers and domesticks' — on 12 December. On this second occasion, Coga received 14 oz. of sheep's blood, and still seems to have been alive five days later when Oldenburg wrote to Boyle in Oxford informing him of the repeat experiment. Apart from a little feverishness, which may well have been occasioned by Coga's 'disordering himselfe by intemperate drinking of wine', he seems to have suffered no serious consequences from the two sheep-blood transfusions.[26] As the distinguished Curator of Experiments at the Royal Society in 1667, and a Royal Society colleague and perhaps something of a vivisectionist pupil of Dr Lower, it is hard to imagine that Hooke did not play a part in this experiment. But after subsequent French transfusions resulted in fatalities, and moral questions were raised about the ethics of animal to human transfusion, these experiments were terminated around 1670.

What is truly remarkable, however, is how any human being could survive the transfusion of around half a pint of animal blood into their bodies with no more than temporary stiffness and pain at the point of injection and some nausea. The rejection symptoms should have been catastrophic. And yet it is clear that both Anthony St Armand in France, who was one of the individuals upon whom Jean Denis had experimented, and Arthur Coga in England not only survived their ordeals, but probably underwent them more than once. What might be suggested by way of explaining this survival is that, in the absence of any anticoagulants, the amount of blood that actually entered the veins of St Armand or Coga could well have been much less than the half-pints being spoken of. Yet even a few ounces, or cubic centimetres, should still have been sufficient to bring on violent and perhaps fatal rejection symptoms. However, we must be cautious in explaining away these transfusion survivals by arguing that the blood probably coagulated before reaching the men, for the seventeenth-century experimentalists were fully aware of the problem of coagulation taking place within the blood-conveying pipe. Indeed, as though to convince contemporary and future sceptics that a substantial quantity of blood really had flowed, the author of the report on the first experiment on Coga stated that a silver pipe as long as three writing quills (probably about 18 inches) connected Coga to the sheep, and when 'we drew the Pipe out of his Vein, the Sheep blood ran through it with a full stream'. Of the 12 or so ounces of blood removed from the sheep, a good nine or ten had entered Coga's veins, so the author thought. These men, after all, fully understood the need to accurately quantify their results.[27]

By the time Hooke had become part, albeit in an 'apprentice' capacity, of the Oxford experimental community by the late 1650s, however, its physicians had already acquired a wider reputation for themselves as a result of an ineptly-performed public execution in December 1650. Anne Greene was a young servant girl from Duns Tew, Oxfordshire, who was executed for supposedly aborting her unwanted child. She was hanged for half an hour at Oxford Castle following her condemnation at the Assizes, after which a soldier struck her body with his musket butt, and suspecting that breath might still be in her, dealt a further blow to her corpse while various men stood upon her, to ensure that Anne had been finished off. Anne's body was then handed over to a company of Oxford medical and scientific men, including Dr (later Sir) William Petty, Professor of Anatomy, the Revd Dr John Wallis, and others, and taken to an unspecified location to be the subject of an anatomical dissection. But when the body was placed on the table, and Petty was about the make the primary incision, 'they perceived some small ratling in her Throat', whereupon the same doctors 'used means for her recovery, by opening a *Vein*, laying her in a warm *Bed*', and 'also using divers *Remedies* respecting her *senselessness, Head, Throat* and *Breast*, in so much that within 14 Hours, she began to Speak, and the next Day Talked and Prayed very heartily'.[28]

Having survived the executioner once, Anne was pardoned, after 'the mediation of the *Worthy Doctors*, and some other Friends'[29] with the legal authorities, and apart from shock and a severely bruised neck and body, she seems to have fully recovered from her ordeal. Needless to say, her survival caused a great stir, and while it owed relatively little to her doctors beyond their timely observation of her faint breathing and generous wish that she should be pardoned, it inevitably drew attention to the Oxford experimental physicians. Indeed, she became the subject of several poems, one of which read:

'Ann Green was a slippery queane,
In vain did the jury detect her,
She cheated Jack Ketch,
And then, the vile wretch,
'Scap'd the Knife of the learned dissector.'[30]

Sixteen years later, by which time Hooke had come to Oxford and then left it for London, the continuing members of this experimental medical community made another study which, at least within the world of the University, was to cause a further stir.

10 May 1666 was an oppressively sultry day, during which a lightning storm had been gradually brewing around Oxford. The storm's increasing intensity and the progressive synchronicity of the bangs and flashes, moreover, were monitored from the top of the Bodleian Library Tower with a 'minute watch' by the same Revd Dr John Wallis who had been present at Anne Greene's recovery, and was Savilian Professor of Geometry, experimentalist, Fellow of the Royal Society and another of the scientific influences upon Robert Hooke.[31] At the same time, a group of Wadham College undergraduates had gone boating for the afternoon up the River Isis to the nearby village of Medley. Then, as the students were preparing to return, and one Samuel Mashbourne was pushing off the boat, a bolt of lightning came down, striking him, in John Wallis's phrase, 'stark dead', and 'stonying' his friend William Harman, who was thrown rigid into the river, 'like a post not able to get himself out', though rescued by College friends. Like the half-hanged Anne Greene, Mashbourne's body was put into a warm bed and spirits poured down his throat in a fruitless attempt at resuscitation. The treatment seems to have worked in Harman's case, however, for he lived to be elected Moderator *Novae Classis* in the University in 1668.

Mashbourne's body was taken back to Wadham and prepared for burial. By some unspecified subterfuge, however, but probably involving the crossing of palms with coins, on a night just before his funeral, Mashbourne's corpse was taken away to an unspecified location, where a large number of medical men gathered to see an autopsy on the remains of a man killed by lightning. It is unlikely that this autopsy was performed in accordance with any legal requirement, for as Mashbourne had not died by any criminal act, was not presumed to have been poisoned,

105

nor, like Anne Greene, was not an executed criminal, the case is not likely to have been referred for forensic investigations by the Coroner's Court. The dissection, therefore, was no doubt motivated by intellectual curiosity rather than by legal necessity, and would probably have been technically illegal.[32]

It is important to remember, however, that in 1666 not only was lightning a very imperfectly understood piece of natural phenomena, but the instantaneous death of a fit young man was also puzzling, though in June 1664 Hooke had reported to the Royal Society the recent case of a man who had been struck by lightning, and spat blood, but who lived.[33] After all, in the seventeenth century death usually came after a person had been worn down by a fever, or suffered a violent injury, so how a bolt from the sky could kill a man and leave nothing but two small round black spots on his neck and some superficial burning on his chest (probably caused by wearing metal buttons in his waistcoat), while blasting his hat apart and damaging his clothes, was a mystery. The obvious element of secrecy and possible illegality surrounding Mashbourne's autopsy was made clear by the dark and congested circumstances of the act itself: though the May evenings are long and light, the autopsy seems to have been performed by candlelight, while the jostle of town and gown doctors around the dissecting table is made clear by the recorded fact that the operator lacked elbow room in which to wield his saw when opening Mashbourne's cranium.

The mysterious effects of death by lightning were made evident from the seemingly healthy state of Mashbourne's internal organs—his brain and its attendant blood vessels were all found to be normal, unscarred and in a healthy state, as were his other major organs. Indeed, apart from superficial injuries to the skin, there seemed to be no obvious reason for Mashbourne's death. Though Hooke was no longer resident in Oxford by 1666 (though he was probably still out of London following the Royal Society's adjournment during the plague year), he must have been intimately familiar with the case, for a detailed account was published in the Royal Society's *Philosophical Transactions*; while the Oxford diarist Antony à Wood (who was a next-door neighbour of Dr Willis in Merton Street) did his own research into the Mashbourne case, and wrote and later published an independent narrative.

One might suggest, however, that the reason why there was so much concern with Mashbourne's brain, and why Willis (who in 1668 left Oxford to reside in London at the request of the Archbishop of Canterbury)[34] had 'addicted [him]self to the Opening of Heads',[35] and even perhaps reformed and stabilized the drunken and debauched Arthur Coga by blood transfusion, derived from a developing fascination with the physiology of perception. Willis's neurological researches were inspired on the one hand by a desire to explore the ramifications of Harvey's circulatory physiology, but on the other by a wish to understand the function of the brain, and its relation to the soul, to the intellective functions of thought and feeling, and to the more obviously mechanical ones of heartbeat and muscle action. René Descartes' well-known mechanistic physiological researches conducted before 1650 and *De Homine* (posthumously published, 1664) had provided an obvious starting-point for serious investigation of how the 'mechanical' body might be controlled by, or else work in synchronicity with, the conscious mind, but Willis's *Cerebri Anatome* (1664) and *De Anima Brutorum* (1672) represented major physiological steps forward. Willis, moreover, a devout Christian whose High Church Royalist principles, it will be recalled, had occasioned his expulsion from Christ Church during the Cromwellian period, saw no incongruity in investigating how the mechanics of the brain operated, for as human souls were immortal gifts from God that temporarily inhabited fleshly structures, what interested him was how this immortal and divine perception reacted with the sense impressions of sight, touch and sound, which the mechanical body collected, brought into focus and stored up as memory.[36]

It is also plain that Robert Hooke shared this fascination with perception, which he discussed on a number of occasions over the years, beginning in *Micrographia*, where he spoke of scientific instruments as the 'artificial organs' which clarified and strengthened the *senses* whereby the human body obtained its facts about the natural world,[37] along with *memory* and *judgement*[38] through which it organized them into coherent schemes of understanding. And other perceptually-related researches gave clear evidences of his skill as a practical anatomist. Hooke's Lecture on Light, for instance, read at Gresham College in June 1681, shows his own dual interest in both the anatomical and the perceptual, for he commences with a basic anatomical procedure for examining the structures of the eye. He begins, 'Having taken the Eye out of its place, or Socket in the Skull, and having taken off carefully all the Muscles that serve for its Motion, as being not now considered, we have a round Ball...' Hooke then proceeds to a detailed anatomical dissection of the eye, showing how it forms images 'to shew the Picture of Objects without', and draws on animal parallels, such as the peculiar features of the eyes of 'Young Kitlings' or cats. After describing in great detail the anatomy of the eye, and the production of focused images upon the retina, he reminds his audience that the capacity to see is prior to all that we might learn from telescopes and microscopes, whereby 'the wonderful Wisdom of the Great Creator [is] more manifestly shewn'.[39]

Then in a subsequent Lecture on Light, undated but given the designation 'Lecture VII' by Richard Waller[40] who edited them for posthumous publication, Hooke tries to grapple with what goes on in the human brain when we perceive. The brain, Hooke argued, supplied the organic foundation for perception, but mental life only really took place when the soul somehow reacted with the brain. Here we see Hooke wrestling with a mind–body problem first addressed by Descartes, and which in many ways remains just as elusive today: exactly how do the optical images impinging upon the retina, or shock waves upon the ear drum, really produce that mental awareness which we think of as conscious perception?

In Hooke's analysis, time and memory were of fundamental importance, for all sensory experience was sequential: 'Time as unnerstood [*sic*] by Man, is nothing else but the Length of the Chain of these Ideas';[41] and if time were a chain of events, then memory was that capacity to retain and reflect upon these impressions. In this whole process, the soul was the *primum mobile*, 'So that Thinking is partly Memory, and partly an Operation of the Soul in forming new ideas'.[42]

It cannot be denied that in his observations on time, memory and perception, Hooke is overtly philosophizing, for there were no techniques available in 1680 whereby anyone could reliably connect mental experience with controlled experiments in brain physiology. Ideas of the kind outlined above, however, had well-established resonances in the writings of Descartes, Gassendi, Willis and Hobbes (whom Aubrey tells us Hooke admired) and would receive their definitive philosophical treatment from Westminster's and Christ Church's own John Locke in 1690. Yet what should be remembered is the great importance of both the physiology and the philosophy of perception to the wider understanding of learned men in the late seventeenth century, and how deeply aware of these issues Robert Hooke was.

In addition to vivisection experiments on respiration and physiologically-related hypotheses about perception, Hooke's concern with flight, and attempts to make 'succedaneous Muscle for flying, and give one Man the strength of ten or twenty if required' also led him to study real muscle tissue under the microscope.[43] On 4 February 1675, for instance, he demonstrated to the Royal Society how muscles consisted of 'an infinite number of exceeding small round Pipes, extended between the tendons of the Muscles, and seem'd to end in them'.[44] On boiling, they became very flexible, like gut. Then on 25 April 1678 he suggested that muscle fibres under the microscope resembled 'a Chain of small Bladders', like a string of pearls, and were probably activated by a 'very agill Matter'[45] to effect their expansion and contraction. One wonders whether he thought this 'agill Matter' was similar to the subtle 'animal spirits' which Willis believed activated the cerebral and nervous systems.

From a reading of the rapid advances that had taken place in scientific medicine and physiology, one might reasonably expect that great strides had also been made in the practical art of healing the sick;

but this in fact was far from the case, for a great gulf existed between what a man might do when wearing the gown of a learned experimental physiologist in the lecture hall, and what he did at the bedside when faced with actual human suffering. Practical therapy, indeed, lagged far behind experimental physiology; and while these physiological discoveries would later bear fruit in the late nineteenth and twentieth centuries, after many more of them had been made, the practical healing capacity of Dr Willis, Dr Lower, Sir William Petty and their colleagues remained pitifully low. It was not for nothing that when the greatest medical trauma of the seventeenth century hit the Capital in the Great Plague of 1665, killing around 100 000 people in a few months, most of the physicians left London and followed their rich patients to the clean air of the countryside for the duration.[46] The Royal Society, belonging as it did to the world of the rich and leisured, left as well, but not before Robert Hooke – soon to depart for Surrey and the Isle of Wight for several months – had drawn some interesting conclusions about the nature of the plague. Hooke, now just installed in his Gresham Professorship, had noticed by 28 June 1665, when he wrote to Boyle, that the plague did not seem to be communicated by some vague aerial *miasma*, by which infectious diseases were then thought to be spread, for he concluded, 'I [cannot] imagine it to be in the Air'. Instead, Hooke's own observations of the spread of the 'infection or contagion' led him to believe that it was communicated only by direct contact with infected individuals or substances. This was, indeed, a remarkably acute observation, telling us much about Hooke's capacity to observe and draw conclusions from the environment at a time of danger and panic, and in the same letter he reported another: in spite of the prevailing summer heat, there 'is a very great scarcity of flys and insects', and Hooke asked Boyle to confirm whether or not this was also the situation outside London. While one may not have been able to cure the Plague, at least one could try to ascertain its vectors of transmission, and the disease's possible connection with wider environmental factors. And this, after all, is the basis upon which all rational and scientific medicine stands.[47]

In spite of their physiological researches, Willis, Lower, Petty and their colleagues, when faced with real sickness, still reverted to that limited repertoire of tried and trusted Hippocratic, Galenic and Arabic therapies that would have been familiar to Chaucer's Doctor of Physick in 1380. In this way of thinking, health was the happy result of a balance between the four humours – blood, yellow bile, black bile and phlegm – and disease the product of an imbalance. A patient who had a burning fever, for instance, would have a vein in the arm opened so that a half pint of 'overheated' blood could be taken out, in a misguided attempt to cool him down. In April 1673 Hooke's erstwhile patron, Sir John Cutler, was found 'newly recovered of an ague by bleeding at the

Jugular 16 ounces'.[48] Likewise, the victim of an 'apoplexy', or what we would call a stroke or heart attack, would also be bled with benefit, along with the drowned and asphyxiated, as in the case of Anne Greene. In Hooke's time, one should not forget, doctors still believed that the blood's quality varied with the seasons. In the invigorating weather of the early summer of 1673, Hooke recorded a personal example: ' ... being the first summer morning and finding my blood to ferment, I walked in the fields with Blackburne', presumably to cool it down.[49] Five months earlier, however, just before Christmas 1672, Hooke had paid half a crown to Mr Gidly the surgeon to 'let me 7 ounces. Blood windy and melancholy.'[50] This would, one presumes, have been heavy, thick, winter blood, that inclined one to depression. Even more strangely, a person who had lost blood during an unanaesthetized surgical operation would be 'phlebotomized', or let blood, some hours after the ordeal of surgery, as happened to Prince Rupert in January 1666, when a hole was cut into his skull which successfully alleviated a head injury.[51]

Yet while phlebotomy was a clinical leftover from classical, pre-circulatory medical days, when it was somehow believed that hot, corrupt or thick coagulated blood had to be 'purged' to make way for new healthy blood, the technique remained in the physician's armamentarium until the early Victorian age because, oddly enough, it could work in certain circumstances. What we now know happens when blood is lost is that the adrenal cortex is stimulated, so that the body automatically secretes 'feel good' chemicals which are carried to the brain, and can help to revive the comatose. However, if the bloodletting followed an already bloody surgical operation, the revival was only likely to have been temporary. So doctors might do right for what we would now consider to be the wrong reason, solely on the ground that this technique had been used empirically, yet successfully, for centuries, and had become a regular part of the therapeutic ragbag. However, the physician had to be careful not to over-bleed a chronically sick person, or else the short-term benefits of the 'feel-good' effect would be fundamentally undermined by the degenerative consequences of continual haemorrhaging.

It was also generally believed in the seventeenth century, moreover, that the red inflammation which can develop around a wound — and which we now know is a healthy response by the body's immune system — was a sign of poisoning. Since bloodletting made the patient anaemic and thereby made the inflammation appear paler, doctors misguidedly believed that they were draining poisons, not repair substances, from the site of injury, when they bled a post-operative surgical patient.[52]

Blood-letting, or 'breathing a vein', however, was also seen as a form or purging, as were most of the other techniques employed by the bedside physician, which were generally used because they had, or were believed to have, a purgative or a shock effect. Humours, after all, could get sluggish, and produce extreme cases of heat, cold, moistness or dryness, and needed subduing or strengthening if the patient was going to get better. 'Jesuit's bark' or crude quinine — probably the greatest single pharmaceutical specific discovered in the seventeenth century — could reduce a malarial fever, Ipeaccuanha shifted bile and phlegm and was used to treat both dysentery and bronchitis, while Jalap was a bowel purgative. Cantharides (or Spanish Fly), made as it was from crushed insects, was an 'irritant' employed to make blisters and keep sores discharging, while various Elixirs — including that of Vitriol, or sulphuric acid — were used to strengthen and invigorate. One might try to purge the eyes for blindness, the ears for deafness, the lungs for bronchial phlegm, or the arm or tongue afflicted with stroke for their 'lameness'. More generally, one could purge the guts with rhubarb or Jalap, for as constipation was a perennial problem in that heavy-meat-eating age, self-induced confinement to the privy or closed stool for a day or two could only do one good, as the now over-active bowels were believed to set off a chain reaction of sympathetic purges throughout the rest of the body.[53]

While he could, with his Oxford education, reputation and Royal Society contacts, have easily obtained a licence permitting him to practise medicine for money, there is no evidence that Robert Hooke treated anyone other than himself. Yet when we look at his diary, correspondence with Robert Boyle and other documents, it becomes immediately clear that Hooke was not only extremely *au fait* with the sometimes bizarre official practical medical treatments of his day, but that he was also a relentless self-medicator.

In Hooke's Diary, for the years 1672–80, one encounters an enormous amount of medical data, much of it relating to Hooke himself. The Diary's opening, in August 1672, finds Hooke thrashing around to find relief from a collection of symptoms variously described as vertigo, headaches, head noises, unsteadiness, swimming eyes, indigestion, sickness, stomach disorders and sleeplessness, to name but a few. They occupy his Diary entries for about a year in a serious way, and to a lesser extent down to 1680, for Hooke, while relentlessly active and rarely bedridden, was hardly ever well. Whether these symptoms were the product of a long-term undiagnosed physical malady, or whether they were the results of constant stress generated within a distinctly manic and obsessive personality that could never relax, is a matter for speculation; but one cannot help wondering how far an unrecognized caffeine addiction may have played a part in these maladies, for visits to Garaways, Childs, Tooley's, Man's and several dozen other coffee houses recorded in the Diary were part of his daily round, not to mention whatever coffee or 'chocolat' he may have drunk at home in his Gresham College rooms.[54] Seventeenth-century coffee, moreover, was drunk thick and strong, and while he rarely records the number of cups drank—'Slept well after 2 dishes of Coffee—hand shook'—one suspects that the alcoholically abstemious Hooke might have used the bean as an alternative to the bottle.[55] English coffee-houses only began in 1650, in Oxford, and unlike alcohol, coffee's effects upon the human mind and body were still very imperfectly understood. It is believed, though hard to prove, that one of coffee's first English habitués, indeed, was none other than the elderly doctor Harvey himself who, according to Aubrey, 'was wont to drinke Coffee: ... before Coffee-houses were in fashion in London'.[56]

Yet while Hooke had many academic physicians amongst his own social circle in London, including the distinguished Fellows of the Royal College of Physicians Dr Jonathan Goddard FRS and Dr Walter Charleton FRS, some of whom he consulted professionally with regard to his own health, one cannot help feeling amazement at the sheer eclecticism of Hooke's personal therapeutic quest: in addition to what Mr Gidly the surgeon or Dr Goddard may have prescribed, Hooke seemed willing, quite literally, to try out any old wives' tale or scrap of quasi-mystical medical lore that came his way. He drank gallons of the reputedly medicinal waters of Dulwich, put great store on the special ale and folk remedies prescribed for him by Mrs Elizabeth Tillotson, wife of John Tillotson, Dean of Canterbury and later St Paul's, then Archbishop of Canterbury, noted that 'a woman in the Tower cured divers [sufferers] of vertigo by stone horse dung',[57] mentioned in the winter of 1674 Mr Wild's belief that 'the blood of a black cat would cure chilblains',[58] and noted down, on the recommendation of Robert Boyle, that *stercus humanum*, or dried and powdered human faeces, was held to be an excellent remedy for removing films and mists from the eyes when blown into them.[59] If a scientist of Boyle's stature was willing to recommend such nostrums, one need not be surprised to find ground-up lice, distilled toads and the mould from dead men's skulls—the ubiquitous 'Usnea'—appearing in the prescriptions of qualified yet perhaps less intellectually exalted medical men, let alone amongst the irregular healers. Indeed, Hooke's Diary specifically mentions 'Sir R. Morays present of Usnea *Cranii humani*' on 12 March 1673, while six weeks later he 'took Goddard's Drops', a famous compound concocted by his friend Dr Jonathan Goddard FRS, which also contained human skull mould and ammonia.[60]

Some substances seem to have been used in medicine for their purgative qualities, while others were valued for their fortifying or building-up power, such as when 'Mr Hedges told me that the flesh of tortoises any ways eat is a certain cure for the Consumption'.[61] Perhaps the rationale behind this theory derives from the ostensible fact that consumptive tortoises seem to have been very rare!

As the late Roy Porter reminds us, early modern medicine was less of a coherent system than a market-place in which monarchs and milk-maids felt no embarrassment in drawing upon whatever expertise seemed to suit the needs of the particular case.[62] As the remnants of humoral physiology still tended to emphasize that cures were not general, but had to be tailored to the 'temperament' of the patient, and the new chemical drugs such as mercury or antimony were just as likely to kill as to cure, the pick-and-mix approach to medicine made as much sense then as it strikes us as ludicrous today. So the greatest experimental scientist of his age, and subsequent possessor of a Lambeth Doctorate in Medicine, could confidently try out a cure for piles recommended by his maidservant Nell,[63] or dose himself with simples from a Dean's wife's medicine chest, with dubious substances recommended in coffee houses, or with revolting human and animal products vouchsafed by Fellows of the Royal Society. Practical medicine, or the business of curing the sick—as opposed to experimenting on the lungs of dogs—had still not yet broken free from its roots in common folk culture, and even an educated man might solicit tips on how to cure his 'melancholy' or constipation, in the same way that he would seek advice about gardening or animal-breeding.

But when one looks carefully at what Hooke believed he was doing when he dosed himself, one can discern a clear rationale, within the context of the medical theories of the age, for virtually all of Hooke's nostrums and procedures were intended to be, in some way or another, purgative. The 'snuffing ginger' which he took on 22 December 1672 was intended to purge his head of dizziness, aching and phlegm, and 'I was much relieved by blowing out of my nose a lump of thick gelly', while the 'Andrews drink' of 16 February 1673 which 'wrought... [several?]... times with me brought much slime out of the gutts and made me cheerful', was intended to cleanse his queasy stomach.[64] Likewise the sundry purges, Venice treacle, and other clysters (enemas) to which he subjected himself were intended to purge his bowels: on 21 November 1675, for instance, Hooke 'Took senna. Stayd within all day.'[65] These purges, bizarre as the clinical theory behind them may appear to us today, were monitored with a scientific eye, and precise record kept of the number of times and the degree to which they 'wrought' with him.

Vomits were, of course, another form of purge, as were sweating régimes, and it is interesting that on two occasions, in 1675 and 1676 respectively, Hooke went to one Tom Hewks (who seems to have been an apothecary) for each of these treatments. In August 1675 he 'Took Hewks Vomit it made me incline to vomit after 1 hour [of] taking. It made me straine after 2 houres but brought up nothing.' A little later, however, he used a feather to encourage further retching, but it made his 'head and eyes much worse'.[66] Then about nine months later he was off 'To Thom Hewks about... sweating box... fluxing err time'.[67] The sweating box seems to have been a form of enclosed Turkish bath. Yet despite each of these therapeutic rigours Hooke's intellectual curiosity won out in the end: after his above-mentioned vomiting marathon he ended his day in his Gresham College turret making astronomical observations, and after his sweat he was off with Tompion and other scientific friends—though he was 'Ill with cold'.[68]

On the other hand, Hooke, like most of his contemporaries who both administered and received this heroic physicking, was all too well aware of how often doctors could kill, as well as cure. When Hooke's old friend and patron, Dr John Wilkins, now Bishop of Chester, died in London in November 1672, his official cause of death was a six-day stoppage of the urine, initially ascribed to bladder stones. During his last six days, the dying 58-year-old Bishop Wilkins had been prescribed cider in which red-hot oyster shells had been quenched by his clerical friend the Revd Dr Joseph Glanvill FRS, a frightful-sounding salt of tartar emetic preparation, while his physician, Dr Jonathan Goddard FRS, recommended blisters with Spanish Fly at the neck and feet, to drain away the supposed poison, as a way of alleviating Wilkins's urinary blockage. Yet when his scientific friends performed an autopsy on Wilkins's body shortly after death, Dr Needham informed Hooke 'of Lord Chesters having no stoppage in his uriters nor defect in his Kidneys...' So that '... 'Twas believed his opiates and some other medicines killed him, there being no visible cause of his death, he died very quickly and with very little pain, lament of all.'[69] What seems so bizarre to the modern reader, however, is the contrast between the hotchpotch of old wives' tale 'therapies' tried out on the dying Bishop, and the precise and scientific search for the cause of death in autopsy.

It is quite clear from Hooke's Diary that Wilkins's death had not only set the scientific community talking about the loss of one of its most respected members, but that the precise cause of his death, and the problem of 'the stone' in particular, had led to an active swapping of nostrums. In November 1672, for instance, Sir Theodore Devaux reported to Hooke that the late and internationally-celebrated royal physician, Sir Theodore Mayerne, used to cure 'stone in kidneys by blowing up bladder with bellows etc.'[70]

In addition to the above therapeutic procedures found in Hooke's Diary, one can also trace, in the writings of Boyle, Willis, the Revd John Ward and his other medical and scientific friends, an attempt to develop new substances and procedures based upon the supposed virtues of the newly-devised metal and mineral pharmacies. We have already seen that, since the sixteenth century, mercury came into widespread use as a prophylactic against the new disease syphilis, and probably did so on little more than the grounds of visual analogy. Mercury caused the patient to salivate heavily—sometimes up to two quarts of saliva per day while under treatment—giving the impression that excessive cold phlegm was being expelled from the body.[71] The metal mercury, moreover, having an ancient parallel with the planet of the same name, was also somehow seen to be linked (by astrological connection) with Venus, the planet associated in classical astrology with sex and lust. As a result, the quip came into being that those profligates who spent 'a night with Venus' would end up spending 'a lifetime with Mercury'.

According to his Diary for 1 August 1672, Robert Hooke seems to have taken both mercury and steel in the form of a drink though, in spite of his recorded amorous encounters with his maidservant-cum-mistress Nell Young, it does not seem to have been to treat a sexual complaint. Mercury, let us not forget, was also believed to have wider therapeutic uses than just the treatment of sexual diseases. Even so, one cannot help but feel intrigued at the unspecified complaint for which Hooke sought treatment on 28 December 1672, at Mr Guidly's [or Gidly's], the surgeon, when 'I made an issue in my Pole' [penis?]. 'Dr Chamberlaine was here and directed. He made it with caustick, I gave him 5 sh[illings]. Slept ill.'[72]

Metal-based drugs had grown in popularity after the exotic Swiss physician alchemist Paracelsus began to advocate them in the years prior to his death in 1542. It is true that much of the reasoning behind Paracelsus's ideas derived from the supposed interconnectedness of the seven metals with the seven planets and with the vital forces of the human body in a series of frankly occult hierarchies, but Paracelsus's stand on the relationship between chemistry and medicine struck a powerful chord in Renaissance thinking. If Paracelsus was the prophet of mercurial medicines, so the *Triumphal Chariot of Antimony* (1604), which was attributed to Basil Valentine, did likewise for antimony, the newly-discovered metal which could, in the right preparation, have a beneficial effect upon fevers and other conditions. Mineral salts that were formed by the action of strong acids upon metals, often through unnecessarily elaborate distillation processes, and which were the products of the rising fascination with manipulative chemistry at this period, were invariably pressed into medical usage as well. Johann Glauber's *sal mirabile* or 'wonderful salt' (sodium sulphate) of 1658 was one of them, entering the pharmaceutical canon as 'Glauber's Salts'. In the early seventeenth century, the Flemish chemist John Baptista Van Helmont not only advocated chemical medicines, but also did highly influential work in developing increasingly complex and refined laboratory techniques, especially in the preparation of metallurgical and mineral substances. It is in addition intriguing to note that the man simply referred to as 'Leonard', whom Hooke had met when staying at Lord Conway's Ragley Hall, Warwickshire, in 1680, had previously been 'an ingenious German mechanick servant to Van Helmont'.[73]

Robert Boyle, Hooke's probable mentor in practical chemistry, had also been deeply influenced by Van Helmont and the earlier metallurgical chemists, while at the same time trying to strip most of the occult and mysterious assumptions from their ideas. While discounting much of the astrology and occultism, therefore, Boyle began to explore the chemistry of the metals, not only in pursuit of a *philosophical* (as opposed to a greed-driven) search for the key whereby one metal could be transmuted into another, but also for metal-based drugs which, in accordance with Bacon's vision, could help to ease and prolong human life. Because of gold's supposed purity as a substance, various gold preparations were devised for use in pharmacy, working on the assumption that a noble metal should produce a noble and powerful medicine. Boyle himself subscribed to the concept, borrowed from Paracelsus, Van Helmont and others, and elaborately-prepared gold salts named *Aurum Fulminans* (flashing or exploding gold), *Aurum Potabile* (drinkable gold) and *Aurum Vitae* (living gold) were administered by physicians to patients — usually for their supposedly special purgative and humour-restoring powers. Unravelling exactly what these *iatrochemical*, or mineral, medicines were in modern physical chemistry terms is difficult, although it is likely that what they contained was gold tetrachloride. Hooke would almost certainly have been acquainted with the preparation of these gold medicines, as would have been many of the more adventurous physicians of the day; while even the Revd John Ward, Vicar at Stratford-upon-Avon, chemist, medical cognoscente and close friend of the Oxford experimentalists, describes the production of *Aurum Vitae* and its chemical relatives in his surviving manuscripts. Sadly, no one who took one of these preparations in the hope of finding the key that would banish all illness from the sheer 'nobility' of the gold would have felt rejuvenated by what would really have been a foul and indigestible emetic.[74]

Not all the new medicines which were discussed at this time were metallic in their origin, and it is interesting to see how burgeoning trade and acquaintance with non-European cultures were introducing into Europe drugs and other substances which were of established familiarity in eastern countries and in the Americas. We have already seen how coffee, probably from the Turkish world, 'chocolat' from the Americas and tea from China had suddenly been absorbed into the social world of seventeenth-century London, and the search for new medicines from both east and west was keen. On 18 December 1689, Robert Hooke reported the effects 'of the Plant, called *Bangue*', or what was probably Indian hemp [*Cannabis indica*] before the Royal Society. Though from the context of his narrative Hooke does not seem to have taken it himself, and was communicating what had been passed on to him by the original and unnamed English experimenter, he reported that it was used by 'mendicant Heathen Friars' in India to produce all manner of visions and states of mind. Hooke's informant said that enough powdered *Bangue* as would fill a tobacco pipe bowl should first be washed down with water, and what followed was better than drunkenness. The substance had the power 'to put a Man into a Dream', from which when he awoke, he felt refreshed and well, with no hangover. Hooke proposed that *Bangue* might be used to treat 'Distempers of the Head and Stomach', to pacify and clear the heads of lunatics, and to impart a sense of well-being to the distressed. It is surprising, therefore, especially considering England's growing commercial relations with India, that *Bangue* or hemp did not really enter European pharmacy until the Victorian period.[75]

Seventeenth-century medicine still largely operated on the assumption that, in this vale of tears and human illness, there was a special medicine or rule of health that when prepared or applied properly would provide a wonderful key which would suddenly allow access to perfect health. The wisdom of philosophers and the sagacity of experienced physicians were assiduously directed at finding these keys (*clavis* in the Latin). The search for this *clavis medicina* is implicit in the perfect purge which will set all humours in balance, the elixir which will rejuvenate the feeble and the ointment that will make the blind see. It also carries a strong indebtedness to Christianity and the way in which physical health was seen as having parallels to spiritual health and to the image of Christ the physician: just as in the Gospels Christ's physical touch or uttered word brought instant cure to both the physically and the spiritually afflicted who had faith in Him, so the perfect medicine, once found, must give immediate and enduring relief from all suffering. The early-seventeenth-century poet George Herbert caught perfectly this sense of both spiritual and physical healing in his medically-allusive poem *The Elixier* (pre-1633). Not only did the physician's elixir rejuvenate the body, but Christ's elixir would rejuvenate the soul, while within the text of the poem Herbert uses the image of the apothecary's 'tincture' which will make all 'grow bright and clean'. Herbert even went on to speak of faith in Christ in the imagery of an instantly potent alchemical medicine:

> 'This is the famous stone
> That turneth all to gold
> For that which God doth touch and own
> Cannot for lesse be told.'[76]

Did not a practical chemist 'tell' the purity of his gold in the art of classical assay, whereby a new piece of gold was scraped across a 'touchstone'

117

upon which its thin golden trace could be compared with a trace from a piece of gold of known purity?[77] Tinctures, moreover, in the medico-chemical language of the seventeenth century, were potent and pure concentrates, one drop of which could bring about wonders of colour change (Latin *tinctura*), healing, cleansing and refining.

To understand the mentality that lay behind the more traditional aspects of seventeenth-century medicine, therefore, one must enter into a world shaped by Christian imagery as well as pagan Greek humour-alism, in which not only physical and spiritual health but also the great truths of nature were hidden from us by the fallen state of mankind, which necessitated our discovery of a hidden key to find relief, under-standing and salvation. It was not for nothing that in the 'Preface' to *Micrographia* Hooke described scientific research as conducing towards spiritual redemption.[78]

Perhaps, in the light of the above, one can see Robert Hooke's relent-less self-medication as bearing all the marks of a man searching for the key to perfect health, which he variously hoped to find in Dulwich water, 'Elixier Proprietalis', Dr Goddard's Drops, painful blisters to the head, and, which for a short period in the summer of 1675, he genuinely believed that he had found in spirit of sal ammoniac. But in this vale of tears none of these substances gave enduring satisfaction, although one wonders to what extent Hooke's genuine medical problems, whatever their cause in modern diagnostic terms, were only aggravated by the endless self-infliction of purges, vomits, clysters, metals, irritants, phlebotomies, scarifications, blisters and opiates which were intended to draw up foul humours and facilitate that sudden disappearance of symptoms which three centuries ago passed for health. Perhaps he would have felt better had he followed the older Hippocratic regimen of diet, exercise, rest and gentle herbal medicines, though this might have seemed less intellectually adventurous.

Ruthless as Hooke's self-medication sometimes was, however, he never seems to have attempted to inflict it upon others. In those incidents recorded in his journal when people close to him went down with grave illnesses — such as his young Isle of Wight relative and pupil Tom Giles in 1677, and his niece Grace Hooke in 1679, both of whom had fevers — he was all too willing to call in professional help and keep out of the business of treatment himself.[79] Tom unfortunately died, but Grace recovered, only to die of another fever in 1687. One suspects that his relentless curiosity might well have made Hooke regard his own body as an experimental laboratory, though he had the good sense and kindness not to extend this appetite for experimentation to others, especially if their lives hung in the balance.[80]

But when looking at the medical ideas and practices of the seven-teenth century, one finds the scientific revolution at its most paradoxical;

for those self-same individuals who were recognizable modern scientists when laying the basis of circulatory, cardio-vascular and respiratory physiology and neurological anatomy inevitably descended into what appears to us like empirical clinical shamanism when dealing with actual human disease. This situation, moreover, would continue for well over 150 years into the future, and when Georgiana, Duchess of Devonshire, suffered from an acutely inflamed eye in 1796, she was bled, scarified and subjected to a bizarre regime of counter-irritants that would have seemed eminently sensible to one of King Charles II's doctors. And all of this was in spite of enormous advances in academic anatomy and physiology over the intervening years. Not until the late Victorian period, indeed, did practical therapeutics begin at last to benefit from what by then had been 300 years of scientific research into the workings of the human body.[81]

CHAPTER 7

SURVEYOR TO THE CITY OF LONDON

The fire which raged through the City of London between 2 and 6 September 1666 was to provide, ironically enough, one of the great opportunities of Hooke's life. By totally destroying many of the most important and wealthy sections of the metropolis, including Guildhall, 52 Livery Company Halls, 86 parish churches, St Paul's Cathedral and 13 200 houses in 400 separate streets, while at the same time claiming only six human lives, it opened up the way for lavish reconstruction work.[1]

Almost before the ashes had stopped smoking, King Charles II, of whom it was said that he had gone personally into his blazing capital to inspire the fire-fighting operations, was considering plans for a complete re-design of his City.[2] With the teeming, plague-breeding medieval town gone, King Charles and a few other influential people briefly dreamed of turning London into a showpiece capital of Europe, with wide streets and great civic buildings that would outshine Bernini's Rome.

Gresham College had been mercifully spared the blaze, and it must have seemed natural for the apparently omni-competent Virtuosi to produce the grand design. Within a matter of days Dr Christopher Wren (he was not knighted until after the fire) had produced a plan for a complete new city. Indeed, it is interesting to think of Wren submitting a design at all, for while he had held an Assistant Surveyorship of the Royal Buildings in London since 1661, he was not a trained architect, but rather, as a Gresham College and then an Oxford Astronomy Professor, a gifted amateur. Surveyorships were often granted as rewards, and Wren's fulfilment of the office so far had consisted very largely of recommending modifications to existing structures, such as old St Paul's and various royal buildings. It is true that he had produced a daringly original plan for Oxford University's Sheldonian Theatre, incorporating a brilliantly engineered unsupported ceiling based on a complex geometrical lattice of short wooden beams first designed by his professorial colleague John Wallis, but it was still unfinished in 1666. In Wren's London design, the ancient street plan was to be radically altered, as the new St Paul's Cathedral would act as the hub of a great 'wheel' of new streets that would radiate from it.[3] Robert Hooke, who so far seems to have had no architectural experience beyond conducting experiments into the load-bearing strengths of wooden beams, submitted, according to his biographer Richard Waller, an equally remarkable street plan wherein 'all the chief Streets as from *Leaden-Hall* corner to *Newgate*, and the like, to lie in an exact strait Line, and all the other cross Streets turning out of them at right Angles, all the Churches, public Buildings, Market-places, and the like, in proper and convenient places, which, no doubt, would have added much to the Beauty and Symmetry of the whole.' Hooke showed his plan to the Royal Society on 19 September 1666, and one cannot help but be struck by the speed with which it had been thought out and produced. Richard Waller, in his *Life* of Hooke, could not explain why such a design had never been accepted by the City authorities, though he did believe 'it is probable this might contribute not a little to his being taken notice of by the Magistrates of the City, and soon after made Surveyor'.[4] It is unfortunate however that Hooke's City plan of September 1666 does not survive, so that we only know its general details from Waller's description.

One must bear in mind, however, that in the seventeenth century the strict geometry of the classical style of architecture, brought into vogue by Donato Bramante and Andrea Palladio in Italy and Inigo Jones in England, made it possible for seeming amateurs to produce designs of brilliance and originality. Architecture, after all, was often studied by gentlemen as a branch of polite learning, rooted as it was in the geometry of cubes, orders, mathematical proportions and carefully-worked-out rules of elegance. A man with a knowledge of Latin and Euclidean geometry could go on to master Vitruvius' *De Architectura Libri Decem* from the first century BC along with sixteenth-century Italian architectural writers like Andrea Palladio. (We know that Hooke was very familiar with Vitruvius through borrowed copies, and his works are referred to several times in his Diary,[5] but it does not seem to have been until 25 March 1675 that he spent 34 shillings, plus 5 shillings for binding, on a copy for himself.) When, feeling unwell in September 1672, he 'read Sirlio took syrup of popys, slept little with sweat and wild frightful dreames', one hopes that the mid-sixteenth-century *Architecture* of Sebastiano Serlio did not occasion them.[6]

An aspiring gentleman architect, therefore, could then go on to make a close study of detailed engravings of Greek, Roman and Renaissance buildings, and if he were also a mathematician who combined a technical grasp of thrusts and forces, arches and beams, with a natural flair for art and design—as was the case with Hooke and Wren—there was no reason why he should not turn his hand to the design of buildings or even cities, when circumstances required. After all, Inigo Jones himself, the architect of the King's Banqueting House, Whitehall (1619) and the

Queen's House, Greenwich (1637), and the man who had first brought coherent classical architecture to England, had originally been a court stage-set designer; while Sir Roger Pratt, a county gentleman and gifted dilettante, had produced influential classical buildings in pre-Civil-War England, and would become one of Wren's fellow Commissioners of the King's Buildings.[7] But England's architectural Renaissance had been halted by the Civil War, as had most of the other arts, though not, curiously enough, the sciences. Indeed, it is interesting to consider why this should have been the case, and one wonders how much it had to do with the religious politics of images. For buildings—especially Italianate classical buildings, just like religious paintings and stained-glass windows—were redolent of 'Popery' to the Puritans and therefore not to be encouraged. Science, on the other hand, was innocent, for its visual images, being the phenomena of nature, were the direct handwork of God Himself, and therefore untainted by human pride or superstition.

Beautiful as the new ground plans of Hooke, Wren and other Virtuosi might have been, they were to remain dream cities, for London was rapidly becoming the great commercial metropolis of Europe, and could not wait two or three decades to be re-designed. It was a close-fisted city of merchants, and proud of its corporate independence from Royal control; and with Oliver Cromwell only eight years in his grave, Charles Stuart had more sense than to behave with the same self-indulgent autocracy—at least outside the confines of the Royal Court—as several of his brother absolute rulers in Europe were accustomed to doing. Regardless of dreams about a new London, therefore, property-holders began to clear their sites and rebuild afresh as soon as the charred embers were cool enough to touch. The best that could be hoped for was that new City legislation to rebuild with wider streets and more fire-resistant materials would make the new London a safer place than the old one. And remarkable as it may seem, considering the appalling devastation of the fire, which destroyed 273 acres of valuable property worth millions of pounds in the money of the day, things got moving again with astonishing speed, and by the beginning of December 1666 most of the fire debris had been cleared away. Then during the winter of 1666–7, the entire city site was staked out—not in accordance with a new master plan, but simply following the lines of the medieval streets—for which service each householder was charged 6 shillings and 8 pence by the City authorities, though larger sums were sometimes paid, perhaps for a more detailed or more complex survey, as when in September 1673 Abraham Jaggard paid Hooke £1 for his Survey Certificate.[8] The Surveyor's plans, however, incorporated not only the surviving old boundary marks of people's property, but crucial new safety requirements for wider streets and firmer foundations capable of carrying stronger walls bearing slate and tile rather than thatched roofs.

By the spring of 1667 Cheapside, Ludgate Hill, and Fleet Street, along with other great thoroughfares, were springing up as living streets again, as money and trade, the real life-forces of London, began to flow once more. Indeed, London's resurgence after September 1666 was a perfect testimony to the energy and vigour of King Charles's Capital which, while loyal to the Crown in this new post-Cromwellian world, was nonetheless proud of its own powers of initiative and self-determination.

Though neither of the official designs of Hooke and Wren was adopted, both men were to play a vital part in the creation of the new London. As we already know from their scientific work dating back to their Oxford days in the 1650s, the two men were close friends (and even distant relatives by marriage), and would come to work so closely together on their subsequent architectural projects as to sometimes make it difficult to know exactly who contributed what particular aspects of certain designs. Not long after the submission of his City plan on 19 September 1666, Robert Hooke was appointed one of the Surveyors to the City by the Aldermanic Council, to work in conjunction with his colleagues Edward Jarman, Peter Mills and, soon after, John Oliver, to supervise and erect the buildings of the City and mercantile community. Exactly who was responsible for securing him this appointment is not clear, though it is possible that Dr Wilkins, who was well connected in the City of London, and Sir John Cutler might have had a hand in it. As Jarman and Mills died within a few years, Hooke soon found himself the Worshipful City of London's Chief Architect and Surveyor. Similarly, in 1669, Christopher Wren, who was already one of the Commissioners for the Royal buildings, along with other 'gentlemen' architects such as Sir Roger Pratt and Hugh May, was appointed Surveyor General to the King in succession to Sir John Denham. As it was Hooke's responsibility to supervise the street plan, safely regulations, and the design of commercial and municipal premises, so it was Wren's to take charge of those premises belonging to the Church and royal government. With over 80 city churches destroyed in the conflagration, Wren's ecclesiastical buildings alone represented a major undertaking, including his great masterpiece, the new St Paul's Cathedral, which replaced the vast medieval Gothic church which the fire had gutted and the surviving walls of which it had weakened.[9] In architecture, both men were relative beginners, though Wren was perhaps the more experienced, with some completed work in Cambridge already behind him, and some unfinished in Oxford. Yet while the City of London never became beautiful after the fashion of Venice or Rome, it was nonetheless destined to become the magnet of eighteenth-century Europe. It formed the backdrop against which such figures as Newton, Handel, Hogarth and Samuel Johnson lived, and a collection of visiting foreigners, from Benjamin Franklin and Canaletto to Casanova, came to gape, admire and watch the world go by. It was the 'flower of cities' where, it was said, one could see every-body and everything; but still a living, practical city, as befitted a pair of Virtuoso architects.[10]

The Halls of almost thirty City companies are noted in Hooke's Diary in connection with building work. He also records undertaking work on the Thames waterfront, which, in the days before the Victorians embanked the strongly tidal river, consisted of a maze of landing stairs and small commercial boat quays west of London Bridge, while east of the Bridge were London's docks, where the wealth of the world was brought ashore from ocean-going vessels and stored in great warehouses. We do not know how many City Livery Company Halls and great commercial premises he designed as such, and how many were other men's projects about which he was consulted, or gave advice to other archi-tects, but he seems to have been closely associated with the Armourers', Dyers', Bakers' and Mercers' Halls in particular. And the Mercers' Company was one of the richest and most powerful in the City, having supervisory interests over Gresham College, whose buildings, before its establishment as a collegiate institution in 1597, had been the home of the great Tudor financier and mercer, Sir Thomas Gresham.[11]

In addition to certificating surveyed plots, and the actual designing of particular buildings, it was Hooke's job as Surveyor to make certain that all buildings erected in the City conformed to the new safety regula-tions, aimed at preventing a second great fire. Timber was to be replaced by brick and stone, thatch with slate and tiles, so that the quaint but dangerous medieval city of timber and white plastered buildings gave way to one of smart red brick. These new measures seem to have been inspired in part by John Evelyn, who had lived in Holland, and had been impressed by the relative safety, as far as fire proofing and healthi-ness were concerned, of the brick-built, tiled, broad- thoroughfared Dutch cities. The streets of the new London had to be wider with easier passage to prevent fire hazards, although this brought roars of complaint from Londoners who now found that increased street widths bit deeply into what had once been valuable commercial frontages. As Surveyor, Hooke had to deal with endless cases of this sort, besides approving new plans, advising on legal disputes concerning building lines, suggesting compensation where appropriate, and making certain that the safety requirements were met.[12] His Diary is full of references to architectural and building work, including City Court attendances and hearings which were probably in relation to disputed land-holding claims or the new regulations. Robert Hooke's contribution to the modern concept of town planning still deserves to be more fully studied and assessed by historians of this branch of architecture, though the masterly researches of Professor Michael Cooper into City of London archives have already opened up a whole new historical understanding as far as Hooke's part in the rebuilding of London is concerned.

Exhausting and time-consuming as his City Surveyorship must have been to a man who was still a full-time scientist, at least it was well paid. Not that the City itself lavished an especially large salary upon him—he received £150 per annum, paid quarterly from Lady Day (25 March) 1667[13] — but rather, it was his right to collect fees and perquisites from all architectural clients, building applicants and others who made demands upon his time. Hooke made 'many thousands of pounds' from his City position, and accomplished the transformation thereby from relatively penurious scholar to a gentleman of considerable estate. Indeed, his Diary during the 1670s records numerous financial transactions relating to surveying, providing building plans, costing up jobs, and the like; and while these sums could also have included amounts due to be paid out to others, one gets a sense of the very substantial sums that passed through Hooke's hands and a healthy percentage of which no doubt entered his own pocket. Rebuilding one of Europe's great capital cities was not, after all, a cheap enterprise, and considering Hooke's very senior role in the business, which saw him courted by rich merchants, shopkeepers and even members of the aristocracy and Royal Court, one can fully understand how a man of his natural energy and acumen could scarcely avoid growing rich.

But it is not just in the realm of civic administration that Hooke's architectural reputation is to be found. Though we have no idea from whom, or how, he really learned his architecture, he began almost straight away to produce designs for major public buildings that were both beautiful and completely functional. This was, after all, the traditional task of the architect and builder from classical times onwards, and it is a sad aberration of much of the twentieth century that it expelled beauty from architecture and made the once reputable word 'functional' synonymous with 'brutalist'. Many of Hooke's buildings, moreover, were comparatively cheap, considering their size and grandeur, and the sudden shortage of materials which the complete rebuilding of London must have occasioned.[14]

So while Hooke never formally studied architecture either at home or on the Continent, he absorbed the main elements of the fashionable French, Flemish and classical styles and used them in several of his best creations. It is quite likely, however, that he gained a grasp of basic style and techniques by studying the pattern books of buildings and detail decorations which were quite readily available, as well as by purchasing published plans of the great buildings of Europe. We know, for instance, that though Sir Christopher Wren worked in an essentially Italian style, he never personally travelled much farther south than Paris, in which city he met the celebrated Italian architect Giovanni

Lorenzo Bernini, and through him procured drawings of classical and Renaissance masterpieces. While we have no record that Hooke ever met Bernini, he may perhaps have visited France in the mid 1660s, although he could easily have gained access to important pictorial material in London. The spoken word, too, played its part in firing Hooke's creative imagination. In early July 1674, for instance, he was clearly fascinated by a verbal account of two new Dutch buildings: 'Mr Story... returned from Holland Saturday last. Told me of the new Lutheran church 70 foot diameter... and 70 foot over Amsterdam. Of the Burghers hiordiage. Of the Jews new Synagogue 100 foot square.— Drank ale with him.'[15] But it is impossible to be sure where Hooke acquired that architectural wherewithal which enabled him not only to be considered for the Surveyorship to the City of London, but also to attract a string of lucrative private commissions that included Livery Company halls, country mansions, hospitals and churches.

One of Hooke's biggest and probably grandest buildings was London's Bethlehem Hospital in Moorfields. This medieval foundation, with its 25 inmates, popularly known as 'Bedlam', was one of the notorious lunatic asylums in Europe, and even today we use the word 'bedlam' to signify a scene of uproar. In the 1670s the old 'Bedlam' Hospital, which had fortunately survived the fire anyway, was completely rebuilt and vastly extended by Robert Hooke between 1674 and 1676 at the behest of its Governors. The new Bethlehem Hospital boasted a magnificent façade 540 feet long, with three pavilions and adjacent buildings. Set out in gardens, it had more of the aspect of a palace than a madhouse, and became one of the showpieces of the new London. When its similarity to the new palace of Louis XIV at Versailles was observed, this gave rise to the remark that what was fitting for a French king was only good enough for an English lunatic, though the wit Ned Ward, in *The London Spy* around 1699, quipped 'I think they were mad that built so costly a College for such a crack-brained society'.[16] Not only was the outside of 'Bedlam' well designed, but in its early days it must have been something of a model hospital for the insane, with individual rooms for the inmates, and light, airy surroundings. These luxuries were not to last long, however, for by the 1730s 'Bedlam' had become the overcrowded, insanitary, prison-like madhouse captured in graphic detail by William Hogarth in the last picture in his series *The Rake's Progress*, and parodied in Alexander Pope's *Dunciad*. It also became a notorious haunt of voyeurs, who paid good money to the keepers for the pleasure of being allowed in to stare or laugh at the inmates. In the belief that the insane, in losing their reason, had also forfeited their right to human dignity, it was held by seventeenth-century physicians and 'mad doctors' that the lunatics should be chained, whipped and confined in darkness as a way of making them regain control over themselves—a regime all too graphically depicted in Hogarth's Bedlam engraving. Furthermore, the increasing condemnation and humiliation of the insane was linked to the medical realization that the growing army of the mad had somehow brought their condition upon themselves as a result of catching syphilis. What came to be called the 'general paralysis of the insane' was a polite term for tertiary syphilitic madness: Hogarth's Bedlam-incarcerated Rake has upon his chained and naked body the dark chancres of the disease which he had caught by letting his lust destroy his reason.

All that survives of Hooke's splendidly-appointed hospital today, apart from an engraving which captures its spacious elegance, are the two badly-worn statues of the 'Crazy Brothers', depicting *Raving Mania* and *Melancholy Madness* — representing what we might call schizophrenia (combined perhaps with syphilitic madness) and depression — which had once flanked its grand entrance.[17]

The popular association between the Royal Society and Bedlam asylum must have run deep in the popular thought of the period, for in Ned Ward's scurrilous account of late seventeenth-century London, portrayed in his journal *The London Spy*, the Londoner who was showing his country visitor around Town went 'from Maggot-Mongers' Hall [the Royal Society, a nickname redolent of Hooke's *Micrographia*] to survey Madman's College [Bedlam]', as though the two institutions possessed a natural association.[18]

But not only was Bethlehem Hospital built in an astonishingly short period of time, considering its size and architectural opulence, but like so many other Hooke buildings, it was finished comfortably within its £17 000 budget; and so pleased do the Governors appear to have been with their new building, that they gave Hooke an extra £100 beyond the £200 fee which he had charged for his services.[19]

One of the historical problems in assessing Hooke's architecture is the frequent confusion of his buildings by subsequent writers with those designed by Sir Christopher Wren: a confusion, indeed, compounded in Sir Christopher Wren junior's life of his father in *Parentalia* (1750) which leaves one to assume that Sir Christopher senior had rebuilt London pretty well single-handed! As the two men were working closely together, often in a similar architectural style, in the same City, the confusion is understandable; and it was not until the discovery of Hooke's Diary in the late nineteenth century that the true provenance of several buildings traditionally accepted as Wren's became certain.

One of these was the famous Royal College of Physicians' building in Warwick Lane, erected between 1671 and 1679 and referred to many times in the Diary: e.g., in December 1672, 'Gave Dr Whistler estimate of physitians theater'.[20] Though it was demolished in the nineteenth century, accurate drawings and even photographs survive, which clearly demonstrate the grandeur of the structure. For a time the physicians could not decide whether to place their anatomical theatre at the front or back of the building, and when they finally opted for the back, Hooke crowned it with an ornate golden ball, which soon earned the nickname of the 'gilded pill'. The finely-proportioned fenestration and levels of the physicians' building are superb, and certainly discredit those modern architectural critics who claim that Hooke's creations were heavy and prosaic. The fact that posterity so frequently confused his work with Wren's is surely proof enough of its quality. It is also likely that his 'View at Chirurgions hall 7 sh[illings]' was a small fee pertaining to work on this building, such as a minor inspection.[21]

Hooke's other public buildings include Bridewell Prison, the great gothic gatehouse of Newgate gaol, and parts of Christ's Hospital, all of which became established parts of the eighteenth-century visitor's tourist trail, the first being an institution for the punishment and reform of prostitutes (the interior of which was depicted in plate 4 of Hogarth's *The Harlot's Progress*) and the second a charitable school that trained up young gentlemen to become mathematically-competent naval officers. The Monument or 'Piller', erected in the 1670s to mark the starting point of the Great Fire near London Bridge and still standing, was an official Wren creation but almost certainly constructed in consultation with Hooke, combining as it did both royal and civic remembrance of the conflagration. Indeed, Hooke's friend John Aubrey, who cannot always be relied upon, listed 'the Piller on Fish-street-hill, and the Theatre there'[22] as Hooke buildings. The 'Piller' or Monument was an elegant stone column 202 feet high, ascended by an internal spiral staircase, and afforded panoramic views of the City until modern skyscrapers closed it in. Even nowadays, on a clear day, the view from the top can be breathtaking. Crowned by the phoenix rising from the flames, the Monument was also intended as a place for making astronomical and physical observations if a zenith sector telescope were set up in its 202-foot interior shaft. While the Monument never became the location for significant astronomical observations, Hooke did record on 16 May 1678 that when a barometer was conveyed from street level to the top of the structure, its mercury fell by one-third of an inch due to the fall in air pressure with altitude.[23] Although it will never be possible to be sure how much of Hooke's personal input went into the design of the Monument, we certainly know from his Diary in October 1677 that he was included with Wren and others in a discussion 'about monument inscription', which remark presumably relates to the Monument's dedication stone.[24]

Though Wren is still officially regarded as the architect of the Royal Observatory buildings at Greenwich, it is now known that he collaborated with Hooke in the work. Hooke had, as was shown in chapter 5, played a well-documented part in designing many of the instruments originally placed there, and no doubt proposed designs for the very specialized buildings intended to accommodate them. The original regulator clocks, with 13-foot pendulums, were almost certainly inspired by Hooke's and Wren's recent experiments with long pendulums (although their escapements were designed by Richard Townley), and themselves dictated the height of the famous Octagon Room in which they were mounted.[25] This elegantly-proportioned Octagon Room forms the centrepiece of the Observatory, and in addition to accommodating the long-pendulum clocks — with which Flamsteed first confirmed the homogeneous rotation of the earth on its axis in 1678 — has tall, narrow windows designed for the handling of the long-focus telescopes of the period, as Robert Thacker depicted in his contemporary engraving. Nor is it by any means unlikely that Hooke collaborated with Wren on many of his other buildings, including his churches, just as Wren was probably invited to make suggestions in the drafts for Bethlehem Hospital or the College of Physicians. The two men, after all, had been friends since their Oxford days, and Wren's name occurs more frequently than that of any other single individual in Hooke's Diary between 1672 and 1680.

As mentioned above, Hooke's contemporary reputation as an architect can also be gauged from the private commissions which he received in addition to his official City work. These included projects as diverse as the laying out of very up-market residences in St James's Square, charitable almshouses, at least one church and even involvement in the design of the defensive Mole or fort when the British briefly held Tangier. In the 1670s, moreover, the connoisseur Lord Montague commissioned a design for his new house, standing on the site of the present British Museum. But the mortar had scarcely had time to set before Montague House accidentally caught fire and was gutted. A French architect was called in to supervise the rebuilding, but it has been thought that most of his work concerned the interior, as Hooke's stone shell had probably survived. Some of the fragments of Hooke's Montague House may well have been incorporated into the British Museum building, which was put up in the eighteenth century. Other private commissions included work for Lady Ranalaugh, and Lord Conway's Ragley Hall in Warwickshire.[26]

Hooke records that he left London in the middle of June 1680 to ride up to visit Lord Conway at Ragley Hall on a journey that lasted well over a week, and for which he received a fee of 30 guineas in gold. One immediately gets a sense of Hooke's social status from his Diary account of the visit to Ragley. He was given 'Lodging in the Best Roome', dined, went to Church, and socialized with Lord Conway and his family,[27] and after making suggestions about the architecture, was prevailed upon to stay an extra day 'to see Sir J. Mordant and Mr Parker who dined with My Lord and Dick Kemson'.[28] Indeed, Hooke's visit to Ragley gives us a delightful glimpse of the life of a great country house of the Stuart age, where, apart from enjoying his own obvious celebrity status, as Virtuoso and gentleman architect, he was interested to make the acquaintance of 'Leonard an ingenious German mechanick servant to Van Helmont, the Butler who had lived 3 years in Constantinople', along with one 'Holbert a Carpenter but a Pap.[ist]', and 'viewed the remainder of H. Stubbs Library to be yet sold'. Hooke held 'Mr Wilson the Parson' in low esteem, he being 'I doubt a Sychophant or worse. Dined with My Lord. Not at Church in the afternoon'.[29] Though altered by Wyatt and subsequent architects, Ragley, which still stands and is occupied, has Hooke's imprint upon it, and in its day was regarded as a most elegant house.

Indeed, these jaunts out of London to direct or advise on architectural work were not only profitable financially, but could also be highly enjoyable and flattering social occasions. In addition to the celebrity visit to Lord Conway's Ragley Hall, the Diary recounts another convivial jaunt out to Shenfield, Essex, with his client Dr Wood, in May 1678, where he ate 'pyes... suppd with Dr. Wood and his Lady. Surveyd his house and directed it', then walked with Wood. But Hooke noted down, like the good professional that he was, that Dr Wood paid all his expenses and fares, though he found the stagecoach journey back to London tedious because of the 'twittle twattle company in the coach'.[30] One suspects that Hooke was not good at small talk with strangers who did not have active intellectual interests.

The intervening three centuries, however, have dealt harshly with Hooke's architectural creations. This has come about through no inherent defects in the buildings themselves, but by accident, changing fashion, and the rebuilding of London in the nineteenth century. One of his best pieces of interior design, the magnificent wooden screen in the Merchant Taylor's Hall, London, was destroyed by a bomb during the blitz in the early 1940s, although photographs of it still survive.

The last fully authenticated, complete Hooke building still standing is the parish church of St Mary Magdalene, in Willen, Buckinghamshire, which is now part of Greater Milton Keynes. This beautiful little church, seating about one hundred people, has come down almost intact, the only alteration being the addition of a small apse to the east end, to increase the altar area. Built in red brick with white limestone facings in a Flemish style somewhat reminiscent of the Royal Observatory, Willen Church was commissioned by Hooke's old Headmaster and friend, Dr Busby, who owned the advowson. From the extremely detailed accounts and documents that survive from the building of the church, and are preserved in Westminster Abbey Library, we can reconstruct a good deal about Hooke's approach to provincial building commissions and the costs involved in 1680, when the fabric was erected.

By the end of March 1679, Hooke had 'made draught of church for Dr. Busby',[31] while on 4 April he 'Drew mouldings for Dr. Busby.' Indeed, Dr Busby and work pertaining to Willen Church make fairly regular appearances in his Diary over the next few weeks, and on 21 April he records 'Drew designe for Dr Busbys church', which could well relate to the drawing which survives in the British Library.[32] Yet Willen Church as it was built and still survives does not correspond to this drawing, and the contrast between the drawing and the one surviving Hooke building gives us a hint of the amount of design modification and negotiation that may have gone into a single piece of Hooke's architecture. On the other hand, Willen was not the only church with which Busby was concerned, and the 21 April drawing could indeed relate to a different ecclesiastical foundation altogether. Even so, the fundamental re-working of initial architectural designs was common: Sir Christopher Wren, after all, went through an elaborate series of modifications with the Dean and Chapter for the much grander Cathedral of St Paul's.

Fortunately, quite a large body of information survives about the construction details of Willen Church. The materials, including 242 613 bricks, timber and labour, cost £2202, which was thought to be expensive by prevailing price levels.[33] Hooke seems to have provided drawings and designs for Willen Church, no doubt sent up some of his London workmen to keep an eye on things, and made at least one visit to Willen, or 'Welling' as he called it, between 3 and 5 May 1680. The 50-mile journey, via St Albans, Dunstable and Newport Pagnell, began at the George Inn in Aldersgate Street at 4 o'clock on Monday morning, and lasted 13 hours, after which he 'Slept well.' But after he had 'measured church' on the following morning, one senses that Hooke's celebrity presence attracted some local note, for after it, he 'Dind at Mr. Stevensons, Dr. Aterbury, Plucknet, Bates, Horn, Tufnell.' Supper and another night of sound sleep followed at *The Swan* in Newport Pagnell. He travelled back to London on the Wednesday, and must have been on his way before dawn, first riding the 16 miles from Newport Pagnell in time to catch the 7.30 a.m. coach from Dunstable. He arrived back in London at 6.00,p.m., and immediately repaired to Jonathan's Coffee House to talk to his friend Sir John Hoskins.[34]

While it is clear that Robert Hooke did leave London on several occasions to undertake architectural consultancies, it was not unusual for an architect of the period to supply drawings for a commission and then, very largely, leave it to a reliable builder to erect, decorate and fit out a fabric. This could well have been the case with some country commissions, for Hooke's commitments meant that he was rarely free to travel far from London.

While Hooke was not an ecclesiastical architect in the way that Wren was, there is no shortage of documentation for the fact that he was involved in church structures. Many of these commissions, no doubt, were undertaken in conjunction with Wren, though it is clear that fees for specific aspects of church building—most likely those that pertained to the new building legislation—were paid directly to Hooke and his approved workmen, as when, in 1673, he 'agreed with Russell to take down St Antholin's Steeple to the water table viz. 40 foot for £8'.[35]

The seventeenth-century building trade, especially considering the new opportunities offered by the post-Fire rebuilding of London, was notorious for what might be called its creative financial practices. Stone, bricks, timber, and other commodities from other parts of Britain that were brought up the Thames by coastal traders became the subjects of all manner of 'graft'. Similarly, the labour market, with its gangs of specialist workers, all with their craft and 'tribal' affinities, operated an elaborate system of perquisites and fee distribution, as did parts of the City of London's own bureaucracy. If these time-honoured practices were not abused, however, they were accepted as part of the natural order of things: in the days before the Victorian concepts of an adequate salary that bought the whole of one's working time and of public account-ability came to change the business ideals of the Western world, taking one's time-honoured cut for the job was not reckoned to be dishonest. Indeed, even Wren himself was perfectly happy to work within this system of perquisites.

While Robert Hooke no doubt took his own time-honoured cuts as Surveyor, and amassed a 'great Estate' from a relatively modest profes-sional income, all the evidence suggests that Hooke was reckoned to be a man of great probity who could be trusted not to cheat his architectural and City clients. It is also likely that a good part of Hooke's 'great Estate' derived 'from Persons soliciting to have their Ground set out, which, without any Fraud or Injustice, deserved a due recompense in so fatiguing an Employ'.[36] Yet when one considers how many individual plots of ground the City Surveyor had to determine and for which he would receive a fee before his clients could legally re-build their premises, one senses the magnitude of the earnings that could legitimately come his way.

Hooke's sense of professional probity in his architectural work, and the pride that he felt in keeping his building projects within the estimate given to clients, was clearly hurt, however, following a controversy which broke out regarding a commission which went well over budget in 1691. This pertained to Hooke's estimates for the building of a charity hospital at Hoxton, paid for out of the will of the late Alderman Ask. As Hooke argued in his defence, the excessive building costs were occasioned not by his own ineptness or dishonesty but by two factors beyond his control. Firstly, the clients had made their own additions and alterations to Hooke's original plans and estimates, thereby driving up the cost; and secondly, Hooke was not able to procure the workmen personally, 'which if he had done, as he said, he would have ingag'd it should have come to little or no more than his first propos'd Sum'.[37] This latter statement in particular gives us a glimpse of how Hooke routinely related to and advised his clients, having, no doubt, his own cadre of trustworthy workmen whom he knew would do a sound and economical job, and feeling annoyed when a client chose to replace them with non-approved men of the client's own choosing. Quite clearly, Hooke exercised a system of quality control on his own building contracts.

Architecture transformed the physical and financial circumstances of Robert Hooke's life, lifting him from the ranks of comfortably-off bachelor dons and academics who lived on a hundred or so a year, and enabling him to become, in the words of his friend John Aubrey, a man of 'great Estate',[38] who would leave almost £10 000 at his death. Professor Michael Cooper has done more than any other scholar to unravel the sources of Hooke's income, architectural and otherwise, from detailed researches, conducted primarily in City of London archives. It seems, for instance, that Hooke earned something in excess of £1500 for the 3000 ground plans, or Foundations Surveys, which he did for property owners who wanted to start rebuilding their premises, though most of this work was already completed by the early 1670s.[39] Then, of course, there was his £150 per annum salary as Surveyor, which he would have received for a good 30 years, not to mention private fees for numerous individual architectural commissions, such as the £300 which was paid to him by the Governors of Bethlehem Hospital.[40] All of these building and architectural monies were quite separate from his scientific and academic incomes, which comprised his Gresham professorial stipend of £50 a year from 1665 to 1703, his £30 Royal Society Curator's salary, running at least until 1678, and the total sum of around £1500 which he received eventually for his Cutlerian Lecturership (£925 had already been paid, and for the outstanding £550 he successfully went to law in 1696).[41] Truly, a 'Gentleman, free and unconfined', who as early as 1676 had seriously considered investing £4000 in the purchase of broad acres at Alvington on the Isle of Wight, though eventually deciding against it.[42]

Yet a life of leisure was not for Hooke: he simply added architecture to his already prodigious list of skills and duties after the Great Fire, and then found he was good at it. For someone apparently lacking previous training or experience in the building trades before suddenly finding himself responsible for re-erecting the richest city in Europe, he manifested remarkable talent, not to mention confidence. Possessed of a natural eye for proportions, an inborn artistic ability, and the obvious knack of getting workmen moving, this man, who a year or two before had been known primarily as a microscopist and experimental scientist, was designing some of the most beautiful buildings in London.

Nor must it be forgotten that architecture and the various bureau-cratic duties of the Surveyorship were in no way allowed to supplant Hooke's *real* career, that of a scientist. During the hectic years when London was rising from the ashes under his direction, Hooke was still delivering Cutlerian Lectures, which he wrote up as the books *Helioscopes*, *Lampas*, *An Attempt to prove the Motion of the Earth* and *Animadversions on Hevelius*, curating the Royal Society, then acting as its Secretary between 1677 and 1682, and corresponding with Virtuosi across Europe, besides equipping the Royal Observatory and accepting private architectural commissions as well. What is amazing is how he found the time and energy to do so many things all at once, let alone struggle against seemingly ever-present, though perhaps to some degree stress-related, illness, prostrate himself with purges, and occasionally fall out with his colleagues. But diversity was the essence of Hooke's life, and as he dashed around the City on foot, 'stooping and very fast', carrying architectural plans or pieces of apparatus, or discussed gravity, blood-transfusions and the formation of lunar craters in a coffee house, one finds the Curator of Experiments at his most irresistible and compelling.

CHAPTER 8

A WORLD TURNED UPSIDE DOWN: HOOKE'S GEOLOGICAL IDEAS

Although, as we have seen, Robert Hooke's scientific bent inclined him more to physics than to natural history, one branch of the natural history disciplines held a lifelong appeal for him: geology. From his early days on the Isle of Wight down to his last 'Discourse' on the subject delivered to the Royal Society on 10 January 1700, Hooke was fascinated by those processes which formed the earth and produced both the 'Ante-Diluvian' and modern habitats for the creatures that lived upon its surface. Of course, one could say even here that Hooke's interests came closer to those of the physicist than of the naturalist, for what really concerned him were the *physical* processes that had formed the earth as an astronomical body, while his interest in surviving organic remains was more by way of their use as physical benchmarks to archaic 'Earthquake' processes. Even so, Hooke's impact on early scientific geology was enormous, and modern scholarly studies by David Oldroyd and, more recently, Ellen Tan Drake have emphasized the quite extraordinary and far-sighted contributions which he made to the science, and which only in the late twentieth century have received the recognition they deserve.[1]

Hooke's geological writings were published in three places. As with so many other lines of research, it was in *Micrographia* that he reported his first studies, the examination of 'petrified' or fossilized wood under the microscope.[2] Between 1667 or 1668 and 1700, he delivered a series of at least 27 lectures or 'Discourses' to the Royal Society on the generic subject of 'Earthquakes', or earth-forming processes, which were published in 1705, though it would appear from a remark made in 1697 that he had lectured on the subject of Earthquakes as early as 1664.[3] Among Hooke's manuscript materials gathered up and published by William Derham in 1726, there were a series of pieces on petrified bones, the 'Nautilus' shell creature, and notes and observations on solidified resinous amber as dug out of deep sand measures.[4] As usual, the ever-informative Diary recorded details about contemporary earthquakes either witnessed in England during the 1670s or recounted to him by travellers from abroad.

One suspects, however, that two events were particularly instrumental in focusing Hooke's thoughts upon a physical, empirical science of geology. The first was the above-mentioned examination of petrified wood, as described in *Micrographia*. The second was the one or more visits which he paid to the Isle of Wight after October 1665. We know that he visited the Isle in the autumn of 1665, after the death of his mother in the preceding June, for he says as much in a letter to Boyle in September 1665;[5] and then, in an undated 'Earthquake Discourse' delivered some time before September 1668, and most likely that of 27 June 1667, he refers to what were almost certainly geological observations made on the Isle: 'I had this last Summer an Opportunity to observe upon the South-part of *England*, a Clift whose Bottom the Sea wash't that at a good height in the Clift above the Surface of the Water, there was a Layer, as I may call it, or Vein of Shells, which was extended in length for some Miles.' Had Hooke's childhood familiarity with the natural curiosities of the Isle of Wight produced new insights when these were re-examined by the 30-year-old Fellow of the Royal Society?[6]

Where Hooke's geological writings are so significant is in the remarkably open-minded approach to the science which he displayed: in his willingness, at this infant stage in geology's development, to review evidence as objectively as possible, bearing in mind the way in which cultural perspectives inevitably mould our objectivity. His aspiration to objectivity, however, lay in his readiness to look at things in their own natural context as far as was possible, and not to attempt to squeeze them into established formats suggested by classical or Biblical literature. While Hooke undoubtedly possessed an acute sense of the Divine Providence and Majesty, he believed that God's gift to mankind of a quizzical intelligence was meant to be used to unravel the wonders of Creation; and while he seemed to acknowledge that the New Testament was the true road to salvation, he was nonetheless cautious about taking the *Genesis* narrative too literally. While accepting that God was the Creator, Hooke was all too aware that whoever wrote down the Creation story could not have witnessed it personally, nor in all of its sublime detail. What we get in the divinely-inspired but clearly subsequent Mosaic narrative, therefore, is not a direct eye-witness account of a piece of phenomena, but a communication of divine awe and power as recorded through the limited and *post-facto* perceptions of mankind. Moses, whom seventeenth-century scholars believed was the recorder of the *Genesis* account, was simply writing the Creation narrative, along with the Jewish Law, as it came to him from the word of God during that period between the Children of Israel's Exodus from Egypt and their entry into the promised land of Canaan. A time, indeed, when it was believed that the earth was already over 2000 years old.[7]

While Hooke's geological writings fully acknowledge the central truth of Scripture, they are not restricted by it on a literal level, and on matters of terrestrial time-scales, continent formation, extinction, fossilization and species change, he was willing not only to look at the physical evidence as objectively as possible, but also to draw ideas taken from classical and Chinese literature, and even from geology's more junior science of archaeology. As an accomplished classical scholar, moreover, Hooke was familiar with tales of transformations and changes in nature, and it is not for nothing that Ovid's *Metamorphoses*, with its own Creation and Flood narratives and underlying message that things have not always been what they are now, is so frequently referred to in his geological writings.[8]

From his observations of a piece of petrified wood originally given to him by 'my highly honour'd friend' John Evelyn in or sometime before 1664, and from sea shells dug out of the ground, Hooke drew a conclusion of enormous geological importance. He discounted the traditional view that such objects were indigenous mineral encrustations made by some *Vis Plastica*, 'Plastick Virtue', or moulding force in the earth, or else some 'Astrological or Magical Fancy',[9] and proposed instead that these substances had once been living, organic things. Under the microscope, he was able to see cell structures within the fossilized wood that were identical to those seen in normal wood, though their burial underground in the right chemical conditions had meant that '*petrifying* particles' had been 'intruded' into them and had effectively replaced the organic structures with mineral ones. Chemical tests with vinegar and other solvents produced the same reactions as they did with normal stony fragments. 'Nor is Wood the onely substance that may by this kind of *transmutation* be chang'd into stone', Hooke reminded his reader, for he had examined similar structures of bone and sea shell all 'petrify'd'. What is more, he had even 'set down Shels, found about Keinsham, which lies within four or five miles of *Bristol*, which are commonly call'd *Serpentine-stones*'.[10] These were probably forms of ammonite, which were a type of fossilized creature to which he would return on several occasions. By 1665, therefore, Hooke had clearly come to realize that, under the right conditions, buried organic materials could be 'metamorphosed' into petrified mineral, before going on to discount the old notion that 'figured' or 'petrified' stones were sports of nature formed by a 'Plastick Virtue' or inexplicable moulding force within the rocks themselves. This concept of the organic origin of fossils (which was to be one of the crucial pre-conditions for the rise of scientific geology) would be developed and applied by Hooke over the next three decades to a variety of geological phenomena as part of his study of the Ante-Diluvian or pre-Biblical-Flood world.

The Great Flood of Noah, as described in the Seventh Book of Genesis, was universally regarded as the primary geological agent in the seventeenth century, and contemporary writers about terrestrial history, such as John Ray and Thomas Burnet, invariably spoke of it as such. Indeed, most writers would continue to do so until the early nineteenth century.[11] This Flood was believed to have taken place some 1655 or 1656 years after the Creation, or in 2349 BC, being sent by God as a punishment to wipe out the wicked people of the world, leaving only Noah and his family to survive in their Ark. According to Scripture, the Flood was brought about by 40 days and nights of ceaseless rain, which produced an inundation that covered the world's mountains and stayed on the earth for 150 days after the rain had ceased. Not only did it destroy sinful mankind, but also those animal species which failed to get into Noah's Ark in breeding pairs.

It was this Flood, according to received opinion, which was responsible for terrestrial stratification, as the waters dried up and deposited their detritus. It was also held to be responsible for erosion features in the landscape, the seams of shells on mountaintops, and the deposition of peculiar skeletal remains — which were believed to be those of lost creatures that had not managed to get into the Ark. What no one doubted, however, was that Noah's Flood was the *last* divinely-inspired global catastrophe, for quite clearly in Genesis VIII:21, God Himself had stated 'I will not again curse the ground any more for man's sake.'

While Robert Hooke had no problem in accepting the historical truth of the Genesis Flood narrative — most clearly seen in his use of the terms 'Ante' and 'Post' Diluvian for the world's population — it would seem that even by 1668 he did not regard the Flood as the first or indeed the most decisive geological agent to change the earth's surface. Quite simply, considering the complex character of the earth's surface and the petrified objects deposited within it, Hooke felt that a mere 190 days of aquatic turmoil was insufficient to account for everything and, as he suggested in his Lecture of 26 May 1697, they had not even been enough to soften the ground to facilitate the major sinking or moving of large objects within it.[12] In the same lecture, he discounted the reported find of a great ship in Switzerland with 'Anchors and sails, tho' torn' and the remains of its crew, discovered some 600 feet underground, as having been an ancient vessel overwhelmed by Noah's Flood.[13] A much more likely explanation, argued Hooke, was that the ship had been the victim of some ancient earthquake which had dragged it into a deep submarine crevice and subsequently buried it, or else it had been on a great and deep lake when the earthquake sank and buried it. Lake Geneva, after all, so Hooke reminds us, is an example of a large and very deep Swiss lake, and the ancient vessel could well have been overwhelmed in one like it.

But either way, Hooke saw no connection between this Swiss wreck and Noah's Flood, and was only willing to ascribe the wreck to a sudden *natural* cause. Indeed, in his numerous writings on earthquakes, Hooke says relatively little about Noah's Flood as a *primary* geological agent, probably because he could find no obvious or completely convincing natural mechanism to adequately explain such a *global* inundation, while the earth-softening effect of a mere 190 days of flooding would have been insufficient to account for all the geological phenomena ascribed to it. In 1694, moreover, Edmond Halley, who as Clerk to the Royal Society after 1686 must have heard a good number of Hooke's Earthquake Discourses, quite apart from the ideas he picked up in conversation with Hooke himself, was to do his own experiments and calculations on the volume and power of 40 days and nights of heavy rain, as will be discussed shortly.[14]

While Hooke may have been reluctant to attribute too much geology to the Noachian Deluge, he was certainly of the opinion that the earth had witnessed many violent disturbances, or earthquakes, and that these had been the agents behind not only the present state of the rocks, but also what we would now call the ecological distribution of living things. It is perhaps not surprising, therefore, that Hooke's recorded investigations into the ancient earth say relatively little about either the supposed geological events of *Genesis* or of the commonly-assumed 4004 BC Creation date. Instead, especially in his earlier 'Discourses' before 1687, Hooke concentrates upon observable *phenomena*: anatomically-interesting fossils owned by the Royal Society or dug up by reputable scholars; strata encountered when sinking deep wells (such as one found near Amsterdam which, at a depth of 99 feet, struck a four-foot-thick layer of marine shells and sea sand);[15] and marine detritus found on mountain tops. Hooke abstains from commenting upon Biblical narratives, one suspects, *not* because of any fear of ecclesiastical reaction (after all the Bishops of Chester, Rochester and Salisbury, along with John Tillotson, Dean of St Paul's and later Archbishop of Canterbury, were amongst his personal friends, along with those eminent Christian laymen, Boyle and Willis), but because, like Galileo, he believed that the Bible was to teach us how to go to Heaven, rather than to teach us 'How the Heavens go' in precise scientific terms.[16]

Central to Hooke's geology was his awareness of the great changes through which the ancient earth had passed, and of which there was no clear written record in Biblical or pagan sources. It was in this respect that his youthful and subsequent knowledge of the Isle of Wight was significant, for the Island, along with the mainland coast across the Solent, such as Purbeck and Portland just to the west, seemed to give proof of major changes in prehistoric sea levels. Near the Needles, to the west of the Isle, only a couple of miles from his native Freshwater, for instance, was a bed of shells over a mile in length, and lying 60 feet above high water mark.[17] Hooke also cited similar cases of shell-beds high in the Alps—not to mention those found 99 feet below modern ground level near Amsterdam.[18] Everything suggested that there had been fundamental changes in land and sea levels that could *not* be simply accounted for by Noah's Flood and its 190-day inundation.

Bearing directly upon these changing levels were discoveries of large 'serpentine' shells, *Cornu ammonis*, or giant ammonites. Reference has already been made to Hooke's *Micrographia*[19] observations where he mentions such stones being found near Keynsham, Bristol. However, the area around the Solent became especially significant once again: 'it has very much been urged upon the Consideration of the Petrifaction or *Cornu Ammonis* taken out of the Quarry of Stone in the Isle of *Portland*, whether it could be reasonably supposed that ever there were in the World a species of the *Nautilus* of this shape, and of so vast a bigness, of which it is supposed the World has not afforded an equal in a living Species'.[20] Yet such giant shell fossils were not unique to the Solent or west country, for Sir Jonas Moore had seen a giant ammonite in Yorkshire 'which was full as big as the fore-Wheel of a coach',[21] while Henry Howard of Norfolk presented one of 2 feet 6 inches in diameter to the Royal Society.[22] And as if to cap them all, Hooke drew attention to 'the concave Impression of one of greater Magnitude, which I found in a Piece of *Portland*-stone'.[23] Whether Hooke found the ammonite *in situ* on the Isle of Portland, just down Channel from the Isle of Wight, when he was geologizing there in 1665 or 1666, or whether in a block of Portland stone which he subsequently examined elsewhere, is not made clear in his statement.

So Hooke's conclusions, by the time that he delivered his third series of 'Earthquake Discourses' in January 1687,[24] were threefold. Firstly, that fossils—which are to the naturalist what 'medals' are to the antiquary when it comes to understanding geological sequences—were the 'petrifyed' remains of once-living creatures, and not just twists in the rock. Secondly, that there had been radical changes of sea level, as indicated by the discovered location of organically-derived remains. And thirdly, and quite momentous in its implications, was Hooke's suggestion that hill-tops in England had once formed the beds of *tropical* oceans. Hooke drew this conclusion from the giant ammonites found around Portland and elsewhere, for as the 'hotter Countries, such as are in the *Torrid Zone*, produce Turtles or Sea Tortoises, abundantly more exceeding the smaller sorts of these colder Regions',[25] so it was more likely to have given rise to gigantic *Nautili* or *Cornu ammonis* than to the modest shell creatures now found in the cold waters of the English Channel. But it was the mechanism Hooke proposed by which all of this came about that marks him out as an imaginative as well as an inductive scientist of genius. This was his 'wandering poles' hypothesis announced in his Royal Society Lecture on 26 January 1687, in which astronomical phenomena combined with stratigraphic to produce the first dynamic theory of continent formation in the history of science.[26]

Hooke argued that while the earth rotated upon its axis each day, and moved around the sun in a year, maintaining both a stable orbital plane with relation to the sun and a stable angle of orbital tilt at $23\frac{1}{2}°$, it appeared to slowly roll about within these fixed coordinates over long periods of time. The result of this rolling was that any one point on the surface of the earth could have been pretty well anywhere else given enough time. The present polar regions of the globe could, for instance, have been in the same latitude as the present Mediterranean: 'I would desire them to consider, whether . . . this very land of *England* and *Portland*, did, at a certain time for some Ages past, lie within the *Torrid Zone*; and whilst it there resided, or during its Journying or Passage through it, whether it might not be covered with the Sea to a certain height above the tops of the highest Mountains.'[27]

In this extraordinary model for geophysical activity, Hooke is proposing that the gradual yet inexorable motions of the earth's mass, within the prescribed parameters of its fixed orbital dynamics (of plane of rotation and tilt of axis), combined with tidal and as yet unclear gravitational drag effects, are constantly re-modelling the face of the planet. Also, while the $23\frac{1}{2}°$ axial tilt of the rotating sphere might itself be dynamically stable, those two areas on the earth's surface which constituted the poles at any one time inevitably 'wandered off', to be replaced by new polar zones in other parts of the planet. Hence, one could find the long-dead denizens of a tropical ocean entombed in the rock of the Isle of Wight 60 feet above the present-day Solent, or ancient periwinkles in beds of Alpine shale.

Yet what really gave a plausible edge to Hooke's 'Polar Wandering' hypothesis in its developed form by 1687 was the late-seventeenth-century realization that the earth was not a simple sphere, but an *oblate* one, wider across the equator than through the poles. An oblate or polar-flattened shape for the earth had first been proposed after the French astronomer Jean Richer discovered in 1672 that when a pendulum clock that had been regulated to beat exact seconds in Paris was set up at Cayenne in South American French Guyana, it lost 2 minutes and 28.5 seconds per day.[28] The suggested explanation for this phenomenon amongst most astronomers, especially in England and Holland (though Giovanni Domenico Cassini in France disagreed), was that Cayenne, which is 5° north of the equator, was slightly farther away from the centre of the earth than Paris, at 48° north. If, therefore, the earth was oblate, then even in this still gravitationally hazy period of scientific history, it seemed that the pull of gravity on a swinging pendulum would be slightly less in Cayenne than in Paris. Then when the young Edmond Halley was at St Helena in 1678, he also found that the pendulum of his regulator clock did strange things.[29] At first, Halley had regulated the seconds pendulum of his clock to keep exact time at sea level on St Helena. But when he transferred his observatory up the mountain, he found that the clock lost, or ran slower, because it was now several hundred feet farther away from the centre of the earth than it had been at sea level. And by the mid-1680s, astronomers had made detailed observations and measurements of the exact length of pendulum needed to beat dead seconds in many parts of the world, including far eastern locations such as Siam and the northern extremities of Europe, so Hooke reminds us,[30] and had found that all observations pointed to the fact that the earth's gravitational action upon swinging pendulums was slightly weaker at the equator, thereby suggesting that the globe was oblate.

This oblateness became a central component of Hooke's 'Wandering Poles' theory, for if the equatorial bulge was geophysically permanent, and the product of a *centrifugal* axial rotation, while at the same time the earth rolled around within the fixed coordinates of its orbit through space, then places on the earth's surface must *rise* and *fall* with relation to the centre of the globe.[31] This must in turn cause continental masses to rise up and sink below the oceans over long periods of time, thereby producing earthquakes, inundations, and related earth-shaping phenomena. In his idea that, as the terrestrial sphere gradually rolled around within its orbital planes and its current equatorial regions were always pulled outwards because of centrifugal force, Hooke became the first scientist to suggest that large rocky masses such as planets might be sufficiently plastic in themselves to behave like viscous liquids or semi-solids.

By 1687 Hooke had developed an all-encompassing mechanism to explain stratification, extensive marine depositions away from the sea, and with it the implication that the present-day continents need not have been the globe's original land masses. And all of this had arisen directly from Hooke's interest in fossilized or 'petrified' remains, for were not many of the fossils dug up in England and Europe quite unlike the creatures living in these habitats today? In addition to the giant ammonites found in southern England, the skeletal head of a hippopotamus had been unearthed at Chatham, Kent, an elephant's leg bone in Norfolk, and other elephant-like remains in Germany.[32] Had it not been fully demonstrated 'by Multitudes of Observations (divers made by my self, and many more by others) that all *England* ... not only the Vales, and lower parts of the Land, have been sometimes the Bottom of the Sea, but even the Tops of the Hills and Mountains'?[33]

Hooke also used his knowledge of amber to add weight to his ideas about the constantly changing nature of the earth's surface as a result of sea level variations and earthquakes, in a series of papers on amber which he delivered to the Royal Society in the early months of 1697. Much of their content seems to have been derived from a Prussian publication, but with his usual ingenuity Hooke incorporated the German observations into his wider hypothesis and drew additional insights from English finds. Amber, Hooke argued, originated in a natural resin which fell from trees to the ground, and could, at this viscous stage, entomb those creatures which got stuck in it, such as the birds which Tacitus said were sometimes found within amber.[34] These lumps of amber could then get carried away into rivers or lake beds, or even into the sea, so that after several great cycles of geological change had taken place, the amber was found in a petrified state in layers of fine subterranean sand which, no doubt, had once been beaches. Then as his *coup de grâce*, Hooke concluded that amber and other petrified objects '[do] seem to me plainly to have been [at] the Bottom of some Sea that has formerly covered all that Country [Prussia]; which Country has, in Process of Time, been rais'd above the Level of the Surface of the Present Sea'.[35]

It is hard, indeed, to see how this mechanism could be easily reconciled with the mere 190 days of the Flood of Noah, for in Hooke's discussion of amber formation, its washing away from its place of formation, its presence upon beaches and sea-beds, and then the raising-up of the same sea-beds to form dry land, he is implicitly dealing with long periods of time.

It is impossible to read Hooke's 'Earthquake Discourses' and related geographical works without becoming aware of two consequences of this way of thinking which seem, on the surface, to have challenged some of the deepest religious and cultural beliefs of that age. The first of these was the implication of species-change, which formed a natural corollary to Hooke's demonstration of the historical fact of the extinction of ancient creatures and their apparent replacement by different modern ones. The second was the tacit possibility that the earth was vastly older than the supposedly 'canonical' date of 4004 BC. I would suggest, however, that these two consequences might seem far more momentous to us today, knowing what we do about the ructions that surrounded the Darwinian controversy after 1860, than they would have done in 1670.

For one thing, 'species fixity' had not acquired the scientifically canonical status that it would under the formative influences of eighteenth-century naturalists such as Carl von Linnaeus. It is true that Aristotle and other classical writers had spoken of things belonging to their own generic types, and contemporary seventeenth-century naturalists such as John Ray denoted species as real categories within nature but, as Hooke amply demonstrates, several classical writers, including Ovid and Pliny, had been relatively open to things 'metamorphosing'.[36] And quite simply, within a Christian context, why should not an all-powerful God modify His own Creation across time in the way that He saw best, either by 'metamorphosis' or by special creations of new species?

On the other hand, one should be careful not to read too much in the way of causal agency into Hooke's statements about ancient or extinct creatures. Hooke was not so much advancing a coherent theory of evolution as making it clear to his lecture audiences, at least as early as 1668, 'That there have been many other Species of Creatures in former Ages, of which we can find none at present',[37] or in other words, that extinctions had taken place. He also reminds them, in the same passage, that there were probably species alive today 'which have not been from the beginning', though they could just as well have been the results of special and subsequent divine Creations as of evolution. By 1689, however, he is perhaps suggesting a clearer line of descent from the extinct to the living, for species might change 'by mixture of Creatures [which] produced a sort of differing in Shape, both from the Created Forms of the one and other Compounders, and from the true Created Shapes of both of them'.[38] Yet there again, as Hooke would no doubt have been aware, any skilled animal stockbreeder could have said something similar by 1689. Not for a further two centuries, in fact, would scientists have acquired a sufficiently thorough knowledge of speciation to really know the difference between extreme variation within a species type (such as that between a wolf hound and a lap dog) and a fundamental change of type.

If Hooke delivered his first Earthquake Discourse to the Royal Society in 1664, as his 1697 remark suggests, then it was only a decade after Archbishop James Ussher's *Annales Veteris et Novi Testamenti* had supposedly established the Creation date at 4004 BC. This date, indeed, was to take on a canonical significance for later generations, especially after Ussher's chronology for the Creation, Flood and other ancient events came to be printed in copies of the Authorized Version of the Bible after 1701. I would suggest, however, that during Hooke's time 4004 BC would have constituted less of a stumbling-block than it did for later generations. For while there was a medieval scholarly tradition that the earth was about 4000 years old by the time of Christ, this was in no way hard and fast. What is more, Archbishop Ussher, who had died in 1656, was far from being the narrow-minded bigot that post-Darwinian and certain other modern writers have made him out to be, but was a scholar of immense learning with a lifelong passion for historically recorded fact. Ussher realized that there was no simple generational chronology in the Bible, for the figures often failed to add up correctly not only when compared with the Greek or other secular records for Biblically-related events such as battles or royal epochs, but even when checked against other internal Biblical dates. Instead, Ussher's chronology was based on sophisticated systems of cross-dating which could hopefully be tested alongside non-Biblical sources. As Ussher was aware, other chronologists had come up with other dates. Julius

Africanus around AD 217, for instance, calculated that the earth was 5500 years old by the birth of Christ, while the Greek *Septuagint* text of the Old Testament made the world out to be older than the Hebrew text. So in Hooke's time dates were by no means set in stone.[39]

Yet while scholars might have differed by the odd few thousand years when it came to reckoning the age of the earth, all seemed to agree that it was relatively young, and that the interval between the Creation and the Flood was still only a matter of two or three millennia at most; but everything Hooke wrote about geology as an inductive science seemed to need vastly greater periods of time.

Hooke was clearly aware of the short span of the Biblical chronologies, and what one finds as the 'Earthquake Discourses' develop over four decades is a move away from an almost total avoidance of any chronological concerns towards a genuine sense of perplexity on the subject. Indeed, the 'Discourses' up to 1668 are very much confined to the discussion of physical evidences in palaeontology, but by the late 1680s and 1690s Hooke is clearly becoming concerned with time scales.

It is impossible to know whether Hooke harboured genuine doubts about the short spans of all the Jewish-derived chronologies—for as he was all too well aware, no one was making a *contemporary* record of events (such as the days of Creation) as they unfolded—and whether he really did believe that forces might have been more intense and far-ranging in the pre-Deluge days than they later became. By December 1687 he is beginning, just like the Biblical chronologists, to ransack the classical authors, in whatever cultural traditions to which he could find linguistic access, as if searching for clues or for inspiration. Once again, he was of the opinion that Ovid's writings, especially his Creation and Deluge stories, contained significant evidences, and that the Egyptians, Phoenicians and other ancient cultures seemed (through the second-hand accounts of Roman and Greek writers) to have had myths stretching back into the mists of time—and did not Plato's *Timaeus*, with its account of the Atlanteans, speak of a time scale of stretching back 9000 years before Solon and 11 000 before Hooke?[40] But were these true histories or 'Romantick Fables'? Even so, the very detail in which Hooke was combing ancient literature (and incidentally displaying an erudition which would have won plaudits for a classical historian, let alone a scientist) is enough of a clue to the fact that he was becoming seriously concerned about time scales. And then, in 1687, in the midst of a discussion of Graeco-Roman myths, he suddenly interjects that 'Mr *Graves* tells us, that the Chinese do make the World 88,640 000 Years Old.'[41] But the Chinese figure is stated baldly, and supplied with no comment or amplification.

I think it is wrong to make the assumption that Hooke was somehow afraid of saying more because of a fear of ecclesiastical reproach. As we have seen, Hooke had friends within the episcopate, let alone among other clergy and devout laymen in the Royal Society, and there is no evidence that the senior dignitaries of the Church of England, of which Hooke was a member, were concerned with enforcing any kind of intellectual conformity in matters of science—though they might well have responded angrily had Hooke suddenly declared himself a Roman Catholic or an Anabaptist. Indeed, it is hard to imagine Hooke's ideas receiving ecclesiastical censure when, in 1691, he was honoured with a Lambeth Doctorate in Medicine by the Archbishop of Canterbury.

It would be mistaken, however, to assume that Hooke's geological ideas did not meet with resistance, and even evoke gasps, though the gasps were more of incomprehension than of shock at religious heresy. His 'Wandering Poles' theory seemed baffling to many who found it hard to grasp Hooke's concept of an earth that was dynamically stable in its orbit, yet whose continental masses not only changed their latitudinal and longitudinal positions with regard to the stars over time, but even their distance from the centre of the earth. Similarly, others were sceptical about the interpretation of Richer's and others' pendulum observations after 1672, and the oblateness of the earth that Hooke deduced (correctly, as was later shown) from them. Surprisingly enough, one point upon which Hooke and Newton seem to have been in accord was the oblate shape of the earth, and Newton was to use this concept, as derived from pendulum observations, as one of his key arguments in gravitation theory in *Principia* in 1687.[42] But when Edmond Halley, in his capacity both as Clerk to the Royal Society and as a loyal Oxonian, communicated Hooke's geological hypotheses to the 71-year-old Revd Dr John Wallis, Savilian Professor of Geometry at Oxford, in February 1687, they received short shrift. Hooke's ideas of wandering poles, global inundations, and 'ye top of ye Alps [being] ever sea; Except in Noah's Floud' seemed to him simply ridiculous.[43] As Wallis was quick to point out (4 March 1687) and as Hooke himself could not help but concede, there was not the slightest evidence in the historical record for the changes occasioned by Polar Wandering, 'For Egypt & Canaan, Arabia, Syria, Chaldea, Babylonia & Mesopotamia ... are mentioned as planted & inhabited very early; & have so continued to be peopled (without ever having been made sea) ever since.'[44]

Indeed, it is probable that some of these ideas were not that easy for Hooke himself to swallow, for as a rigorously evidence-based experimental scientist he was all too conscious of the sweeping claims and numerous weak spots in the hypothesis. As Hooke fully realized, with his lack of direct observational evidence of dynamic processes to back him up, as well as his patent indecisiveness when it came to the matter of geological time scales, the whole system was based on a sort of ingenious inference. As he announced in his 'Discourse' of 23 July 1690, as if in response to Wallis's criticisms, 'The greatest Objection, I say, against this, I find hath been, that there were wanting Instances to confirm its History. For that, all Places, Countries, Seas, Rivers, Islands & Co., have all continued the same for so long time as we can reach backward with any History'.[45] Indeed even those parts of the world for which the most ancient written cultural records survive, such as Greece, Italy and Egypt, make no reference to earth-changing cataclysms, leaving us to suppose that the earth's surface has remained substantially unchanged since the first glimmerings of civilization. Hence, the paradox of which Hooke was all too acutely aware.

Yet not only was the written historical record silent about long-term continental or sea-level changes, but Hooke realized that astronomical evidences for polar wandering were also absent. After all, if every place on earth was inexorably wandering into or out of a polar region, as part of a gradual process, then there should be discernible changes of measured latitudes of places between antiquity and the seventeenth century. Hooke suggested in his 'Discourse' of 2 February 1687 that the alignments of ancient buildings, or latitudes carefully measured in antiquity, might be used for discerning polar movement over time, and to this end consulted John Graves's [Greaves] *Pyramidographia* (1646), which contained his detailed survey of the Great Pyramid of Giza, made in the 1630s. Hooke was annoyed to find that while Greaves's survey was archaeologically detailed within itself, it paid little attention to the pyramid's astronomical alignment, and hence could not be made to show whether the latitude or orientation of Cairo had 'wandered' over several millennia.[46] On the other hand, Wallis pointed out that both ancient and modern measurements of the latitude of Marseilles indicated that the astronomical position of that city had not changed between the solar shadow observations made between Pytheas in *c.* 300 BC and those of the contemporary French astronomer Pierre Gassendi.[47]

Hooke responded, somewhat petulantly, to Wallis's criticisms, and one suspects that he was all too sensible of important points which he could not answer. Instead, one might say, Hooke fell back on an argument from historical analogy, for just as the now generally-accepted Copernican hypothesis had once seemed to fly in the face of all the physical evidences, so would his ideas of wandering poles and global inundations be some day shown to have taken place.[48]

But as an empirical scientist, Hooke was assiduous in collecting precise information about modern-day earthquakes and their power wherever he could find it. His Diary records such references, as when on 9 October 1674 he was 'At Garaways. Mr Lodowick gave me Relation from India about an earthquake',[49] while on 12 January 1676 he received 'Letter from Cole about earthquake in the West'. Then on Sunday 6 April 1690, at about 5.00 p.m., a violent earthquake took place on the

Caribbean island of Nevis. It was felt on other islands in the Leeward group, and reports of it were being published in London by June 1690, so that Hooke came to discuss it at length in his 'Earthquake Discourse' of 23 July of that year. What fascinated him was this modern quake's power and damage capacity: its occasioning of temporary sea level rises, and the 'burying of thousands of Trees'.[50]

Then there was the destruction of the mercifully uninhabited 'Isle of Rotunda' which blew up so that part of it 'tumbled down and sank into the sea', accompanied by massive artillery-like bangs and clouds of white smoke.[51] This information put Hooke in memory of the explosion of gunpowder mills in Hackney, 'which I heard, being within a Mile of it, in the Fields'. It led him to wonder whether certain types of local earthquake activity—with its deep blasting noise and smoke such as that at Rotunda—could proceed 'from such Subterraneous inkindling as resembles Gun-powder'.[52] What one sees in the period following the news of the Leeward Islands earthquake is a growing interest in what the causes of such quakes could be, and Hooke was now coming to think of them as powerful subterranean explosions capable of generating what later scientists would call tidal waves.

Indeed, by his 'Discourse' of 23 July 1690,[53] only a few weeks after the account of the Leeward Islands quake had reached London, Hooke was speculating about volcanic and earthquake activity being the product of the reaction between 'the saline Quality of the Sea-Water [which] may conduce to the Producing of Subterraneous Fermentations with the Sulphureous Minerals there placed', for which idea he gives credit to 'Signor Bottoni' in his *Pyrologia Topographica*.[54] This subterranean chemical reaction leading to an explosion theory of earthquakes was developed by Hooke over the next decade, receiving what was really its definitive expression in his Discourse of 30 July 1699.[55] Herein one finds one of the persistent themes in Hooke's broader chemical understanding: axiomatically combustive, 'sulphureous' substances being excited to release their fire and force through contact with a dissolving agent. In the early nineteenth century the Oxford chemist and geologist, Charles Daubeny, would return to the idea of invading sea-water reacting with beds of unoxidized rocks and causing a sort of fermentation, in his proposed mechanism for volcanic activity.[56]

The most powerful earthquake—albeit mild by Caribbean standards—of Hooke's time to strike England and northern Europe was that of 8 September 1692. It was felt very strongly at Deal, Sandwich and Canterbury in Kent, where pewter rattled on kitchen shelves, buildings shook, chimneys fell down and there was a 'great Terror and Affrightenment of many'.[57] It was also felt in London about 3.00 p.m., and across most of southern England where the shaking of the earth 'affected most persons with dizziness for the time and shook most houses in and about London', as was recorded by the Oxford diarist Antony à Wood.[58] Wood also mentioned that the sulphurous smell generally thought to come out of the ground during earthquakes (and hence, to confirm their supposedly explosive status, but which were more likely to have been produced by the release of sulphuretted hydrogen from stagnant ponds) put a pack of hunting dogs off the scent on Enfield Chase. But perhaps most interesting of all was a macabrely amusing incident that the quake occasioned in the Anatomy School in Leyden University in Holland where, according to Wood, 'some persons viewing the skeletons in the anatomy chapter... perceived them to move, and not dreaming of the cause were so affrighted that they all ran away'.[59] It is not known whether Hooke himself heard this story of the Leiden anatomists, for he leaves no record of it, but if he did one assumes that it would have produced some laughs as it circulated around the London coffee houses.

Yet as with so many other areas of research in his life, geology too involved Hooke in priority disputes. In particular, he claimed that Henry Oldenburg had transcribed his early 'Earthquake Discourses' and circulated them around Europe.[60] On the face of it, one might take this as a compliment of the highest order. However, when Nils Steno, the Danish Roman Catholic scientist living in Italy, published his own far-reaching researches in which he argued that certain 'Tongue stones' found in the Tuscan hills were similar to the teeth of modern sharks, and that fossils were probably organic in origin, Hooke's hackles rose. While Hooke never seems to have blamed Steno himself, whose work he respected, he was always quick to smell a rat in any project in which Henry Oldenburg was involved, especially when Oldenburg translated Steno's *Prodromus* (1669) into English and began to openly praise the Dane. Hooke was not so much angry at Steno, therefore, with whom he had a great deal in common when it came to geology, as at Oldenburg for advancing Steno's name and reputation before the Royal Society, while seemingly playing down Hooke's own originality and priority.[61]

Yet while it is true that many of Hooke's contemporaries such as John Wallis found his geological ideas beyond belief on evidential grounds, these ideas certainly seem to have inspired several younger scientists within the Royal Society, and none more so than Edmond Halley. Halley's interests in the natural causes of climate, the atmospheric circulation of moisture, and the importance of both rainfall and condensation in supplying the waters to the world's great river systems went back to the mid 1680s, or just before Halley's communication of Hooke's geological ideas to Wallis in Oxford in 1687.[62] But what Halley was to do over the next few years was to look at the formative influences that could have moulded the ancient earth, and develop a theory that they may have acted somewhere between the Creation of the World and the

Six Days described in Genesis. While Halley was not denying that God had made the earth, what the 1694 paper which he delivered to the Royal Society suggested was that an unspecified period of time had passed between God's *ex nihilo* Creation of the world and the re-ordering of the 'Old Chaos' in the Six Days recounted at the beginning of the Book of Genesis. In this way, Halley was creating what might be called a creative cosmological space within which a variety of physical agents moulded the face of the earth, and animal species came and went, and were entombed within the rocks. While Halley is explicitly going beyond Hooke, especially in regard to time scales, one finds so many Hooke components in the theory he developed over the next few years that it is impossible to exclude the likelihood that the adventurously young scientist had found inspiration in the ideas of the brilliant experimentalist of the older generation.[63]

Within the vast period between the Creation *ex nihilo* and the six-day Adamite Creation suggested by Halley, sea levels could have changed, continents emerged and sunk back into the deep, and all parts of the earth's surface inundated, dried out, and inundated again. As the detritus of the continents ended up in the seas, through river action, Halley even postulated that one might be able to calculate the age of the earth from the slowly increasing saltiness of the seas, and calculating backwards, find an epoch when all the waters on the newly-created earth had been fresh. He further argued that violent cataclysms in the earth's past might have been brought about as by cometary impacts, and even went so far as to suggest that the Caspian Sea, with its fresh water, was a scar on the earth's surface resulting from a glancing cometary blow. The Caspian Sea's fresh water, moreover, perhaps indicated that it was still sufficiently recent, in geophysical terms, as not yet to have accumulated salt as a result of river action.[64]

The reason why Halley proposed a cometary impact, or a series of such impacts, to keep re-modelling the globe, however, was because he was not convinced by Hooke's 'Wandering Poles' idea. And this resulted from his own researches into the history of observational astronomy, for Halley too realized, just like Wallis and Hooke, that if indeed the continents were slowly wandering round the globe, then the fact should have left its trace in the astronomical record. As good-quality observations of star positions had been made in Alexandria for some 2000 years, from the time of the ancient Greeks through that of the medieval Arabs and down to John Greaves's observations of the 1630s, Halley argued that it should be possible to use them to extract a good value for the latitude of Alexandria, and hence the elevation of the pole as observed at Alexandria, over two millennia. Yet when he compared ancient and modern Alexandrian latitude values, he found an error of only 4 arc minutes, which was well within the range of observational error for the pre-telescopic astronomers. From this fact, therefore, Halley concluded that the poles did not seem to have wandered at all in 2000 years, which they ought to have done, even if only to a small extent, if polar wandering was a gradual process.[65] As a consequence, Halley began to cast around for other likely earth-transforming agents, and came thereby to propose cometary impacts.

Of course Halley, who like Hooke was an accomplished classical scholar, was aware that no human record or memory survived from these colossal impacts and terrestrial disruptions, for all these things had happened in the old chaotic times of global history, *before* God had created Adam and Eve and commenced human recollection. Indeed, only *one* such terrible event had probably happened during human history, the Great Flood of Noah, which as Hooke also knew seemed to have some collateral record in the myths of the Greeks and other peoples, though the complexity of geological finds suggested that this Flood had been far too brief in its duration to be able to account for every stratum, bone and mountain-top shell known to science by 1690.

Then in 1694, Halley put forward what could be considered as a new piece of physical evidence for the Noachian Flood's relative transience and weakness: its relative lack of water. Using evidence from the Royal Society's weather records, Halley took the rainfall for the wettest place in Britain, at around 40 inches a year, and multiplied this figure to obtain the amount of water that rainfall would produce if it came down continuously for 40 days and 40 nights across the entire surface of the globe. The figure that he came up with suggested that the Genesis Deluge could have delivered only enough water to raise global sea levels by 132 feet.[66] While this volume would have flooded the low-lying parts of the continents, it would have been entirely insufficient to lay down beds of sea shells in the Alps.

The cometary impact idea also appealed to William Whiston who in 1696 proposed that the Noachian Deluge could have resulted from a near miss by a massive comet with a water-vapour envelope or tail. While in Whiston's hypothesis the earth was saved a destructive collision, it nonetheless received a ducking when it passed through the comet's tail.[67]

What Hooke's extensive 'Earthquake Discourses' clearly did was stimulate a great deal of interest in and discussion about the *physical* history of the earth. While they did not dispute the events of the Old Testament, and sought at every turn to lend substantiation to such perceived milestones as Noah's Flood, they tacitly implied that the early books of the Bible should not be reckoned as a comprehensive and internally sufficient history of the terrestrial globe, although they in no way challenged the belief that the Bible did provide such a history for both the spiritual and bodily origins of mankind. In his theories about the organic origin of fossils and, in particular, the constant re-modelling of the planet's surface—and consequent ecology—with reference to its wider astronomical environment, Hooke displayed originality of the highest order. One can also appreciate how Hooke's ideas went on to influence seventeenth-century figures as diverse as Nils Steno, who became a Roman Catholic Bishop, and William Whiston, who was expelled from his Cambridge chair for holding Socinian, or Unitarian, beliefs, not to mention James Hutton a century later and even early-nineteenth-century stratigraphic geologists such as the Revd William Buckland.[68]

Yet one of the most remarkable things which Hooke's 'Earthquake Discourses' bring home to us is the sheer range of his erudition. While no one has ever doubted Hooke's brilliance as an experimental scientist and as a deviser of ingenious mechanisms, one cannot read his published 'Discourses', covering as they do a creative span of almost 36 years, without being impressed by his knowledge of ancient history, literature, mythology and linguistics. This secular learning, moreover, was matched not only by his mastery of the Bible, but also by sophisticated insights into contemporary chronological studies. When Hooke tried to find equivalents in world literature for events such as Noah's Flood, he was attempting to use a technique of collateral cross-dating which was similar in principle to those employed by Renaissance chronologists such as Joseph Scaliger or James Ussher to date more obviously historical Biblical events from parallels in ancient Greek or middle-eastern annals narrated in the Greek language.[69]

Seventeenth-century learned culture prized broad-ranging, virtuosic and polymathic scholarship, and perhaps nothing shows us better than the 'Earthquake Discourses' how much at the heart of that scholarly world Robert Hooke really was.

Plate 1. *Wenceslaus Hollar's heroic depiction of the founders of the Royal Society.* The central bust on the pedestal is that of the founder, King Charles II, original patron of the Society. To the left, and pointing to the pedestal, is Viscount Brouncker, first President, and to the right sits the spirit of the late Francis Bacon, whose writings up to 1626 supplied the philosophical foundations out of which the Society grew. The Muse crowns Charles with a laurel wreath, while in the background landscape one sees telescopes, thermometers, balances, clocks, chemical apparatus, and numerous other devices signifying the Society's intention to measure and quantify 'to a common stock' the whole phenomenon of nature. The engraving is dated 1667, and while it was popularly said to be associated with Thomas Sprat's *History of the Royal Society* (1667), I have still to find a copy of that work with the engraving bound in. It is likely that the engraving was issued independently of the book.

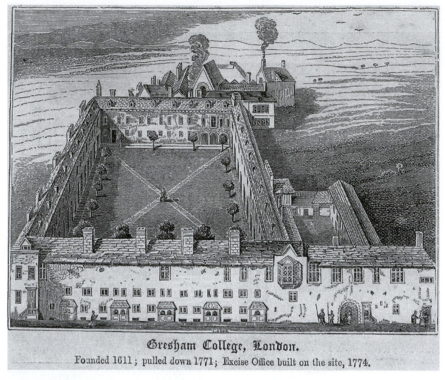

Gresham College, London.

Founded 1611 ; pulled down 1771; Excise Office built on the site, 1774.

Plate 2. *Gresham College, London.* Hooke's rooms — a house really — are in the far right corner of the quadrangle. Author's collection.

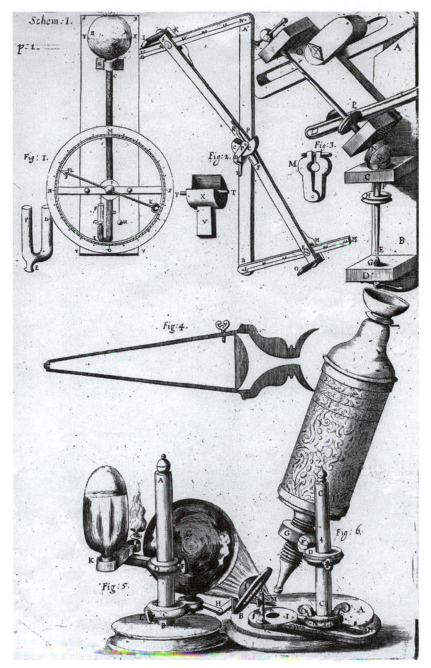

Plate 3. *Robert Hooke's microscope and illumination system.* The instruments at the top of the plate include his Wheel Barometer, lens-grinding machine (see present book, pp 156, 158), and other Hooke inventions. *Micrographia* (1665), frontispiece, Scheme I.

Plate 4. *Hooke's drawing of the Blue Fly: Micrographia*, Observation XLII, Scheme XXVI.

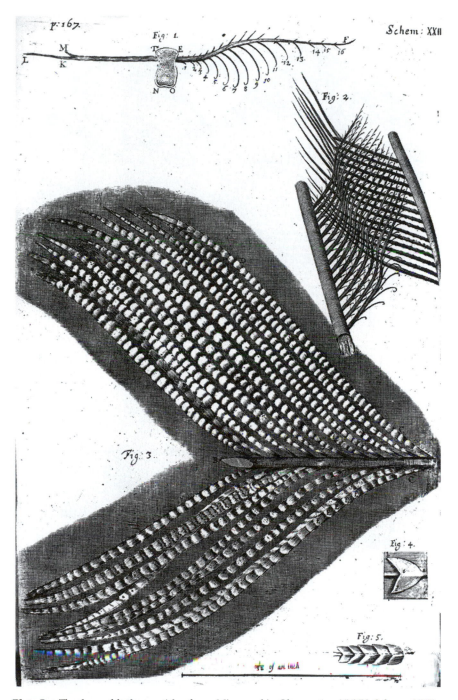

Plate 5. *The shape of feather particles,* from *Micrographia,* Observation XXXV, Scheme XXII.

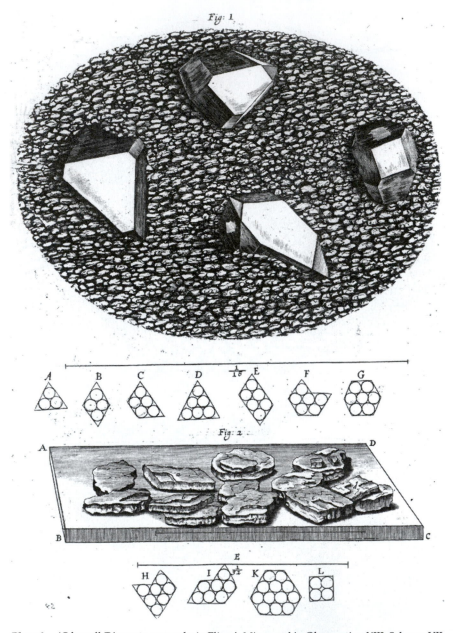

Plate 6. 'Of small *Diaments*, or sparks in Flints', *Micrographia*, Observation XIII, Scheme VII.

Plate 7. *Blue mould, such as grows on cheese.* This is the first ever drawing of a micro-organism. The scale '32' beneath the mould drawing represents 1/32 of an inch. Hooke tended to think less in terms of a precise optical magnification than of trying to measure specimens as tiny fractions of an inch. *Micrographia*, Observation XX, Scheme XII.

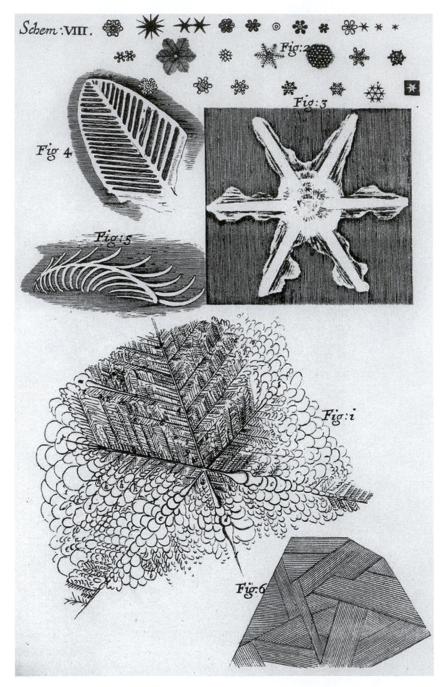

Plate 8. *Ice crystals*, including frozen urine, hoar frost, and snowflakes. *Micrographia*, Observation XIV, Scheme VIII.

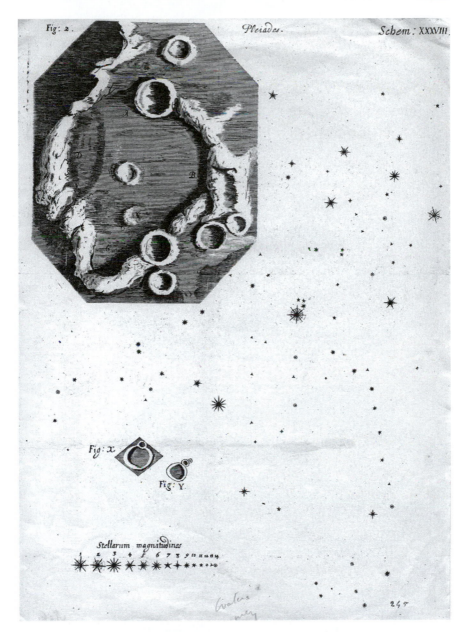

Plate 9. *Lunar craters and the Pleiades Cluster. Micrographia,* Observations LIX and LX, Scheme XXXVIII. See present book, pp 77–9.

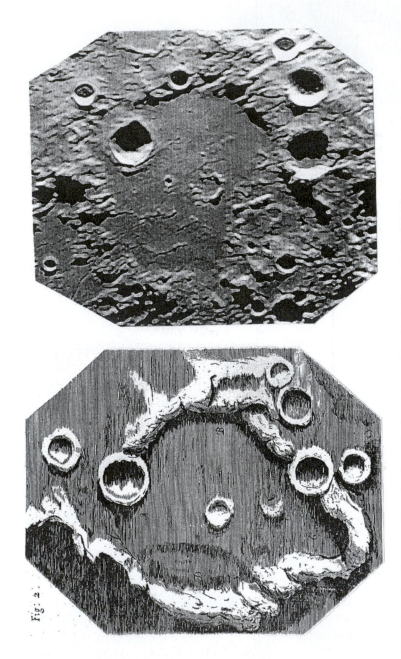

Plate 10. Attempting to assess Hooke's telescopic draughtsmanship objectively. (Left) Hooke's 1664 drawing of the lunar crater Hipparchus; *Micrographia* (1665), made with a 36 foot focal length telescope with a $3\frac{1}{2}$ inch aperture object glass. The drawing encompasses about 90 arc seconds of the lunar surface. (Right) Photograph of the same lunar formation at the same phase illumination. Taken by Gain Lee of Huddersfield, Yorkshire, 28 March 2004, after 8.30 pm.

Plate 11. *Sir Christopher Wren.*
Author's collection.

Plate 10. *Dr John Wilkins.*
Hooke first met him around 1655.
Courtesy of the Warden and
Fellows of Wadham College,
Oxford.

Plate 13. John Wilkins' *Discourse Concerning a New World* (1640). This book popularised both the Copernican theory and also Galileo's telescopic discoveries among English readers.

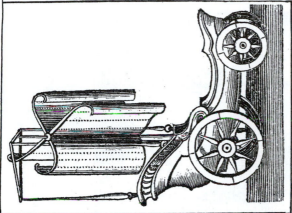

(a)

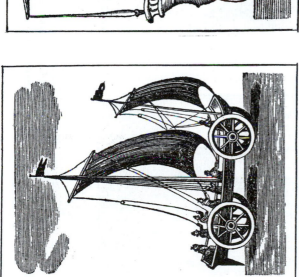

(b)

Plate 14. The wonders of applied mechanics. The possibilities of wind and gears, from John Wilkins' *Mathematical Magick* (1648). It is easy to see how the ideas contained in this book, not to mention his close personal friendship with Wilkins, had such a profound influence upon Hooke, and helped to mould his approach to science.

(a) Land ships. Wilkins claimed that such vehicles were already in use across the flat land of Holland, and that between Sceveling and Putten, it had sometimes been possible to cover 42 miles in *two* hours using similar machines.

(b) Wilkins' idea for a vertical windmill or aerofoil, which—especially if connected to the back axle of the chariot by some sort of gearing—should enable the vehicle to be driven in any direction irrespective of the wind's direction.

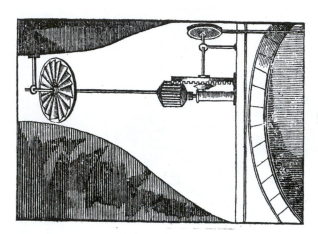

(d)

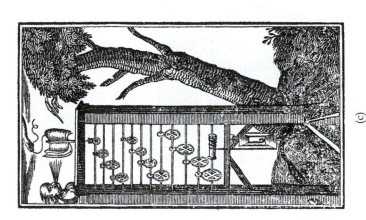

(c)

(c) The increase of mechanical energy by gear trains. The man's puff of breath into the upper windmill will be magnified by gears so as to be sufficiently powerful to uproot a tree at the bottom. Sadly, little was known about inertia and resistance in the 1640s.

(d) An horizontal 'windmill' or aerofoil in a chimney. The heat of the fire, ascending from the fireplace, would cause the aerofoil to rotate at a speed relative to the heat. The heat of the fire, ascending from the fireplace, would cause the aerofoil to rotate at a rate that was related to the heat of the fire. And when connected by cords and pulleys to the cooking spit, it would turn the meat at a rate that was related to the heat of the fire. Was this the first 'smart' invention in so far as it was a mechanical system that responded automatically to the energy being put into it?

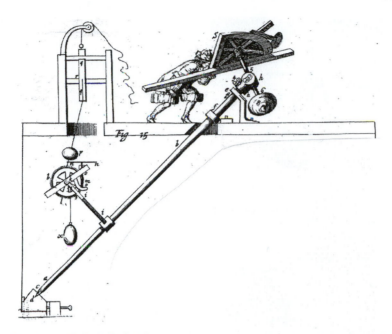

Plate 15.   Robert Hooke's clock-driven equatorial mount for his telescopic quadrant. Though Hooke did not invent the polar-axis equatorial mount, his use of geared clockwork to make the apparatus track celestial objects across the sky was his own invention. *Animadversions* (1674).

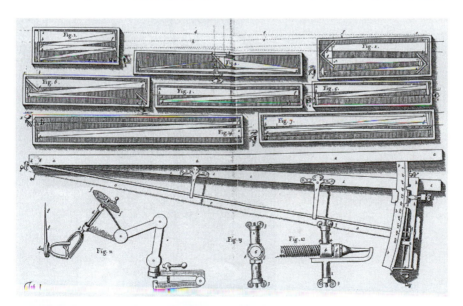

Plate 16.   Hooke attempted to shorten the long focal length of contemporary telescopes by the use of mirrors in periscope-like configurations. Note also his use of the Universal Joint. From Hooke's *Helioscopes* (1676). See present book, p 95.

Plate 17. *The Royal College of Physicians' Building,* Warwick Lane, London, 1673, with the Anatomical Theatre, surmounted by its famous 'Gilded Pill'. It was demolished in the nineteenth century. *Pharmacopoeia Londinensis* (1721).

Plate 18. *Church of St. Mary Magdalene, Willen, Buckinghamshire.* Hooke's only surviving building. Author's collection.

Plate 19. *Bethlehem Hospital*, commonly known as 'Bedlam'. This was without doubt Hooke's grandest building, with a main façade over 500 feet long. Author's collection.

Plate 20. *Montagu House*, built by Hooke in Bloomsbury for Ralph Montagu. The site is now occupied by the British Museum. Author's collection.

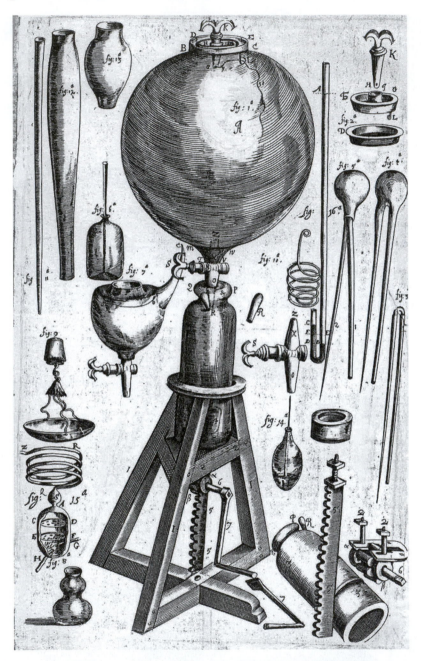

Plate 21. *Boyle's first airpump, 'Machina Boyleiana'*, designed by Boyle and Hooke, and probably constructed under Hooke's direction in 1659. It was used to conduct the researches described in Robert Boyle, *New Experiments Physico-Mechanical Touching the Spring of Air and its Effects* (Oxford, 1660).

ROBERTVS BOYLE NOBILIS ANGLVS MI

Plate 22. *Robert Boyle, c. 1664,* from an original portrait by William Faithorne, re-engraved by Diodati.

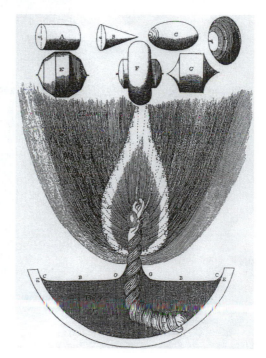

Plate 23. *Studies in the structure of flames,* depicting the rising of the oil in a lamp wick, and the resulting variations in incandescence. *Lampas* (1677). See present book, p 27.

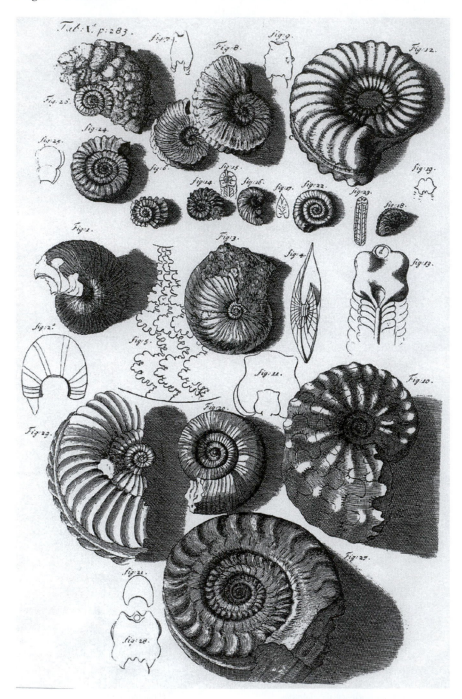

Plate 24. *'Snake stones' and ammonites*, drawn by Robert Hooke. *Posthumous Works* (1705).

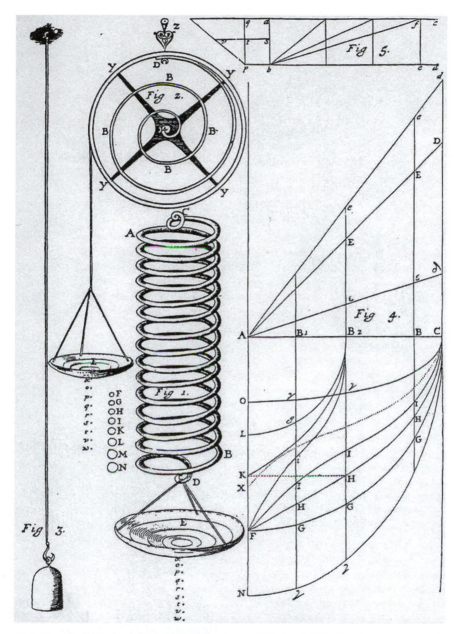

Plate 25. *Illustration from Hooke's Cutlerian Lecture Of Spring* (1678). The plate shows experiments on springs to establish the ratio between a spring's extension and the weight placed upon it. See present book, pp 174–5.

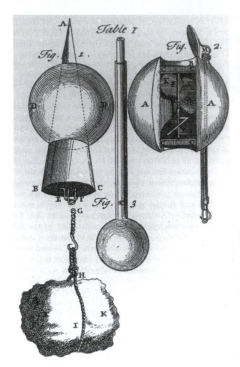

Plate 27. *Sections through hailstones,* drawn by Hooke in 1680. In the late morning of 18 May 1680, a violent lightning storm broke over London, and hailstones as big as pistol bullets and even pullet's eggs fell from the sky. Hooke collected several, and cut sections through them; he described them thus: 'the small white globule in the middle, about the bigness of a pea, was the first drop that concreated into hail; this, in falling through the clouds beneath, congealed the water thereof into several coats or orbs, till some of them came to the bigness of pigeon's eggs, some white, some transparent, according to the several degrees of coldness it passed through, whilst they congealed'. *Philosophical Experiments and Observations* (1726), p 50.

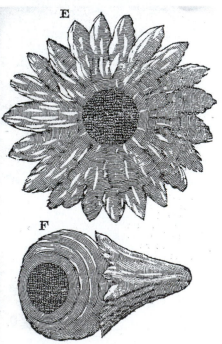

Plate 28. *Hooke's drawing of the comet of 1677.* The plate also includes a telescopic drawing of the planet Jupiter. Note (bottom right) Hooke's use of a distant weather vane as a standard of angular measurement against which to compare the comet. See present book, pp 81–2.

Plate 29. *Hooke's portable camera obscura,* 1694, described as 'An instrument of use to take the draught of picture of anything.' *Philosophical Experiments and Observations* (1726), p 295.

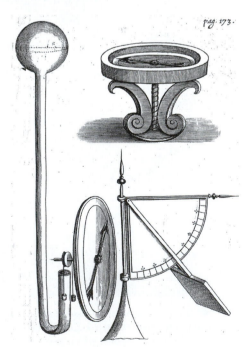

Plate 30. *Meteorological instruments* suggested for Hooke's 'History of the Weather', included in Sprat's *History of the Royal Society* (1667), p 173. The instruments include Hooke's Wheel Barometer, a Wind Gauge, and a Hygrometer with which to measure the dampness of the air.

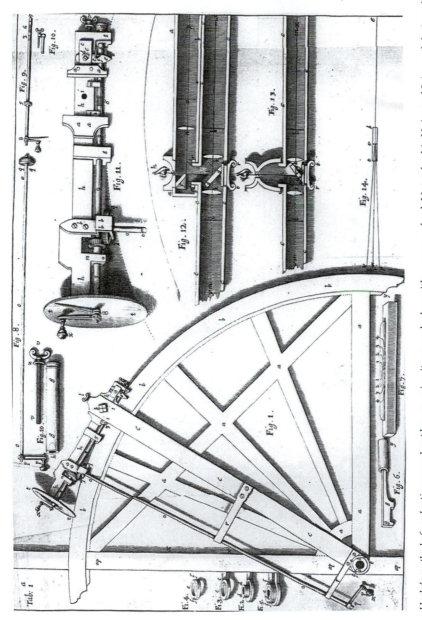

Plate 31. *Hooke's method of graduating a quadrant by engaging its geared edge with a worm-wheel. Note also his Universal Joint and designs for spirit levels. Animadversions on Hevelius (1674).*

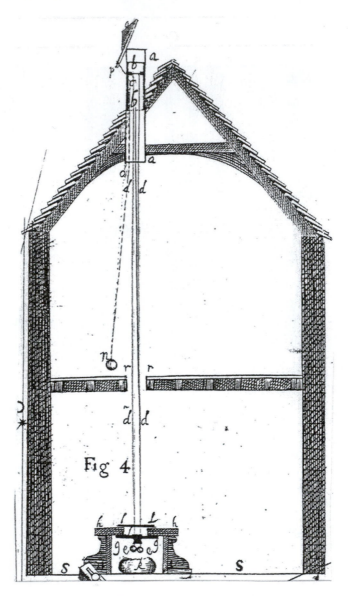

Plate 32. *Robert Hooke's Zenith Instrument of 1669.* The engraving shows a cross-section through Hooke's rooms in Gresham College. A 36-foot-focus telescope object glass was mounted just below the roof, beneath an aperture which could be opened and closed to give access to the sky. Beneath the lens cell, a set of threads or strings supported an eyepiece lens cell, exactly 36 feet below. When Hooke opened the roof shutter, and lay on his back to look up through the lenses, the telescope pointed exactly to the zenith. With this instrument Hooke tried, unsuccessfully, to measure a parallax for the zenith star Gamma Draconis over a six-month period, and hence to demonstrate the earth's motion in space. Hooke, *Attempt to prove the motion of the Earth* (London, 1669). See present book, pp 92–3.

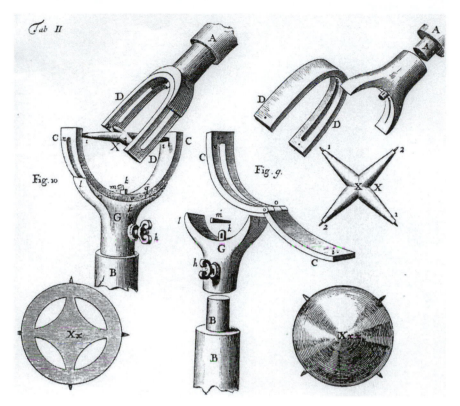

Plate 33. *Hooke's Universal Joint. Helioscopes* (1676).

Portrait 1. Sandra Bates (née Kingsworth).

Portrait 2. Rachel Chapman.

Portrait 3. Rita Greer.

CHAPTER 9

A WORLD OF MECHANISM

Everyone who has written on Robert Hooke, either in his own time or since, seems to have agreed on one point, and that is his genius as a deviser of mechanisms. Richard Waller in his 'Life' of Hooke attributed to him one hundred inventions, whereas Hooke's friend John Aubrey — in a confessedly hurried estimate — says that a complete catalogue would come to around 1000. These inventions, as alluded to by Waller, mentioned in Hooke's Diary, his own published works, or his posthumously published and unpublished manuscripts (one instance being the list of about 106 inventions dating from around 1667 now preserved in the Royal Society Hooke manuscripts) cover an extraordinary range, though most of them would have fallen within the broad topics of theoretical and applied optics, experimental physics, horology and mechanics, along with a few excursions into chemistry and physiology. They would not have been confined to the realm of labour-saving devices, which is what we tend to mean by the word 'invention' today, but would have included improvements to experimental technique or else new elucidations of nature, although Hooke's *c.* 1667 list does, indeed, deal predominantly with useful devices. Even so, it is important to bear in mind that a contemporary would have used the word 'invention' in the sense of the Latin *invenire* ('to come upon'), to encapsulate innovations as diverse as the devising of the spring watch balance, the experiments and reasoning which led Hooke to develop his two-colour wave model of light and colours, and even his insufflation experiment which demonstrated the primary role of air (as opposed to a thoracic reflex) in the respiratory action of a dog, as well as 'a digging Enging', 'A Picture "box"', and 'a new way of printing'.[1]

And when contemporaries described Hooke as a '*great mechanic*', we must not forget that this appellation was not being used in 1660 simply to describe Hooke's undisputed skills at a workbench. Instead, they were referring to his extraordinary capacity to combine *mekhane* ('contrivance', 'device') with *invenire* — to make discoveries using carefully-thought-out practical procedures — and use them as both intellectual and physical tools, or 'artificial organs', whereby he could see yet deeper and more

153

profoundly into nature than mankind's natural and unaided faculties would permit. Starting from the premiss that the whole of creation was a vast divine invention, in which motion was not mysteriously innate — as it had been for Aristotle — but was rather the manifestation of endless trains of interconnected impacts and vibrations, Hooke saw the universe in terms of mechanical systems. In the development of this way of thinking, moreover, one can see Hooke's intellectual debts to Descartes (to whose writings he was introduced by Boyle) and the other continental mechanists, and also to Thomas Hobbes, whose ideas on mechanism he 'loved', and about whom he corresponded with Aubrey.[2] But Hooke went beyond all of these men, and turned mechanism, or the 'mechanical philosophy', from an all-encompassing yet entirely theoretical body of epistemological deductions based on vortices or resisting mediums to a set of ideas and practical techniques through which one could 'put nature to the torture' and go on to uncover entirely new realms of physical knowledge. *Micrographia*, and the rationale that underlay it, was his first and greatest essay in this respect, displaying as it did that perceived interplay between observation, instrumentation, experiment, inspired hypothesis and bold conclusion which themselves were often no more than invitations to further researches.

To Hooke's understanding, vibration, impact, and 'mechanism' were not just the forces through which light, gravity and the other primal agents of nature made themselves manifest, but dovetailed perfectly with the very mechanisms of human perception as he understood them, whereby nervous impulses, responses and memories were able to grapple with and make sense of the realities of the external world. As one might expect from what we have seen so far, Hooke had no time for solipsism and displayed no philosophical qualms about the physical reality of the relationship that existed between the perceiver and the perceived.

Hooke's interests in mechanism and invention, therefore, derive from a broad and coherent philosophy of instrumentation. This instrumentation, moreover, he saw as being historically specific to his own time, only becoming possible after key inventions in optics and mechanics such as the telescope, microscope, magnetic compass and airpump had enabled contemporary scientists to lay firm foundations of fact and experience after millennia of hypotheses and suppositions which had largely gone around in circles. In *Micrographia*,[3] indeed, he put this point over in a pithy analogy of marine navigation, where he equated the new instrument-based sense knowledge to a guiding star suddenly glimpsed, with great relief, by a mariner sailing under a cloudy sky and uncertain of his course. Hitherto the navigator had been forced to lay his uncertain course by the guesswork of dead reckoning, but after the eventual appearance of a familiar star in the sky, he could use a mathematical instrument to fix his position with certainty. Dead reckoning, needless to say, represented the conjectures of a wordy philosophical argument, whereas the sighting of the star represented the sure foundation of clearly observed reality.[4]

Hooke's earliest and some of his longer-term or 'serial' inventions concerned the behaviour of air and its compression. His work on flight with Wilkins at Oxford had been of this kind, as wings were used to compress buoyant cushions of air, while his experiments with the airpump with Boyle, also performed in Oxford, were to constitute some classic examples of the modern method. Hooke's independent researches into the nature of capillary action in 'thin glass canes' or tubes, which he discussed in *Micrographia*, were also studies in air compression.[5] But his earliest and perhaps most significant invention in this respect was his wheel barometer, which he described in the 'Preface' of *Micrographia* and illustrated with a fine engraving. Ever since Wren, Boyle and the early Oxford researchers first realized that the Torricellian column of mercury was sensitive to the pressure changes taking place within an 'elastic' atmosphere, and that these changes seemed to be related to variations in the weather, the barometer, or baroscope, had been an instrument of great scientific and practical potential. Between 1658 and 1664, for instance, it had been discovered that 'during the time of rainy weather, the pressure of the Air is less, and in dry weather, but especially when an *Easten Wind* (which having past over vast tracts of Land is heavy with Earthy Particles) blows, it is much more, though these changes are varied according to very odd Laws'.[6]

One of the problems with the simple mercury column, however, was the relatively narrow spatial range over which the top of its mercury column normally moved. Hooke's Wheel Barometer was an early example of an instrument that was intended to magnify small changes so that the 'height of the *Mercury* will be made exceedingly visible by the motion to and fro of the small Index' or pointer. Hooke's Wheel Barometer—taking the word 'Wheel' from the large calibrated dial against which the index pointer moved—consisted of an upturned or vertical glass tube 25 inches long. At the top of this tube a ball had been blown, to form a 'bolthead' or long glass flask. To the lower end of the tube was attached, by a sealed joint, a glass U-shaped bend, of the same internal bore as the bolthead tube. After Hooke had introduced the 'clear and well strain'd Quicksilver' into the inverted tube, through a subsequently sealed hole at the bottom of the U bend, the volume of mercury was adjusted by means of a siphon, so that when the whole apparatus was returned to the vertical the mercury fell to a point where it filled exactly the lower half of the glass ball, leaving the upper half empty. At the opposite end of the apparatus, the mercury, when perfectly adjusted, would stand an inch or so *below* the upper lip of the

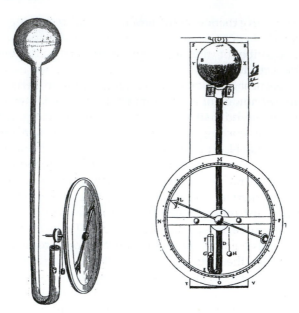

Hooke's 'Wheel' barometer.

upwards-pointing U bend. Only at this point, in an otherwise sealed system, was the tube open to the air.[7]

This arrangement of glass ball, U tube and mercury was firmly secured to a strong vertical board for wall mounting, as became standard in later barometers. Hooke could now calibrate the highest and lowest positions of the mercury as it approached and receded from the open end of the U tube with the variation in weather conditions, and use them to further calibrate a circular scale — the Barometer's 'Wheel'. This wheel was placed in turn with its centre slightly offset above the open aperture of the U tube. At the wheel's centre was a carefully-turned metal cylinder that was able to rotate on a precision needle-axle bearing, so that a lightweight index pointer attached to the cylinder could move around the wheel. All that Hooke now needed to do was place a counterpoised steel-ball plunger, suspended from a fine thread, within the U tube to make it sit on the surface of the dense mercury. The thread was then wrapped around the cylinder and the slack taken up by its matched counterpoint weight, so that the thread always applied tension to the cylinder.

A tiny change in the mercury level, caused by a variation in atmospheric pressure, would now be greatly magnified by the thread and cylinder, to produce a significant movement of the index across the wheel or scale. Such an instrument, Hooke hoped, would enable scientists to study and quantify the 'very odd Laws' which seemed to lie at the heart of the weather.[8]

I have described the design and construction of Hooke's Wheel Barometer in such detail because it provides such an excellent example not only of his approach to instrumentation, but also of his sheer practical understanding of how objects were made. Hooke never tells us whether or not he built the whole apparatus with his own hands, but even if he employed a professional glass-blower to make the bolthead ball and U tube, which is likely, the rest of the device is described with such an intricate familiarity as to suggest that he put it all together at his own workbench. His narrative on the Wheel Barometer, which was clearly written as an instruction and encouragement to others who might wish to make similar instruments, even explains precisely how to introduce mercury into the tube, equalize its pressure with a siphon, make good seals between the conjoined glass tubes with warm craftsman's cement, and then calibrate the mercury column. Very clearly, in addition to being a learned 'mechanical philosopher', Hooke was an accomplished *practical* mechanic.

It is obvious that Hooke and his contemporaries saw the Wheel Barometer as an instrument of considerable importance both in the broader instrumentalization of science and, in particular, within the growth of the new science of meteorology. Improvements and modifications to the basic design were announced in June 1666,[9] while it was illustrated in Sprat's *History of the Royal Society* in 1667.[10] While the significance of the barometer in the early development of scientific meteorology has already been mentioned, it is important in the present context to be aware of the importance of Hooke's Wheel Barometer as a significator of his wider approach to invention, for by 'mechanizing' the rise and fall of mercury by means of a plunger counterweight and dial he was, in a way, connecting atmospherics with horology (as he did most conspicuously in his efforts to perfect the Weather Clock), and suggesting that the whole of nature could perhaps be comprehended through readings displayed on graduated scales. Thus the invention of a refined mechanism, it was hoped, would lead to invention in the intellectually-potent Latinate sense, 'coming upon' or discovering new realms of nature.[11]

One of Hooke's most enduring interests, as far as both practical and intellectual invention were concerned, and to which he returned time and again throughout his career, was light. While Hooke's ideas on the nature of light itself will be dealt with in chapter 11, it must not be forgotten that the practical optics of lens and mirror production occupied a great deal of his time, especially during the 1660s and early 1670s, for it was these optical devices that revealed 'new Worlds and *Terra-Incognita's* to our view' and led us to fresh realms of knowledge.[12]

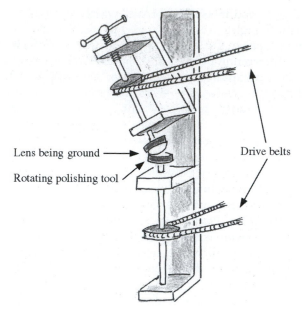

Lens being ground

Rotating polishing tool

Drive belts

Lens grinding and polishing machine.

Hooke's first substantial discussion of practical optical manufacture plunged him into controversy with the French astronomer Adrien Auzout in 1665. This discussion centred around the effectiveness of the telescope object glass grinding machine or 'Engin' which Hooke had described in *Micrographia* and which Auzout believed was impracticable for making large-aperture, long-focus lenses.[13] Of course Hooke was by no means the first person to devise a lathe or engine for figuring telescope lenses, and in recent times he had been preceded by Hevelius, Huygens, Campani, Wren and others. Hooke, however, believed that his 1665 machine would be more versatile in the focal ratios of the lenses which it could generate, and in particular, that it would be able to produce the exact, almost 'flat', curves that were necessary for very long focal length telescope object glasses.

Hooke's 'Engin' was quite straightforward in its design. A stout iron shaft or mandrel about two feet long was made to rotate in the vertical supported by a strong wooden frame, its power imparted by a pulley and cord. On the top of the mandrel was a brass plate or collar of the same size as the lens to be ground and into which the abrasive sand, used to grind and figure the glass, would be placed. The piece of glass that would become the lens was to be cemented onto the collar of a second mandrel that was fixed into the upper part of the same strong wooden frame, though various adjusters made it possible for this upper mandrel which carried the glass to be brought to bear on the sand-abrasive at different angles. When power was supplied, by a cord and pulley, to this upper mandrel at the same time as the bottom one was made to spin, the spinning glass could be figured into an optical curve by contact with the spinning abrasive tool. By adjusting the angle of contact between the glass and tool, moreover, one could impart optical curves of different radii, resulting in long- or short-focal-length lenses as required.

On the surface, it was another piece of Hooke brilliance, but Auzout, who clearly possessed a firm practical understanding of what was needed to figure a long-focal-length telescope object glass, complained in a letter to the Royal Society that the device would never work in practice.[14] The mechanical pressures that were necessary to make a six- or eight-inch-diameter piece of glass bear down upon a spinning abrasive plate of the same diameter were enormous, argued Auzout—and would not Hooke's thin-looking iron mandrels bend or the wooden frame distort under such pressure? (One also wonders how Hooke's three or so inch diameter pulley wheels on these mandrels, clearly capable of carrying only relatively light transmission cords, could ever have been able to transmit the sheer brute force that was necessary to make the precision grinding possible.) If any distortions of the mandrels or other parts really did occur, how could the abrasive tool impart the necessary perfect curve to the lens?

Then Auzout really touched a nerve when he pointed out that, as far as he had been told, Hooke's lens-grinding 'Engin' in *Micrographia* had never actually been tried in practice, and was 'mere Theory'. Auzout then chastised the Royal Society for giving their imprimatur to Hooke's untried machine, in the light of their claim that 'they publish nothing but what hath been maturely examin'd'.

Indeed, Hooke's claim for the capacities of his lens-grinding machine does have a facile quality, suggesting expectation rather than proven reality. If the machine were fitted with the correct mandrels and abrasive wheels, he says, could not it figure lenses with spherical curves of 'a thousand foot long [i.e. in focal length]; and if Curiosity shall ever proceed so far, ... ten thousand foot long'? Quite simply, Hooke is giving the impression that his machine could be scaled up *ad infinitum*, and it is on this point that Auzout is very rightly calling him to heel.

Central to this debate, however, were the ratios of object glass diameters in inches to focal lengths in feet which astronomers and practical opticians of the 1660s believed held good when it came to producing the best telescopic images. Working on the assumption that the best images were produced by lenses ground to perfect spherical curves, and that aberrations were minimized and magnification increased by using the largest attainable focal lengths, Auzout and others published tables of these optimum parameters. In June 1665 Auzout published such a table in *Philosophical Transactions*, stating that the maximum apertures that telescopes will bear 'are in about a *sub duplicate proportion* to their [focal] *Lengths*'.[15] Auzout suggested as examples that a three-foot-focus telescope should have a one-inch-diameter object glass, a 40-foot a three-inch, a 150-foot an eight-inch, and so on.

But Auzout was entirely sceptical about the prospect of making lenses of 300 or 400 feet focal length, let alone 10 000 feet, as Hooke was optimistically predicting. Surely, asked Auzout, a 1000 or 10 000 foot focal length glass would need a clear aperture of 15, 18 or 21 inches, and such plates of glass free from 'Veins' and 'Blebbs' were impossible to manufacture. And even if such large glass plates *could* be manufactured, Hooke's 'Engin' would never have the power to impart accurate spherical curves upon them. As far as Auzout could make out, while Hooke's 'Engin' might have been capable of grinding little eye-glasses and moderately-sized object glasses for telescopes, it would never have been able to figure '*great* ones'.[16]

Quite clearly, Adrien Auzout had major reservations about the future perfection of long refracting telescopes deriving, it would seem, from his own perception of the limitations of materials and technology. 'Whence it may be judged (continues he) that yet we are very far from seeing *Animals* &c. in the *Moon*, as Monsieur Des Cartes gave hope, and Mr. Hook depairs not of', and from seeing the moon as if the observer were no more than ten or twelve leagues (or between 30 and 36 miles) above its surface.[17] Indeed, it is interesting to find the usually practical Hooke being corrected for technological flights of fancy, not to mention for his association with one of Descartes' more outlandish fantasies, by one of Descartes' fellow-countrymen. On the other hand, one must not forget that a belief in the possibility of advanced life on other worlds was not necessarily outlandish in itself during the seventeenth century, for as late as 1695 the elderly Christiaan Huygens wrote a book on the subject.[18]

Hooke's reply to Auzout is fascinating on many levels, going well beyond the rather lame casuistry of its opening, in so far as it casts light upon Hooke's research circumstances and the inner world of practical optics in the seventeenth century. Hooke began by assuring Auzout and the readers of *Philosophical Transactions* that what he had written about his lens-grinding machine 'was not meer *Theory*... but somewhat of *History* and *matter* of *Fact*: For I had made trials, as many as my leisure would permit'. In short, the Curator of Experiments lacked the time and resources to perfect his lens machine, but wished to publicize his design so that those who *did* possess them might 'have a description of a way altogether *New* and *Geometrically* true' whereby they could produce object glasses of a superior quality.[19]

Hooke next took issue with Auzout's remark about the impossibility in practice of obtaining pieces of glass plate sufficiently large and strong to be able to withstand the grinding pressures necessary to make lenses of 15 or more inches in diameter: 'For, as to the possibility of getting Plates of Glass thick and broad enough without veins, I think *that* not now so difficult here in *England*, where I believe is made as good, if not much better *Glass* for *Optical Experiments*, than ever I saw come from Venice'.[20] Unfortunately, Hooke does not tell us the source of this high-quality English glass, though he could well have been referring to George Ravenscroft's contemporary experiments to produce a superior English flint glass with the aid of Italian craftsmen.

Hooke then went on to take Auzout up on specific points of glass-grinding technique, and to explain how his machine really *could* impart the controlled spherical strokes necessary to produce the promised lenses. But as Hooke never cites examples of excellent lenses actually figured on his machine—as opposed to the future possibility of their manufacture—one cannot avoid thinking that Auzout's point about 'meer *Theory*' had struck home. One senses, moreover, that when Hooke insisted to Auzout that lenses of greater focal length than 300 or 400 feet really were practicable (with their 15–21-inch diameters and perfect spherical curves), and that his 'Engin' really could grind them, he is doing little more than whistling down the wind.

The Auzout–Hooke correspondence also brings out some interesting home truths about the object glasses used in the 1660s, at least in England. While one of Auzout's criticisms of Hooke's 'Engin' was its inability to impart perfectly symmetrical curves to lenses—because of the probable bending of the mandrels when working under great mechanical pressure—Hooke simply responded by saying that perfectly symmetrical lenses were not actually necessary! While, ideally, an object glass should be at its thickest at its geometrical centre, a lens may still be 'very good, when it is an Inch or two out of it'.[21] Hooke possessed, so he said, a 36-foot focal length glass (probably the glass of that specification by Richard Reeves with which he did his *Micrographia* lunar drawings and 1666 Mars drawings, and which he would use for his 1668–9 motion of the earth zenith sector observations) which was a 'good one'. When working at its full un-stopped aperture of $3\frac{1}{2}$ inches, this 36-foot glass furnished excellent views of the moon and Saturn at twilight, and yet this lens was not really symmetrical for 'yet the thickest part of the Glass is a great way out of the middle'.[22]

Much of this discussion about the manufacture of object glasses, however, came about in the wake of Christiaan Huygens' first long-telescope discoveries described in his *Systema Saturnium* (1659), which included Huygens's elucidation of the physical nature of the ring around Saturn, followed by Giovanni Domenico Cassini's 1675 discovery that the ring had a division within it, which posterity named after its discoverer. Cassini worked with long-focus refractors in Bologna and in Paris, some of which instruments were over one hundred feet, produced by the Roman optician Giuseppi Campani, who could make 'great *Optick Glasses* with a *Turne-tool* without any Mould',[23] as well as matching them with eye pieces which could, allegedly, produce sharp images without 'Rain-bow Colours'. It was these Campani glasses which had made possible Cassini's discovery 'that Jupiter might be said to turn upon his *Axe* [axis]', thereby providing substantiation for the Copernican theory, as well as the sighting of belts and satellite shadows upon the surface of Jupiter.[24]

Very clearly, Hooke's own astronomical observations with long telescopes (as noted in chapter 5) must be seen in the context of these continental discoveries, which is why discussions about possible techniques for making bigger and better object glasses proliferated in the 1660s. Indeed, not only did they give rise to Hooke's 'Engin' but also to several papers in *Philosophical Transactions* which were concerned with that process whereby an invention in optical technology could give rise to the *inventio* or the 'coming upon' of new discoveries. It is clear that, while all of Hooke's telescopic observations were made using lenses ground not on his 'Engin', but by professional opticians using their own undisclosed methods, the relationship that existed between manufacture and discovery both fascinated Hooke and occupied a very great deal of his time.

In his 'Life' of Hooke, Richard Waller recorded that 'He first produc'd his *Reflecting* Telescope' on 28 February 1674.[25] Exactly what Waller meant by 'produc'd', however, is not clear, for on the fifth day of keeping his Diary, on 5 August 1672, Hooke first recorded having 'Polisht an object speculum of 7 inches'.[26] One presumes that he 'produc'd' it, and subsequent speculae, for the inspection of Royal Society Fellows and other friends, for Saturday 28 February 1674 was spent in the company of a host of scientific friends at Lord Brouncker's, in the Temple, at Garaways coffee house, and elsewhere.[27] But either way, it is clear that Hooke was actively involved in casting and polishing metal mirrors for reflecting telescopes from at least the summer of 1672.

The origin of Hooke's interest in reflecting telescopes, whether from an acquaintance with mirror telescopes of Gregory, Cassegrain or Newton (whose subsequently famous reflecting telescope was shown to the Royal Society in late 1671), or from his own independent invention, is not clear, but during the early 1670s his Diary indicates a serious and sustained interest in the practical business of perfecting such an instrument. The use of a mirror as the primary light-gathering optical surface of an astronomical telescope, was, after all, full of possibilities for both optical invention and scientific discovery that went well beyond the perceived limits of the long-focus refracting telescope. For one thing, reflecting telescopes seemed to hold out the prospect of being cheaper and easier to make than refractors, for as they needed only *one* concave optical surface to produce an image, it was not necessary to grind a pair of matching curves as it was with a lens. And since only the mirror's reflecting surface mattered, the practical optician was not worried about the presence of internal 'veins', variations in refractive density, or 'blebbs' within the composition of the mirror as he was when dealing with transparent glass.[28] In theory at least, it also seemed that there was no natural limit to the size of these 'speculum' metal mirrors, cast as they were from a bright tin and copper alloy, and hence when figured and brilliantly polished capable of being 'charged' with eyepieces that would give much higher magnifications than those that were possible with long refractors. It is hardly surprising, therefore, that an inventive practical optician and astronomer like Hooke should have been strongly attracted to reflecting telescopes and their hoped-for perfection.

During the late summer and autumn of 1672 in particular, Hooke was clearly involved in a hands-on capacity in attempts to produce a workable reflecting telescope. On 6 August 1672, he 'Tryd polishing all the morn', and two days later 'Polisht a concave well'. Then on 23 August, he 'Pollish 9 foot [focal length] speculum well. Saw Moon at night through it very big and distinct', while on 1 September he 'observed Mars with speculum, but not so good'.[29] What is not clear from these Diary entries, however, is how many individual mirrors he was polishing and observing with. While he will occasionally specify a focal length, he scarcely ever mentions the *diameters* of his mirrors, or the optical configuration — such as Newtonian — in which he was using them. On one occasion, however, he was working on a mirror that was clearly well over three inches in diameter, as when in January 1673 'With Harry I wrought on the Specular metall for telescope, and polisht it pretty trew, but not perfectly having a cloud in the middle about the bignesse of 3 inches over'.[30]

Hooke does not tell us where he obtained the round speculum metal castings, or blanks, from which he figured his optical mirrors, but it is by no means unlikely that, with the Gresham College workshop facilities that his Diary suggests were available, and his extensive connections within the community of London metal workers, he could have cast them on the premises. On the other hand, he could have commissioned the blanks, with their specified proportions of tin and copper, from a London brass founder.

One of the main problems with reflecting telescopes, however, was the tendency of their tin and copper alloy metal mirrors to tarnish, especially in the sulphur-polluted air of London. Whilst back-silvered flat looking-glasses were commonplace on the domestic level by the

1670s, it would be almost 200 years before a technique was discovered whereby one could *surface*-silver a piece of glass, without the coating peeling off. In consequence, there was no way in which Hooke or any of his contemporaries could have ground and polished a concave curve upon a piece of plate glass or other optical material and surface-silvered it to make a concave mirror. All of the mirrors with which Hooke, Newton and their contemporaries worked, therefore, were ground and polished metal, and if the reflecting telescope was going to have a serious future it was essential that an alloy be developed that would hold its initial lustre for at least several months.

One sees in Hooke's Diary how this problem acted as a stimulus to practical innovation. In early October 1672, Sir Robert Moray, who had first proposed Hooke as Curator of Experiments to the Royal Society, 'told me of a metall of excellent Reflection that would not tarnish', which seems to have been of a French origin.[31] Indeed, reflecting telescopes were clearly attracting the attention of practical opticians across Europe, for as Hooke recorded, also in October 1672, 'With Mr Boyle, who told me of the improving of Mr. Newton's way of telescopes in Italy'.[32] For a brief period, mainly in 1672 and 1673, after Newton had announced the design and performance of his reflecting telescopes, English and European astronomers hoped that the brighter and less aberrated images produced by freshly-polished optical mirrors might make the awkward long-focus refracting telescopes redundant. But it was but a flash in the pan, and not until the time of Sir William Herschel a century later did the speculum metal reflecting telescope enter serious astronomical research.[33]

What is important in the present context is the sheer quantity of practical and experimental energy that Hooke invested in attempts to develop viable reflecting telescopes, especially at a time when, during the winter of 1672–3, he was also suffering considerable ill health, as his Diary makes clear. But the exercise of the breadth of his invention, it seems, was Hooke's best cordial against infirmity.

Yet if those optical, meteorological, astronomical and horological inventions which Hooke saw as intimately bound up with the mechanical investigation of nature constituted the great themes in his creative vision, it must not be forgotten that he also produced a stream of lesser devices that were intended to simplify, clarify or speed up a variety of tedious daily chores. And in this respect, one can perhaps best see the legacy of Wilkins's *Mathematical Magick* in Hooke's imagination, as Wilkins's contraptions for calculating numbers, roasting meat, measuring distances or travelling faster then horses could run all aspired to ameliorate the drudgeries of daily life in true Baconian fashion. Indeed, one such archetypal Wilkins device was the 'wheele for one to run races in and other mechanical inventions',[34] which Hooke had been trying out with Wilkins at Durdans near Epsom in the summer of 1665.

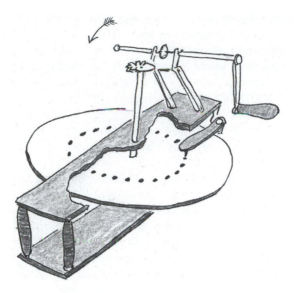

Clock wheel cutting engine.

One of Hooke's more conjectural contributions to practical horology was either his invention, or improvement, of the wheel-cutting engine for the rapid and accurate manufacture of clock gears. Either way, the attention which he paid to the development of a horological bench tool intended to be used in clockmakers' workshops clearly indicates his detailed familiarity with the practical processes of clock manufacture and the awareness of a need to make those processes more exact.

Unfortunately, Hooke never left a properly-documented account of his wheel-cutting engine, and most of what we know about its origins derives from attributions to Hooke by later horological writers such as William Derham, from surviving 'Hooke-type' engines in museum collections, and from brief references in his Diary.[35] Hooke's first cryptic reference to the device comes at the very start of the Diary in August 1672 when he recorded seeing or dealing with a 'Lancashire watchmakers son about wheel cutting engine'.[36] From this, one might guess that not only was such an engine already in existence by that date, at least as an idea, but that it was already sufficiently refined to be capable of cutting watch gears. The Lancashire connection, of course, most likely refers to the already well-established watch-parts manufacture and movement assembly trade in the Prescot district of south-west Lancashire, near Liverpool.[37] These watch components were made on a domestic industrial basis, the basic movements being brought down to London for finishing and setting in gold and silver cases with London names upon their dials. (It is interesting to recall, moreover, that according to anecdotal record Jeremiah Horrocks, who in the late 1630s made some of the first major telescopic discoveries after those of Galileo, and who was seen by Hooke and Flamsteed as the founder of British research astronomy, was first taught geometry by an uncle who made watches at Toxteth, Liverpool.)[38]

Then in mid March 1673 Hooke cryptically recorded that 'Harry cleansd lathe. began wheel cutting engine',[39] though only two days later he 'borrowed Bells engine for cutting wheels', which may have referred to the loan of some sort of dividing plate from a London artisan.[40] But by May 1675, Hooke seems to have been disclosing the operation of his machine to none other than the subsequently celebrated horologist Thomas Tompion himself, when Hooke had 'Much Discourse with him about watches. Told him the way of making an engine for finishing wheels, and a way how to make a dividing plate; about the forme of an arch; about another way of Teeth work...' and such.[41]

The engine in question was intended not only to produce horological gears of a fairly standard accuracy, but also to make them more quickly than the older hand-filing techniques. At the technological heart of the engine was a brass or iron plate about a foot in diameter. Around its centre were engraved a series of concentric circles, perhaps 20 or more in number. (A wheel cutting engine in the store of the Science Museum, London, and dated from internal evidence to *c.* 1672, has its brass plate divided into 25 concentric circles.)[42] Each of these circles was divided in turn into anything between 20 and 100 perfectly equal parts, the largest radial circle being divided into the largest number of parts or digits. And each part or digit was marked by a small hole drilled through the plate. These courses of hole-divided circles were designed to correspond to the familiar gear ratios used in clockwork, so that, for example, if a clockmaker needed a gear wheel with 30 teeth upon it, he could generate it from the circle divided into 30 equidistant holes.

To make this possible, it was necessary to take the hole-divided plate and mount it upon a stout metal axle in a strong iron frame. The circular plate would be mounted to rotate in the horizontal, with its axle pointing upwards. Now when a blank brass or iron disk that was intended to become a clock gear was bolted on to the uppermost projection of this axle, it could be made to rotate in digital fractions of a circle, depending upon which circular course of holes a spring-loaded pin was made to engage in the plate. One could thereby stop down the rotation of both the circular plate, and its attached brass blank wheel, into whatever number of fractional parts the clockmaker desired his brass blank wheel to be divided into. All that was necessary to complete the gear wheel and to incise the teeth was to spin a circular serrated cutter — a circular file, in fact — that was mounted in an iron frame, and bring it to bear upon the edge of the blank wheel. The exact shape of the circular cutter's abrasive edge would then govern the width and depth of the groove that would be incised into the blank clock wheel, so that when the engine wheel had made a full rotation, stopping at each hole in turn, a set of serrations cut into the intended clock wheel would result in a beautifully even set of gear teeth.

Judging from surviving 'Hooke-type' clockmakers' wheel-cutting engines surviving in museum collections, it would appear that the serrated, file-edged tooth-cutter was powered by a small pulley wheel, perhaps attached to the tensioned cord of a bow-drill.

The wheel-cutting engine attributed to Hooke was destined to become an ancestor of one of the most invaluable of all precision machine tools: the engineer's dividing engine. In this form, it would be used in clockmakers' workshops for as long as timepieces continued to be hand-made, and after a denticulated edge and an endless screw worm gear (an arrangement itself taken directly from Hooke's screw-edged quadrant of 1674) had replaced the rows of holes drilled through the plate—in the hands of Henry Hindley, Jesse Ramsdem and Edward Troughton in the eighteenth century—one had the modern engineer's dividing engine.[43]

Hooke's wheel-cutting engine is, of course, a pure piece of ingenious mechanism, depending in no way upon new physical discoveries, such as those in pneumatics or optics. Over the course of his career, Hooke produced a crop of such purely mechanical inventions which were either intended to make life easier through the facility which they provided, or else operated as important adjuncts to more 'philosophical' inventions. Hooke's calculating machine, for instance, was a device for performing routine calculations with greater ease, though it seems to have been one of several similar devices put forth by the Virtuosi of the period. It is not clear exactly how Hooke's arithmetic engine worked, but we do know that when he saw the rival geared calculating machine of Sir Samuel Moreland in January 1673, he regarded Moreland's as 'very silly'.[44]

A similar mechanical invention was the universal joint whereby it was possible to produce a smooth and even motion around bends and corners, which he first illustrated in one of the plates of his *Animadversions* in 1674, and which he had probably invented some years before; while his apparatus for collecting specimens from the ocean depths provides an elegant example of how a metal vessel actuated by ropes and springs might be used to lay the foundations of the science of oceanography.[45] Indeed, these 'useful' inventions seemed to teem from Hooke's scientific and practical imagination. And even by the mid 1690s, by which time his exploration of the great forces of nature seemed to be on the ebb, he was still throwing off ingenious devices. His 'Instrument of Use to take the Draught or Picture of Any Thing' of 19 December 1694, for instance, is an ingenious adaptation of the artist's Camera Obscura. Its novelty lay in the fact that its lens and the glass screen against which the projected image under observation was traced were all enclosed within a hood of thin wood or other such lightweight material which the artist could place above his head and shoulders to produce a local dark environment in which the projected image would be very distinct. By placing a piece of thin tracing paper upon the glass screen inside the hood, one could project an image of the outside world, then copy around that image to obtain an accurate picture with proper perspectives.[46]

Hooke's Cutlerian Lecture *Lampas* (1677) aimed at solving the problems experienced with conventional oil lamps when changing volume pressures within the oil reservoir caused variations in the brightness of the flame. In addition to simply re-designing the fluid mechanics of the oil supply system, *Lampas*, like so many Hooke inventions from his brilliantly creative period of the 1660s and 1670s, then went on to investigate the physical and chemical structure of flames and their relation to combustion in air.[47]

The branch of experimental research which was to lead to several inventions of enduring utility was Hooke's work on spring. Not only did *De Potentia Restitutiva, or of Spring* (1678) develop Hooke's famous Law of Spring, but it also made a major priority claim for the invention of the spring balance in watches,[48] outlined his broader physics of a cosmos activated by vibrations, and contained a description of his spring balance for weighing. Hooke realized that if a spring acted in a predictable mathematical ratio to the tensions placed upon it, then if one suspended a pan from the end of a spiral spring and worked out a calibration scale, it could provide a more compact and portable device for weighing objects than a conventional pivoted balance. Since the late seventeenth century, moreover, Hooke's spring weighing balance has been adapted to a myriad of practical tasks.

While Hooke left no documented claim to having played a part in the development of Thomas Savery's atmospheric steam pumping engine of 1698, such a claim was made for him both by Richard Waller and by John Harris. Waller alleged that Hooke's work on spring and elasticity, most notably of the air and of steam in air, had been important in Savery's development of a suction engine, wherein the suction force was generated by the expansion and condensation of steam in a pressurized container. Hooke's work on spring, says Waller, 'is one of the Principles upon which Mr. Savery's late Engine for raising Water is founded. See *Lexicon Technicum* under Engine',[49] though John Harris's *Lexicon Technicum* (1704) in fact makes no mention of Hooke's contribution.[50] Like Hooke, Savery was fascinated by the application of discoveries in pure science to practical technology, and himself became an FRS in 1705. It has also been popularly believed that Hooke corresponded with

Thomas Newcomen of Dartmouth, Devon, and played an early part in the development of Newcomen's eventual reciprocating steam engine of 1712, which was the next development in steam engines beyond Savery's, though modern scholarly research has found no firm evidence for the belief.[51]

In an undated paper which he presumably read to the Royal Society in the 1690s under the title 'Dr. Hooke's Answer to some particular Claims of Mons. Cassini's, in his Original and Progress of Astronomy', Hooke takes upon himself the task of defending what he perceived as the English priority for many discoveries and inventions in astronomy, and in particular the assertion of his own claims.[52] His tone is often pugnacious, and he is not infrequently reluctant to credit the French with anything original, but the paper tells us quite a lot about his attitude towards invention.

Present in the paper are, of course, Hooke's familiar assertions of his priority of invention of the spring balance of watches. Yet he also claims that the realization that the pendulum length 'would not be the same all over the World ... was discovered by me to this Society, 32 or 33 Years since, as will appear by the Registers of the Society'.[53] While Hooke's paper is not dated, so that we do not know precisely when 32 years previously would have been, his discovery nonetheless, according to his own internal reckoning, must have been in the late 1660s at the latest, and hence *before* the geophysical observations of Richer and Halley. As we saw in chapter 5, Hooke had certainly grasped the significance of Richer's and Halley's pendulum observations by the time of his writing to Newton in 1679. On what grounds Hooke came to this conclusion before 1670, or whether his assertion of 32 years was a mistaken claim for 22 years, we do not know, but it certainly brings home Hooke's awareness of himself as a man whose devices and discoveries had brought about fundamental changes in the understanding of nature across several decades.

One of the most long-standing and influential of Hooke's claims to change the course of astronomical practice in particular was the following: 'A 6th thing is the Application of Clock-Work to keep the Glass [telescope] directed to the Object; but who contrived this Application, will appear in my Animadversions on the *Machina Coelestis* of Hevelius' of 1674.[54] Under this item on his list of invention priorities against Cassini's claims is Hooke's description of the equatorial clock drive for telescopes. Realizing as Hooke had by 1674 that meticulous micrometric measurements of moving astronomical bodies were impossible if the observer's telescope was stationary, he made the inspired suggestion that the telescope and micrometer should be mounted upon a polar axis, and that the axis should be powered by a mechanical clock, so that it rotated once a day, at exactly the same rate as the stars, and hence made it possible to track, measure and observe them in a single smooth motion. Hooke made no claims for having invented the equatorial mount itself, for as he well knew Christopher Scheiner and Tycho Brahe had used and illustrated it before him. However, he was unequivocal in his assertion that he had been the first inventor to drive such an equatorial by pendulum-regulated clockwork, and mount telescopes and measuring micrometers upon it.[55]

Invention was, of course, germane to the very world-view of the seventeenth-century Virtuosi, not to mention a partiality for presenting the new discoveries wrapped up in parcels of one hundred. Sir Francis Bacon's *Sylva Sylvarum* had begun the fashion, while Edward Somerset, Marquis of Worcester, had published his *A Century of... Inventions* (London 1663) which, in the tradition of Wilkins's *Mathematical Magick*, had included descriptions of secret gun mechanisms, perpetual motion devices and — of course — flying machines.[56] And while Robert Hooke's *c.* 1667 list enumerates just over one hundred useful devices, his friend John Aubrey nonetheless reckoned that Hooke's inventions ran into 'many hundreds: he believes no fewer than a thousand'.[57] Richard Waller also claimed that even by 1678 Hooke was 'affirming he had a Centry of the like useful Inventions' that seemed to relate, from the context of Waller's sentence, to spring alone.[58]

Yet what one sees when Hooke or his friends speak of his 'Inventions' is something that goes well beyond the familiar realm of automata or ingenious speculations. To Hooke, inventions were part and parcel of a whole new approach to nature and to practical utility, which aimed to understand the world in terms of mechanism, revealed through a progressive and increasingly sophisticated range of instruments, whereby one could go on to apply nature's inner principles to the Relief of Man's Estate.

CHAPTER 10

A REALM OF VIBRATION: OF FLIGHT, SPRING, WATCHES, AND MUSIC

An active interest in the nature of spring, elasticity and vibration and their practical application to a variety of seemingly self-acting contrivances, goes back to Hooke's earliest years, and certainly to his Westminster School days where, as was seen in chapter 1, 'he... invented thirty several ways of flying',[1] which must have amazed Dr Busby. He found his ideal *métier* at Oxford where, encouraged by Dr Wilkins, Hooke 'made a Module, which, by the help of Springs and Wings, rais'd and sustained itself in the Air'.[2] But finding soon afterwards 'by my own trials, and afterwards by Calculation, that the Muscles of a Mans Body were not sufficient to do anything considerable of that kind', Hooke set about trying to devise a mechanically powered machine that could fly.[3] This led him into the serious exploration of spring and elasticity, which was a topic to which he would return many times in his subsequent career. In addition to thinking of elasticity just in terms of ingenious automata or flying machines, he came to see the concept of shock-wave action and reaction, pulsation and vibration, as lying at the heart of virtually all natural phenomena, from the beating of an insect's wings to the isochronal vibrations of the pendulum, the behaviour of air and the terrestrial atmosphere, the transmission of sound and light, and even gravity itself. To Hooke, the thoroughgoing mechanist, just like Descartes and the notorious Thomas Hobbes before him, the whole business of understanding nature lay in understanding motion.[4]

After abandoning hope of using human muscle power to propel a flying machine, Hooke attempted to devise an artificial muscle substance, though we do not know what this was. As no rubber-based elastic materials were known in the seventeenth century, it is likely that his principal experimental agents were metal springs bent by screws, winches or perhaps the force of gunpowder, as when in October 1674, he 'Discoursed ... of the way of bending springs by gunpowder for flying.'[5] But one Diary entry not only makes a distinctly boastful claim for the invention of an artificial muscle, but indicates that his quest for such a material went back to his Oxford days, when in the company of

Wilkins and Boyle his ingenuity first found expression, for on 11 February 1675, he recorded:

> 'Dr. Croon at Royal Society read of the muscles of birds for flying. I discoursed much of it. Declared that I had a way of making artificial muscule to command the strength of 20 men. Told my way of flying by vanes [wings] tryd at Wadham.'[6]

Hooke's work on spring-powered flight is one of the most tantalizing pieces of research that he undertook, for all that survives are a series of brilliant insights, mixed with extraordinary claims. Neither he, nor his posthumous editors, ever published anything on the subject, though one only wishes that Richard Waller had printed some of the 'several Draughts and Schemes upon Paper' upon flight which he claimed were amongst Hooke's papers at the time of his death. These included

> ' ... some contrivances for fastening succedaneous [substitute] Wings, not unlike those of Bats, to the Arms and Legs of a Man, as likewise of a Contrivance to raise him up by means of Horizontal Vanes plac'd a little aslope of the Wind, which being blown around, turn'd an endless Screw at the Center, which help'd to move the Wings, to be manag'd by the Person by this means rais'd aloft: These Schemes I have now by me, with some few Fragments relating thereto, but so imperfect, that I do not judge them fit for the Publick.'[7]

What would one now give to know what they contained in more detail!

While Robert Hooke left no surviving drawings or even descriptions of either a complete flying machine or of any kind of 'artificial muscle', one of his most detailed treatments of actual flight is to be found in his descriptions of insects in *Micrographia*. In his account of the blue fly, one finds Hooke's genius for observation, experiment and mechanical analysis at its finest. What especially interested Hooke was the mechanical importance of the changing wing-velocities of flies and bees on the one hand, and butterflies and moths on the other.[8]

When reading Hooke's accounts of insects and of flight, one cannot help but be struck by a clear parallelism, at least in terms of thought processes, with the researches of Leonardo da Vinci, conducted around 180 years earlier. In his Note Books, Leonardo left detailed sets of observations of insects, and of the motions of their wings, and studies of the different flight behaviour of creatures such as flies and bees (which Hooke styled as having 'glassy' wings) and moths and butterflies (which had 'feathered' or downy wings).[9] Leonardo also described the flight behaviour of different species of birds—which Hooke does not—though we certainly know that he and Wilkins were interested in bird flight from the few scraps on the subject that have survived from them.

Robert Hooke, however, came to his studies on insects, feathers and flight quite independently of Leonardo's work, for in the 1660s Leonardo's manuscript Note Books were still effectively lost to European scholars, and would not be published for centuries to come.[10] But the intellectual, experimental, model-devising approach to research which Leonardo and Hooke shared and which is most conspicuously seen in their independent researches into flight and flying machines is striking. Of course, what made things so different for Hooke was the new armoury of instruments that had become available by the 1660s, especially the microscope, with their ability to take nature apart, and to examine otherwise invisible structures.

One also suspects that Hooke's innate musicality gave him a further fruitful insight into flight, as he saw a physical, measurable parallel, in the fact that as a fly's wing and flight pattern changed with its speed and flying direction, so did the musical pitch of the buzz. Hooke suggested that a stringed instrument might be tuned in unison with this pitch, and that the ensuing note or vibration could be expressed in mathematical terms.

Very frustratingly, *Micrographia* takes this equation of physical vibration with musical pitch (which was two centuries before the definitive modern researches of Hermann von Helmholz in 1863) no further, other than to conclude that a fly's wing beat 'many hundreds if not some thousands of vibrations in a second minute of time',[11] and that it was the most rapid mechanical action in nature. But Hooke does appear to have come up with a relatively precise value within 18 months of *Micrographia*'s publication, for Samuel Pepys, who was still very much of a scientific beginner, recorded in his Diary a meeting with Hooke on 8 August 1666:

> 'He did make me understand the nature of Musicall sounds made
> by strings, mighty prettily; and told me that having come to a
> certain Number of Vibracions proper to make any tone, he was able
> to tell how many strokes a fly makes with her wings (those flies that
> hum in their flying) by the note that it answers to in Musique
> during their flying.'[12]

Though Hooke does not explicitly state how he believed the wing-action of a fly kept it airborne, he seems to have related it to the spring, or elasticity, of the air, as the down-beat of the wings created a somewhat compressed pocket of air upon which the creature rode, while the up-beat created a form of suction that lifted it up. One can see how these concepts related to his earlier work, conducted when he was Boyle's assistant, into the 'spring' of air.

It has already been mentioned that vibration was widely used as an explanatory mechanism by Hooke. Vibration was intimately related, in

Hooke's thought, with elasticity, spring and resonance, while these in turn were seen as mechanical responses to more fundamental agencies such as 'force', light, weight and gravity.

While Hooke possessed no coherent idea of how springy bodies differed in structure from inert ones, it is clear that his scientific interest in elasticity went back to his Oxford days if not before. Indeed, these experiments, performed with Dr Wilkins and his Club during his undergraduate years, left an indelible imprint upon Hooke's scientific imagination, and at various times in his subsequent career he would recollect incidents of things seen or undertaken during the 1650s which bore a direct relation to what he happened to be discussing decades later. Just as he recalled the Wadham flying experiments in 1675, so in the early 1680s, when discussing spring and elasticity, he mentioned a curious pneumatic fountain that Dr Wilkins had built in the gardens of Wadham College, in which:

> 'the Spring of the included Air [was made] to throw up to a great height a large and lasting stream of water: which water was first forced into the Leaden Cistern thereof by two force pumps which did alternately work, and so condence the Air included in a small Room'.[13]

As this device had been operational before Hooke had built the airpump for Boyle, one wonders at the range of experiments with compressed air that had been going on in the 1650s, before it was possible to work properly with vacuums.

Like many of Hooke's researches, his work on spring and elasticity was done at scattered intervals stretching over several decades, and certainly from the mid-1650s to the early 1680s. He subsequently claimed that in the late 1650s he had been experimenting in Oxford with spring-regulated timepieces, and his 'pendulum watches' attempted to apply the isochronal principle of the pendulum clock to a portable timepiece by 1660, as will be noted presently.[14]

It was in his *Helioscopes* in 1676 that Hooke followed the popular seventeenth-century conceit of announcing the discovery of his 'Law of Spring' in an anagram, or more correctly, in a logogrith: *cediinnoopssttuu*. He published its key two years later, in his most complete treatment of elasticity, *De Potentia Restitutiva, or of Spring*. Here Hooke enunciated the original formulation of the law that bears his name: *Ut Pondus sic Tensio*, or as it might be construed 'the weight is equal to the tension'.[15] As the tension was seen as the product of an increasing series of weights in pans suspended upon coiled springs, it is easy in this pre-Newtonian-gravitation age to understand how Hooke spoke of the *pondus*, or weight, as acting upon the spring. The formulation of 'Hooke's Law' with which we are more familiar today, however, is his alternative expression, *Ut Tensio, sic Vis,* or 'the tension is equal to the force'. In the seventeenth century, we must not forget, scientists had not yet developed a coherent concept of 'energy', and often spoke of *pondus,* or weight, force, pressure and such when trying to define the powers of nature.

In his treatment of vibration and elasticity in *De Potentia Restitutiva,* Hooke discussed the concept of an aether that pervaded 'the whole Universe' and in which particles continually moved. The theory of a vibrative agency that could express the motions of a fly's wings and also the propagation of light, or gravity, in space was the nearest that Hooke came to a 'Grand Unified Theory' in the Mechanical Philosophy. Considering his fundamental concern with *mechanism* in nature, one can understand the resentment that Hooke felt when Newton presented the very different unifying theory of Universal Gravitation to the Royal Society in 1686. But before considering Hooke's ideas on gravity, it is important to look at his work on other vibrating phenomena, such as sound and light, along with the way in which he tried to grapple with such phenomena by means of instrumentation and mechanism, which from the mid 1650s onwards had led him into the physics of horology.

One of the most far-reaching horological inventions ever devised was the spring balance in watches. While the portable clock or watch, powered by a mainspring rather than a falling weight, had been around for almost two centuries, by the mid seventeenth century these instruments were still notoriously inaccurate, lacking—just like the pre-pendulum clock—any natural physical regulator with which to govern the release of the mainspring's energy. Though Renaissance clocks and watches had escapements, the swings of which released the driving weight's or mainspring's energy in short bursts, or 'ticks', the motions of these escapements depended solely upon their tendency to swing and recoil under the direct mechanical pressure of the mainspring. Even by 1650, there was no external natural force that had been harnessed to make the 'ticks' perfectly equal, or isochronal.[16]

While Hooke, at the suggestion of Seth Ward, may have been experimenting with pendulums as early as 1656, it was Christiaan Huygens's successful application of Galileo's pendulum researches to clockwork in 1657, especially after Huygens had devised a pair of cycloidially curved metal cheeks to control some of the secondary harmonic vibrations in a swinging pendulum, that revolutionized the time-keeping capacity of fixed clocks, suddenly making them serious scientific instruments.[17] What the pendulum utilized was the constant force of gravity, expressed in terms of the pendulum's own oscillations, turning it into a natural external regulator, to ensure an even rate of time-keeping for a given point on the earth's surface. The challenge now came to lie in devising a way whereby the seeming isochronal properties of the pendulum swinging with a fixed clock could be applied to a watch to make accurate timekeeping portable and thereby produce a 'pendulum watch'.

Hooke's motive in developing 'pendulum watches', just like that of his perceived rival Huygens, or 'Monsieur Zuliechem', was not simply to improve the efficiency with which gentlemen could keep appointments. It was rather to produce a clock that was sufficiently accurate to enable the finding of the longitude at sea, for it had been known for well over a century by 1658 that if a ship during the course of its voyage could have on board a reliable timepiece that carried the local time of the port of departure, a capital city, or of any other agreed location, then that ship's longitude could be obtained by a simple procedure.[18] All that one needed to do to fix the vessel's longitude, therefore, was to compare local noon, when the sun reached its highest daily elevation, with the time back at home, as carried by the ship's clock. As the sun moved through 15 degrees of sky in one hour, a clock which displayed 3 p.m. when the local mid-Atlantic time was 12 noon made it easy to calculate that the ship must be $3 \times 15°$ (or 45°) west of London, Bristol, Amsterdam or whichever was its port of departure; and if one could also reliably note minute and second divisions of whole hours, then one could fix the ship's place with great exactitude.

But no clock in the world could do anything other than go haywire when placed aboard a pitching vessel in the days before 1658, having no natural isochronal regulator to govern its going. With the announcement of Huygens's pendulum, hopes ran high that a truly reliable sea clock might be developed which would solve the longitude problem.

Sea trials of specially adapted pendulum clocks were made in the early 1660s. Lord Kinkardine, for instance, took a pair of Huygens-type clocks, fitted with short 'half second' pendulums, to sea in 1662, and from his subsequent remark, 'at which I (*i.e.* Dr *Hook*) was present', it seems as if Hooke himself sailed on the test voyage.[19] The clocks were mounted below decks in specially-weighted, ball-joint-suspended cases that hung from the cabin's overhead beams, to maintain their stability. What is more, the two clocks were set up at right angles to each other on the ship to see if it was possible to neutralize the effects of the vessel's pitching and rolling, and thereby obtain a consistent reading; and while the timekeeping rates of the two clocks were found to vary, they nonetheless did keep going. Sir Robert Holmes (later to become Governor of the Isle of Wight and to have involvements with the Hooke family) seems to have been the officer who was in direct naval charge of the clocks and of their sea trials, and after they gave him the position of *Fuego* in the Cape Verde Islands more accurately than other officers in the squadron did by dead reckoning, he reported favourably upon them.[20] Even so, Hooke was of the opinion that a pendulum clock at sea was 'useless' and 'would do no harm' to the spring balance watches, which he was striving to perfect.[21]

All of these experiments, conducted within a few years of 1658, would have been made with clocks using very short $9\frac{3}{4}$ inch 'half-second' swing, or shorter, Huygens pendulums connected to their going wheels through a verge escapement. As a verge escapement required a very wide arc of swing for the pendulum to operate in the first place—sometimes more than 45°—it perhaps explains how they were able at least to keep going at sea, though the shipboard clock's very instability would have introduced an erratic component into the pendulum vibrations which would have resulted in poor timekeeping. The much more accurate 'anchor escapement' which Hooke later claimed, when in a priority dispute with William Clement, to have 'shewed before the *R. Society* soon after the Fire of London',[22] with its 5° arc of swing, could only work, however, in a clock that was completely stable—unless, perhaps, its anchor was attached to a balance wheel which made 'use of Springs instead of Gravity for the making a Body vibrate in any Posture',[23] and was capable of functioning accurately at angles that were out of the vertical. On the other hand, there is no evidence to suggest that Hooke ever used any escapements other than verges in any of his 'pendulum watches', irrespective of whether they were intended for use at sea or simply to be carried in a gentleman's waistcoat pocket. Indeed, verge escapements remained the norm in watchwork for over a century after 1660, in spite of the invention of George Graham's cylinder escapements after 1725 which were, however, difficult to construct. Yet verge or non-verge escapements apart, it was Hooke's claim for the devising of a spring-controlled balance wheel for sea clocks and watches which not only manifested itself in one of his first great outbursts of creativity in applied physics, but also plunged him into one of several cycles of alleged betrayal and plagiarism which punctuated his almost 50-year career in science.

Lying at the heart of Hooke's spring balance was that physical concept which he later came to formulate and publish in the mid-1670s as *Ut Pondus sic Tensio*, and *Ut Tensio sic Vis*, though he might well have grasped its principle two decades previously while still at Oxford.[24] If the energy put into the tensioning of a balance spring bore an exact mathematical relationship to the energy released by the same spring, then one could have, at least in theory, an isochronal action that could be just as exact as that present in a pendulum escapement, with the added advantage that such a spring mechanism, by not depending upon gravity for its regulation, could be portable. The point of controversy that ensued, however, was not about the physics of the mechanism, but about the date and priority of its invention.

In 1674, Christiaan Huygens described an arrangement for a spiral spring balance for which Hooke claimed priority, and which Hooke believed had been plagiarized from him by certain Fellows of the Royal

Society who had been privy to his invention.[25] Hooke attributed the plagiarism to his particular *bête noire*, Henry Oldenburg, Secretary of the Royal Society. Hooke began to make noisy accusations, though Oldenburg warmly vindicated himself in *Philosophical Transactions* and went on in turn to accuse Hooke of being pushy and making excessive claims.[26]

It is very clear that the announcement of the Huygens watch ruffled Hooke's feathers in a big way, and caused him to set in print his own priority claim for the invention of spring-balance watches extending back over 16 or 17 years. Hooke alleged that even by the time of 'His Majesties happy Restauration' in 1660 he had not only developed a viable vibrating spring-controlled watch escapement, but was party to a 'Treaty' with honourable and 'Noble Persons', including Lord Brouncker, Sir Robert Moray and Robert Boyle,[27] to secure his rights for '... measuring *Time*, which was the way of applying *Springs, to the arbor of the Balance of a Watch, for the regulating the vibrations thereof in all postures*'. This 'Treaty', moreover, would have secured Hooke 'several Thousand pounds', although Hooke claimed that he suddenly withdrew from the bond upon discovering that once he had divulged the secret of the mechanism, he stood to gain not a penny if someone else came along and improved upon his original design. For as Hooke shrewdly realized, once the physical principles of the invention were out in the open, ''twas easie to vary my Principles an hundred waies',[28] and hence for him to lose the reward to someone else. If this was, indeed, the condition of the Treaty of 1660, one can fully understand why Hooke rejected it. Then in 1664, having delivered his Cutlerian Lecture 'in the open Hall at *Gresham* Colledge' on the 'application of *Springs* to the *Ballance* of a Watch for regulating its motion', and being prevailed upon 'at the earnest desire of some Friends', he 'cause[d] some of the said watches to be made', though he was reluctant to divulge their precise mode of operation.[29] As Hooke was still relatively poor in 1664, moreover—still being owed his wages by the Royal Society, and not yet having obtained his Gresham Professorship and City Surveyor appointments—it would be interesting to know who would have paid for the construction of such watches.

And here things seem to have rested, relatively speaking, until the Huygens announcement, when in 1675, and to reinforce his prior claim, Hooke now had a watch incorporating a spiral balance spring made by Thomas Tompion for presentation to King Charles II, carrying the engraved inscription 'Robert Hook inven. 1658. T. Tompion fecit 1675'.[30] Whether Hooke really had already come up with a workable isochronal spiral spring balance as early as 1658, which must have been only months after Huygens announced his original pendulum clock escapement, or whether he had perhaps contrived some other balance wheel configuration using straight or zig-zag springs, we do not know.

178

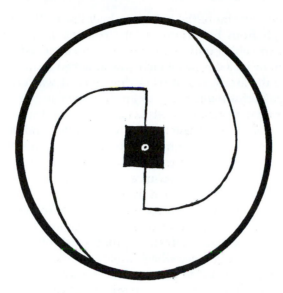

Double spring watch balance.

But what undoubtedly resulted was that Hooke suddenly became very active in the promotion of 'his' spring balance, especially as, as he recorded in his Diary on 12 May 1675, Francis 'Lodowick here about Longitude. Afirmd [*sic*] £3000 premium and £600 from the States General of Holland.' By this date, and again in 1676 when he mentioned a cash premium for finding the longitude at sea in his *Helioscopes*,[31] Hooke was concerned not just with an inventor's prestige, but also with what he perceived as an inventor's just reward.

In the years 1675–8, Hooke's writings, published and unpublished, are full of 'his' spiral spring balance, and his clear wish to stake out his own prior claim over and against that of Huygens. His Diary entries for the months March to August 1675, for instance, bristle with balance spring references, describing a variety of experimental configurations, and often involving the horologist Thomas Tompion. A characteristic reference was the entry for 8 March 1675, when he recorded 'At Garaways. Tompion. I shewd my way of fixing Double Springs to the inside of the Balance wheel. Thus.'[32] [See above drawing.]

It cannot be denied, therefore, that by the end of 1674 Robert Hooke had devised a spiral, arched, or part-spiral double balance spring, though how the design had been modified and improved since 1658 or 1660 is hard to say for certain. In the configuration of March 1675, however, the inner ends of the two spiral curves of the double spring balance seem to have been anchored into the opposite faces of a square box, 180° apart, while their outer edges connected with what appears to have been the inner rim of the balance wheel itself — 'inside the Ballance wheel' — though from his cryptic Diary passage and sketch, it is not completely clear which end of the spring was attached to the rotating balance wheel, or its axle, and which was attached to the watch case.[33]

To make the arrangement work, however, it was necessary for the much more powerful mainspring to apply power to the balance wheel, causing it to rotate, from an impulse communicated through the escapement. In this act of rotation, the balance wheel — one end of which had to be fixed securely to the watch case (to provide something to 'push' against) while the other could rotate with the balance wheel — stored up energy in the delicate spiral balance spring, or 'hairspring'. When a measured amount of energy had been allowed to build up, it was released. At this point, the balance wheel suddenly rotated in the *opposite* direction, until it was checked by the escapement. Now, the cycle of expansion and contraction of the spring, and of the rotation and counter-rotation of the balance wheel, was complete, each rotation producing a characteristic 'tick', so that a new impulse from the mainspring could be imparted through the escapement, and the cycle could begin again. The ingenuity of the balance spring arrangement, therefore, derived directly from Hooke's perception that the amount of energy put into tensioning a spring was directly proportional to that given out when the spring was released. It was an elegant application of a piece of pure physics to practical technology.

Hooke's obviously forceful promotion of his invention led more people than the King to want a Hooke watch. As early as 1 January 1675, he was impressing Sir Jonas Moore, Master of the King's Ordnance (and soon to become one of the early patrons of the Royal Observatory, Greenwich, and supporter of John Flamsteed, the first Astronomer Royal) with his watches. On that night both Hooke and Moore observed a lunar eclipse from Sir Jonas's house in the Tower of London, using an 8-foot telescope and timing it with 'my pocket-Watch, whose ballance is regulated with springs'. Whether by 'my' Hooke meant his own personal watch or one of his design owned by Moore is not clear, but if he was out to impress, he had excellent company around him, for his Diary recorded that in addition to Moore, Lord Brouncker, Captain Sherburn and Mr Collins were also present.[34] Then on 13 June 1675, Hooke recorded that 'they fitted the Deans watch well', which one assumes to refer to the fitting of balance springs, probably by Tompion, to John Tillotson's perhaps already existing watch.[35] On 12 August 1677, he records 'Scarborough here resolvd on a watch with two springs bended about cylinders and a 3rd spring regulator movd by strings'.[36] Whether the gentleman in question was Sir Charles Scarborough FRS, a Royal Physician and a friend of Hooke, or another man of that name is not known. But whatever the case, it is plain that Hooke's balance wheel spring watches, with their superior accuracy, were being widely talked of in the most influential circles.

Exactly when did Hooke come up with his balance wheel principle, and how authentic was his 1675 claim that it went back to 1658? In addition to the already well-known seventeenth- and eighteenth-century records, A Rupert Hall's recognition in the Library of Trinity College, Cambridge, of a manuscript by Hooke describing a proposed marine chronometer has provided valuable new insights to scholars, and led to an attempted reconstruction of the chronometer by the scholar and practical horologist Michael Wright.[37] This undated Trinity manuscript seems from internal evidence such as watermarks to have been produced in two stages, the first part perhaps in the mid 1660s and the latter in the 1670s. Its main points of innovation arise from Hooke's concern to develop a constant driving force for the clock, combined with a constant force escapement, which would minimize energy wastage and conduce to accurate time-keeping. On the other hand, the Trinity manuscript deals with balance spring escapements, some of which employ straight or zigzag springs and some curved ones. (William Derham, for instance, when reviewing the various developments through which the Hooke balance mechanism had passed, mentioned that some were controlled by a 'tender strait spring... one end of which played backward and forward with the Balance. So that the Balance was to the Spring as the Bob of a Pendulum, and the little Spring as the Rod thereof.')[38] Irrespective of the exact shape and configuration of Hooke's early spring balances, what the Trinity College manuscript does make clear is his long-standing concern with isochronal horology and his attempts to apply natural physical principles to the design of portable timepieces.

After 1679 Hooke's preoccupation with a marine time-keeper for finding the longitude at sea seems to have waned, perhaps because he recognized the formidable problems involved in making a reliable marine chronometer, demanding as it did both a physical knowledge and a technological capacity that outstripped the resources of the age. It would not be until over 30 years after Hooke's death in 1703 that John Harrison would be successful in addressing himself to these physical problems, such as the effects of temperature and position change upon a ship-borne timepiece, using his own technologies, such as counter-rotating balances and bi-metallic temperature compensation, with which to counteract them. It would not be until 1764 that Harrison's fourth chronometer proved beyond all doubt that a self-compensating mechanism could be used to find a ship's exact longitude at sea.[39] Yet at the heart of the Harrison machine would be the isochronal action of a spiral spring-governed balance wheel and escapement, and for this he had Robert Hooke to thank.

By the late 1720s, however, when Harrison first applied himself to devising an efficient marine chronometer, the spiral, 'hairspring' balance had long since been established in practical horology, although the future of the spring balance would not lie in the counter-actions of a pair of short curved springs, but in one long piece of very fine spring, shaped to form a volute or spiral. One end of this spiral was to be attached to the axle (known technically as the arbor, or 'staff') upon which the balance wheel rotated, while the other was anchored to the brass plate upon which the movement was built, and was snugly fitted into the gold or silver case. In its mature form, the long single spiral spring would control a balance wheel which rotated much more rapidly and passed through more axial revolutions per 'tick' than the original Hooke mechanism, for in this eighteenth-century arrangement, the escapement could be regulated with much greater accuracy. Indeed, this configuration of spring balance wheel was, with a variety of modifications of detail, to lie at the heart of chronometers, portable clocks and watches down to the 1970s, after which time the superior isochronal properties of a quartz crystal replaced mankind's reliability upon clockwork when it came to measuring time. What is so very evident, however, is the fundamental importance of those experiments into the physics of spring and elasticity which led Hooke—and also Christiaan Huygens—to the development of a mechanism which would dominate isochronal horology for the next 300 years.

In the wider assessment of Hooke's interests in vibration, elasticity, and their practical applications one must also ask how far his inclination towards these inquiries—from studies of the oscillations of a fly's wing to the use of hairspring balances in watches—was influenced by his sensitivity to music. Of this sensitivity there is no doubt, judging from the speed with which he learned to play the organ at Westminster School, his Chorister's place at Christ Church, and his scientific interests in sound, pitch and harmony.

As early as July 1664, as the newly-appointed Curator of Experiments at the Royal Society, 'he produc'd an Experement [*sic*] to shew the number of Vibrations of an extended String, made in determinate time, requisite to a given Time or Note'. From his experiments 'it was found that a Wire making two hundred seventy two vibrations in one Second of Time, sounded *G Sol Re Ut* in the Scale of Musick'. He also demonstrated further sound frequencies and their mathematical relationships on a Monochord.[40] Then 17 years later, in July 1681, he was experimenting with the sounds emitted by rapidly rotating serrated brass wheels and noting, under the right circumstances of adjustment, their equivalence to human voice sounds.[41]

What must be remembered, however, is that not for some time after Hooke's death in 1703 was there anything resembling an agreed or standardized system of pitch or tuning used in European music. While individual consorts and choirs would have had their own scales of tuning amongst themselves, these pitches would not necessarily have been interchangeable with those of other groups of instruments elsewhere, so that a 'G' in the Chapel Royal at Windsor could not be relied upon to have been the same note as a 'G' played in a London theatre band, let alone in the Court of Versailles. But what one finds in these experiments is the beginnings of a *physics* of sound, as a given pitch was seen in relation not only to a particular performer's ear, but also to a set pattern of physical and quantifiable motions in the world.[42]

In what are probably the surviving manuscript notes for a lecture on music given at some unspecified date, and communicated to the Royal Society by William Derham in 1727, one gets a good idea of Hooke's approach.[43] As in the manuscript versions of Hooke lectures published by Waller in 1705 and Derham in 1726, one immediately recognizes the teeming richness of Hooke's intellect, as he first casts historical and anthropological insights at the reader or listener, before moving on to the physics of sound, and proceeding to analyse the psychological impact of different sounds upon the listener.

As part of his wider discussion of sound, Hooke suggested that music was a sort of proto-language that precedes speech. While in the seventeenth century no one had any concept of early or semi-ape man, Hooke pointed out that infants were responsive to different sounds long before they could understand words. He then went on to affirm the vibrative source of all music 'because all sounds are generated by Motion, Sound being nothing else but a tremulous motion of the Drum and Organ of the Ear, excited by the like Motion of the Sonorous Medium, which receiv'd its motion from the Sounding Body'.[44] Indeed, not only can the strings of a musical instrument 'tremulate' and thereby cause vibrations that excite the inner anatomical structures of the ear, but vibrating strings can induce sympathetic vibrations in other strings and, as Athenasius Kircher noted in the earlier seventeenth century, so Hooke reminds us, even sounds produced by vibrating glass vessels can set off sympathetic vibrations in adjacent glasses.

Yet no matter whether the 'tremulations' were generated by strings, pipes or glasses, what mattered to Hooke when it came to producing any given musical note was an exact mathematical motion, such as the 272 vibrations necessary to produce the note 'G'. He saw these vibrations as wave-like in character, in which harmonic sounds were the products of sympathetic, or mathematically synchronous, vibrations, while dissonances resulted from asynchronous waves. It was the harmonious or synchronous sounds that produced the pleasurable responses in humans, for, as Hooke was no doubt aware, a succession of classical and medieval writers took it as axiomatic that mathematically harmonious sounds appealed to our rational faculties, whereas inharmonious ones made our natural reason flinch.[45]

For Robert Hooke, therefore, vibrative motion was one of the great wellsprings of physical truth and hence a primary subject for scientific investigation. And as he saw sound and the applied principles of elasticity as central to this line of research, so likewise did he see light, as we shall find in the next chapter.

CHAPTER 11

'A LARGE WINDOW... INTO THE SHOP OF NATURE': HOOKE AND LIGHT

The generation, propagation and colours of light fascinated seventeenth-century scientists. As Robert Hooke reminded his Gresham College lecture audience in 1680, coherent ideas about the nature of light went back to the pre-Socratic Greek philosophers, and had constituted a serious area of research for such moderns as René Descartes. Also there was Hooke's own early mentor and continuing friend, Robert Boyle, whose *Experiments and Considerations Touching Colours* (1664) reported a diverse body of researches into all aspects of colour, and which stimulated great interest in the subject, though, as a close colleague of Boyle, one assumes that Hooke had been privy to these ideas and researches long before they appeared in print.[1]

Light had always held a high place within Hooke's own list of intellectual priorities, as is evident from his surviving writings, and it is easy to understand why. In an age when religion and science were seen as intrinsically related, light was perceived as one of Creation's divine absolutes. It had been summoned into being by God on the First Day of Creation at the beginning of time, and, as Hooke reminded his listeners and readers, was the agent whereby darkness was driven out of the world in the *Genesis* narrative.[2] On the other hand, this divine absolute was clearly amenable to rational and experimental investigation which, as Jewish, Christian and Arabic philosophers in the middle ages had all argued, filled mankind with delight and brought him closer to the contemplation of his Creator. The study of optics, therefore, was not only a part of terrestrial physics, but also deepened religious faith.

In another respect, Hooke responded to light as a man in love with lenses, microscopes, telescopes and all the optical devices which were currently bringing new realms of knowledge out of the darkness; but on a less obvious level, perhaps, Hooke's artistic sensibilities were aroused by light, for it is hard to read those experiments described in *Micrographia* wherein he investigated the spectrum, and not be aware of the fact that here was a man who was enchanted by the sheer physical beauty of what he was finding, whom the 'lovely reds', 'rich greens' and other colours simply transfixed with delight.[3] It is here that we also glimpse the Hooke who, as a youth, had once worked in Sir Peter Lely's studio, and who in manhood displayed an expert familiarity with painters, limners, and the grinding and mixing of mineral compounds for the artist's 'pallat', which attempted, in their way, to reproduce the 'pallat' of God Himself. It was also the study of light which was to drive Hooke along the first of several collision courses with the young and as yet un-knighted Isaac Newton, and reveal a personality clash that would last a lifetime and do enormous damage to Hooke's posthumous standing as a front-rank scientist.

Like so many of Hooke's scientific investigations, his first articulated researches into light found expression in *Micrographia*. Indeed, by 1665 these ideas had already reached a sophistication which partook of the definitive, so that his subsequent published writings on light, such as in *Helioscopes* (1676), *Lampas* (1677), the manuscript essays and lectures published posthumously by Waller in 1705 and Derham in 1726 and those that still survive unpublished were only refinements and amplifications of them.

While accepting the Biblical origins of light as a divine agent, Hooke was fascinated by its observable and measurable effects in nature. While God was the primary Creator of light, which was an *'Anima Mundi'* or active soul or agent within the world, 'Its Action being so near of kin to that of Spirit',[4] the light that we see about us is made manifest through the *secondary* causes of reflection, refraction and illumination, and is amenable to study as a piece of physical phenomena. And while Hooke did not regard it his business to say what light actually *was*, any more than he could say what the other great forces of gravity and magnetism actually were in an absolute sense, he did see it as the natural philosopher's job to consider how individual rays of light might be generated and how they came to produce colours.

Light rays, argues Hooke, are always generated by motion of some kind. The sun and stars were generally accepted to be burning fiery bodies by 1660, while even seemingly stationary objects that were known to emit weak sources of light, such as rotten wood, fermenting organic matter and glow-worms, clearly contained 'internal motions' brought about by the very act of fermentation or decomposition.[5] All other objects, such as the earth, moon and planets, simply reflected light generated by the sun or some other primary source. This primary light, moreover, whether it came from a star, a candle, or a glow worm, had extension in all directions and an ability to fill space by radiating in straight lines.

Where Hooke was perhaps unusual in the 1670s, however, was in his belief, based on detailed examinations of their nuclei with telescopes, that comets were also self-luminous, light-emitting, as opposed to light-reflecting, bodies. In this respect, he differed from no less eminent a contemporary than Johannes Hevelius, who believed their tails to be nothing more than sunlight refracted through a nebulous mass. Hooke, indeed, was one of the first scientists to think of comets as possessing solid nuclei which emitted some kind of particle stream.[6]

Yet how could the sun, stars and comets actually burn? Hooke provides a solution from his wider researches on combustion, which he first investigated as Boyle's assistant and then independently in *Micrographia*, Observation XVI. All combustion, in Hooke's thinking, was due to the interaction of nitrous and sulphurous bodies, wherein the nitrous *menstruum* or dissolving agent surrounding the sun or a star was believed to act upon it in the same way that 'aerial nitre' attacked heated wood to dissolve it in what seemed like a spontaneous body of flame. So while light itself was divine in its metaphysical source, its local causes were chemical action.

Hooke says nothing about the velocity of light in *Micrographia*, seeming to assume that its transmission was instantaneous, and he indicates as much in his 'Lectures on Light', delivered in Gresham College early in 1680, where he asserts 'the Very Instant that the remotest star does emit Light, in that very Instant does the Eye upon Earth receive it'.[7] But in 1675, the Danish astronomer Olaus Römer, who was working under the patronage of King Louis XIV, drew the conclusion from the apparently delayed motions of Jupiter's satellites at certain places in Jupiter's orbit around the sun that light travels at a finite speed. This led Hooke to return on several occasions to the topic of the velocity of light in his subsequent writings, most notably in his 'Lectures on Light'. Here, Hooke takes issue with Römer, arguing that his explanations for the delays in the Jovian satellite transits could have derived from a variety of factors, such as mistakes in observational technique, and our ignorance of the nature of the satellites' reflecting surfaces. But quite simply, Hooke could not credit that light could travel a distance equivalent to the diameter of the earth, or 8000 miles (as the velocity was then believed), in a single second of time. For, he suggested as an analogy, if one compared the speed of a cannon ball with that of a snail, then the cannon ball's speed was but a snail's pace compared to the velocity of a ray of light![8] Why he should have had any objection to light having an extremely fast, yet finite, velocity, as opposed to an *instantaneous* transmission, is not made clear, beyond the fact that he found 8000 miles a second to be simply unbelievable. One suspects, however, that being a mechanist in his physics, he had no viable physical model for a finite velocity. If he saw space as full of some kind of corpuscular ether, then energy released in one place—such as light-generating motions within a star—would be instantaneously transmitted from it in much the same way as a blow applied to one end of an iron bar results in an instantaneous blow being felt at the other end of the bar. But as he never spoke of the cosmological ether as being elastic, in the way that the earth's atmosphere is elastic (and hence can transmit sound at a finite speed, as particles bounce into each other), it seems that he really could not conceive of any workable velocity slower than instantaneous movement through space. On several occasions in his subsequent writings, indeed, he returned to Römer's work and to his own assertion that light simply *must* be instantaneous in its transmission. One wonders how much further Hooke's scientific imagination would have been stretched had he realized that Römer's early estimates for the speed of light were in themselves snail-like when compared to the values agreed upon today!

What really fascinated Hooke about light, however, was the colours into which it broke down, and why. His obvious historical source was Aristotle (in spite of Hooke's misgivings about Aristotelian science), and those who adopted a broadly Aristotelian theory of colour, through Ptolemy, Alhazen, Roger Bacon and even Descartes. In the mid-seventeenth century light was still seen as naturally white in its purest form, and as taking on its colour tinges because of corruption or — as in Descartes' explanation — by aerial droplets in rotation and mechanical collision.[9]

Yet what actually *was* colour? As something which, in this pre-photographic era, could be neither objectively recorded nor quantified in clear physical terms, and in which the peculiarities of individual perception were pre-eminent, coloured light was seen as a *secondary* quality of nature, in contradistinction to the quantifiable properties of weight, velocity and spatial relationship which constituted nature's supposedly *primary* qualities. The distinction between what appeared to be the personally-perceived attributes of an object, such as its colour, taste or smell, and its obviously measurable ones, such as its physical dimensions, fascinated Renaissance philosophers and scientists, especially after Galileo's ground-breaking experimental researches into the properties of nature. The two apparent sets of properties, *primary* and *secondary*, also gave rise to searching philosophical discussions about the reliability of the human senses, and about what might really be an intrinsic part of nature, as opposed to something which might somehow have adhered to it, like paint on to a neutral ground. So what was colour as seen in nature? Was it a product of decomposing white light that our senses imagined was an intrinsic part of objects — such as the green of grass or the red of apples — or were '*primary* objects' possessing measurable dimensions merely neutral blank shapes upon which the human senses somehow projected the illusion of colour, as a result of a concatenation of mechanical impacts between the outer world and the perceiving eye or brain of the beholder?[10]

The same conjectural physical status could also have been extended to the equally unquantifiable properties of smell and taste, though Hooke was, as we have seen, already coming to associate *sound* with physical waves or pulses transmitted through the air from some impact — such as a ringing bell — and setting off sympathetic vibrations in our ears. And as Boyle's and Hooke's experiments with the vacuum pump had shown that sound needed a medium to pass through before it could impact upon the beholder's ears — as demonstrated by the surprising silence of a bell set swinging in a vacuum — so it seemed logical that light needed some kind of vibrative ether through which it could cross space.

Another question which baffled seventeenth-century researchers was the actual *location* of colour in light. Was the colour somehow extracted by the light from a latent state in the air, glass or water through which it passed, or was colour latent in the light itself, and needed contact with glass or water to release it? Perhaps most puzzling of all was the apparent invisibility of colour in air, which seemed to contrast sharply with the ravishing hues which resulted when it was directed by a prism onto a white surface, or when falling rain supposedly caused it to be refracted and reflected within a cloud to form a rainbow. One can fully understand, therefore, how colour and coloured light became subjects of intense fascination to scientists, and how Hooke could speak of it as a 'Phantasm'.[11]

Robert Hooke made his first serious sallies into the discussion of colour and coloured light in Observations IV and V of *Micrographia*, where he examined delicate silks and silk threads. Could it be that the beautiful colours that could be seen shimmering in silk were the product of the microscopic structures of silk threads themselves? For these threads, under the microscope, appeared as 'small horney, *Cylinders*', perfectly transparent, so that the internal curve of the 'horney' solid structures of the fibres resembled concave mirrors, in much the same way that the inner curve of a glass cylinder filled with water acts as an internal reflector. Could silk's exotic colours, therefore, be the product of this cylindrical structure, especially when the fibres were dyed and their sheen increased?[12] Yet this still did not tell us what colour actually was, or how it was generated. This question was to receive more detailed attention in *Micrographia* Observations IX and X, where Hooke began to examine the coloured rings (subsequently and unjustly termed 'Newton's Rings') found inside the 'thin plates' of mica, and then moved on to a series of brilliant experiments on the properties of coloured fluids.

Under the microscope, Hooke noticed that the thin leaves into which 'Muscovy Glass', or mica, cleaves contained little pits around which coloured rings appeared to form. These rings displayed predominant hues of blue and scarlet, though other colours could be discerned between them in their rainbow-like sequences. These rings or fringes of colour, moreover, not only always fell in precisely the same sequence that one finds in the rainbow, but sometimes reversed secondary fringes were visible, which in turn were analogous to the reversed sequences of a secondary rainbow that arches over the first and brighter rainbow in the sky.[13]

Suspecting that the colour sequences in mica originated from the numerous cleavage planes within a slab of the mineral, Hooke went on to devise a controlled experiment. Taking two pieces of good-quality looking-glass plate, each about the size of a shilling coin (about an inch across), and grinding and polishing each of the four surfaces to perfect optical flatness, he pressed the two round pieces of glass tightly together, and found that he got a colour fringe sequence just like that seen in mica. This suggested to him that the four optical planes of the glass, and their four *inner* reflecting planes within the glass, produced multiple refractions and reflections that, he proposed, caused the colours.[14] He then went on to consider whether the coloured sheens sometimes found on the surface of heat-tempered steel blades were caused by a similar process. Did the 'case-hardening' of the steel result in the formation of 'vitrious laminae', or a thin glossy coat or coats on its surface, that reflected, refracted and broke down the light that struck it from a particular angle? Could a similar sort of transparent surface medium be responsible for the colours seen in mother of pearl, or floating upon the surface of greasy liquids? These experiments were to lead Hooke to draw two conclusions, one right, the other wrong.

As the physical properties of light seemed to originate in some sort of motion, such as combustion or even vigorous rubbing, Hooke concluded that light was 'vibrative' in nature, and travelled from its glowing source to the eye of the beholder in stronger or weaker 'pulses' or 'waves'. This, indeed, is one of the profoundest and truest conclusions that Robert Hooke ever drew from any course of experiments during his 40-odd-year scientific career. He next went on to suggest that one side, or peak, of this wave, when striking obliquely upon any optical surface, produced red light, while the other produced blue. Red and blue, therefore, were simply the products of the inevitable angles and changing planes and forces of impact at which the extreme axes of a sinusoidal wave must penetrate a transparent optical surface, being thereby the result of curves and planes in collision.[15] The intervening yellows, greens and oranges were what might be thought of as vibrative resonances between the two, as each individual wave pulse penetrated the glass or water from an ever-changing or snaking angle of impact.

Hooke then went on to develop his incorrect conclusion that refracted and reflected light consisted of only two pure colours, red and blue, and that the rest of the intermediate oranges, yellows and greens were therefore no more than a proportionate intermixing of these two

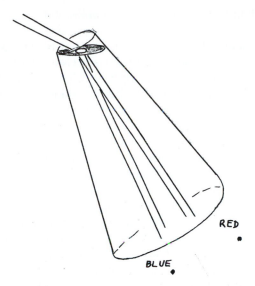

Flask and water experiment (bolthead).

primaries. In framing this conclusion, one wonders how far his own knowledge of contemporary oil painting techniques, and the ingenious artist's ability to replicate all the colours of nature on a canvas from a handful of primary colours, influenced his thinking.

Robert Hooke seems to have been brought to this two-colour theory of light as a result of an ingenious series of experiments, wherein he passed thin pencils of sunlight and lamplight through glass balls, prisms and, most notably, through a two-foot-long conical laboratory flask, or 'bolthead', filled with pure water. No matter how he varied the parameters of the experiment, he always found red and blue light at the extreme ends of the resulting spectrum, with an intermixture of colours between them.

In his flask experiment, Hooke filled up the two-foot conical vessel almost to the brim. He then placed inside the flask's neck, on the surface of the water, an opaque disk or ellipse, with a small hole cut into it, through which was admitted a narrow ray of light. On a white surface at the flat bottom of the flask, he found his familiar red and blue extremes, with their intermediary colours. And from this he concluded that the colours of light resulted when the oblique beam of sunlight with its myriads of rays hit the water surface, so that as each individual wave snaked its way into the water, the internal angles of impact between the red and blue curved ends of the wave constantly changed within a given geometrical range. Coloured light, therefore, was demonstrated to be, so he thought, the product of vibration, as stronger and weaker

191

parts of the radiating force impacted upon the denser refractive medium of the water.[16]

Having concluded by 1665 that all coloured light resulted 'from the two sides [or peaks] of an oblique pulse' or wave,[17] Hooke next went on to see if he could use these two pure optical colours to *build up* a spectrum like that seen in nature. He did this by taking a pair of specially-constructed hollow glass prisms or 'wedges', made like his above-mentioned coloured ring, using plates of perfectly transparent and optically ground mirror glass carefully cemented together. Each wedge measured 7 inches in length and $3\frac{1}{2}$ inches in width at its base. One of these glass wedges he filled with a copper salt solution which produced a fine blue colour, while the other contained an aloes tincture. When he examined the blue wedge sideways on, starting from the sharp apex and going down to the $3\frac{1}{2}$-inch-wide base, he observed 'all manner of Blues', as he looked through a deepening body of liquid. And looking similarly through the aloes-dyed wedge, he found that at the pointed edge was a fine yellow, which changed at the broad end to 'a deep and well colour'd Red'.[18] It is quite evident from his use of language that Hooke's artistic sensitivities were deeply touched by the contemplation of these deepening colours.

The purpose of this experiment, Hooke tells us, was 'in [the] good hope by these two [wedges], to have produc'd all the varieties of colours imaginable', by sliding them alongside each other in strong sunlight and thereby varying the dye densities *vis-à-vis* each other. Hooke found that when he held up each coloured prism in two separate parallel beams of sunlight passing through holes in a screen, and cast their colours on to a white sheet of paper, he could, by manipulating the angles at which the now red or blue light emerged from each wedge, 'make the Paper appear of what ever colour I would, by varying the thicknesses of the Wedges, and consequently the tincture of the Rays that past through the two holes'.[19] Yet when he looked through one prism placed behind the other he saw nothing: only darkness. Indeed, he wrote, 'I found myself utterly unable to see through them when placed both together', though each was perfectly transparent on its own. It was these particular experimental interpretations in *Micrographia* which, as we shall see shortly, were to spur the young Isaac Newton both to undertake his own first prism experiments around 1666, and then to go on to draw his own radically different conclusions about the nature of light.

In addition to the working out of his theory that coloured light was the product of an interrupted geometrical ray, and that all colours were admixtures of blue and red, Hooke's third major investigation into the nature of light concerned its progressive refraction or, as he called it, *Inflexion*, through media of increasing density. His interest in the subject first seems to have been aroused by the making of astronomical observations, for he recorded 'I have seen the Dog-Starr to vibrate so strong and bright a radiation, almost to dazle my eyes'.[20] Why, then, did the stars twinkle, the sun become egg-shaped on the horizon, and red colours appear around objects low in the sky? In some respects, one might think, Hooke was in very familiar optical territory, as writers such as Aristotle, and especially medieval researchers such as Alhazen, Vitellius and Theodoric of Freiburg, had written on these topics, and realized that the earth's atmosphere, just like a lens, bends the sunlight that strikes it.

On the other hand, Hooke approached these familiar optical problems from the new perspective of the post-telescopic and post-barometric age. Why, for instance, did a star stop twinkling, and shine 'plainly and constantly' when viewed with the extra magnification made possible with a telescope? To some extent, one can see Hooke wrestling with the as yet confusing concept of telescopic resolving power here, as object glasses of wider aperture enabled one to see dimmer stars and to collect more light than smaller glasses or the unaided eye. But on a more theoretical level, he is trying to find a consistent model whereby he could explain the apparent reddening of astronomical objects as they approached the horizon. *Inflexion*, or the progressive bending of a ray of light as it passed deeper into a medium of increasing density, was his answer. His idea of optical *Inflexion*, moreover, constitutes an elegant example of Hooke's scientific method, backed up as it was by a set of carefully devised and conducted experiments interpreted in accordance with the geometry of curved and plane surfaces. After all, he was not the Gresham College Professor of Geometry for nothing!

Being stimulated to undertake a particular investigation after observing a piece of phenomena in nature—such as the reddening and distortion of low-altitude astronomical objects—Hooke next repaired to the laboratory in an attempt to investigate the phenomenon under controlled conditions. When directing rays of sunshine through glass or water-filled spheres, for instance, he noticed that those rays which had passed through the densest part of the optical medium produced the reddest light. Likewise, when light was passed around an *opaque* sphere, such as a solid ball, in the laboratory, red light was also seen in the fringes of its shadow. This led Hooke to conclude, in conjunction with his other experiments mentioned above, that the red rays were produced when white light struck an optical surface, be it a glass ball or the earth's atmosphere, at a low or glancing angle. Rays that came in less obliquely, such as that of zenith stars, would travel in straighter courses and not become inflected, and appear as white or blue. But the more oblique the angle at which rays entered the refracting medium, and passed, by geometrical necessity, through the densest part of that medium, the redder they became.

While earlier writers on atmospheric distortion, such as Alhazen, had realized that an oblique ray of light from space would get bent to somehow produce the reds of dawn and dusk, Hooke took the idea several stages further. He argued that there was no clear cut-off line at which the earth's atmosphere stopped, and at which an unchanging angle of refraction began, for recent barometric experiments conducted from the top of Old St Paul's Cathedral and other elevations suggested a gradual diminution of air pressure into space. Yet if this was the case, at what point did a ray of light from an astronomical body entering the earth's atmosphere start to get bent? From a series of figures for diminishing air pressure in inches of mercury in the barometer, Hooke tried to calculate a series of integer numbers for air density, and hence potential refractive power, from the earth's surface into what might be called space. Now if we could measure the refractive power of each integer, Hooke argued, then we could compute a curve of 'multiplicate refractions', to obtain the deepening curve of *Inflexion* as the light ray came down to the observer's eye.[21]

In these experiments and calculations, one cannot help but be struck by the parallel between a ray of light being *inflected* by the earth's atmosphere, and the calculation of a planet's changing orbital velocity and position with regard to the sun, as it moved in an ellipse under Kepler's Laws. Indeed, one sees Hooke transferring the skills of the astronomer (in the astronomer's attempt to quantify the changing velocities of a planet moving through specified sections of an ellipse) to the problems facing the experimental optician in his handling of pressure and refraction gradients. Of course, Hooke must have been aware that one of his mentors, the Revd John Wallis, Savilian Professor of Geometry at Oxford, had already made and published tables and developed calculation techniques to handle integer units within curves:[22] all of these physical and mathematical models for forces acting with curves would eventually find a unifying solution in the Calculus.

As a man who always sought concrete physical models upon which to build his ideas, Robert Hooke set upon a course of laboratory experiments in which he tried to co-mingle perfectly transparent liquids of different densities in the same glass vessel. Pure water, salt water, alcohol and other liquors were tried, as a way of producing a transparent fluid mass that (unlike homogeneous quantities of these individual liquids) became progressively dense with descent. Through this mass, Hooke passed rays of sunlight in an attempt to observe an inflexed curve which 'I imagine to be nothing else but a *multiplicate refraction* caused by the unequal density of the constituent parts of the medium'.[23]

The progressive inflection of light, moreover, would be of the greatest importance to astronomers endeavouring to make exact physical measurements of celestial bodies, for unless all aspects of refraction could be exactly quantified and corrected, such measurements must inevitably contain random errors. As a natural philosopher of wide-ranging and even all-encompassing aspiration, Robert Hooke was not content simply to investigate light as a natural phenomenon. He also strove to apply his researches to the improvement of telescopes and, in his lecture of June 1681,[24] even addressed himself to the physiology of the eye, displaying therein a familiarity with anatomical dissection which one suspects he first acquired when working with Dr Willis during his Oxford years.

It is sad for Hooke that the wealth of observation, ingenious experimentation and discursive optical theory that he produced was to have a stimulating effect upon a man whose subsequent work would not only supersede it, but who himself, unfortunately, would do his best to discount and discredit it. For as the 29-year-old Isaac Newton told the Royal Society early in 1672, he had procured a glass prism around 1665 or 1666 'to try the celebrated *Phaenomena of Colours'*.[25] After producing the well-established spectrum of colours, he then made his famously self-confessed *'Experimentum Crucis'* of passing a single colour from the first prism through a second, and finding that the colour stayed the same. Newton went on to draw the conclusion upon which all modern optical knowledge is based: namely, that white light is a *mixture* of absolute and irreducible colours, rather than being an absolute in itself. In this respect, of course, he undermined Hooke's own idea of light as a blue–red mixture deriving from white light, though Newton's own fundamental mistake was to reject Hooke's *wave, pulse* or *vibration* model of light, and go on to develop a *corpuscular* or particle model. Even so, Hooke claimed that when he and a group of friends were discussing the nature of light and colours at 'our New Philosophicall Club' formed within the Royal Society on 1 January 1676, and Hooke and Wren were proposing different wave forms for light, 'I shewd that Mr Newton had taken my hypothesis of the puls or wave'.[26] But this was in the wake of Hooke's optical battle with Newton fought out after 1672, and when his tendency to see plagiarism at every turn was becoming more pronounced.

While no-one is in any way denying the radical and enduring importance of Newton's own researches into light, it is perhaps not for nothing that these researches were conducted immediately after his reading of the optical sections of *Micrographia*. Indeed, Newton admitted as much in his 1672 paper where he refers to *Micrographia*, and said that the reason why Hooke's experiment with hollow glass prisms filled with tinted water failed to produce the range of colours which Hooke had expected and produced only darkness was because 'it one transmitted only red, and the other only blew, no rays could pass through both'. It is also interesting that Newton referred to the single and irreducible colours produced by his glass prism as his '*Experimentum Crucis*',[27] for in *Micrographia* Hooke used the very same term—with due acknowledgement to Sir Francis Bacon—as 'serving as a Guide or Land-mark' when drawing his own optical conclusions about the seemingly fundamental nature of red and blue light.[28]

Although Hooke's theory of coloured light as deriving from red and blue fundamentals was in many ways a major advance upon the older theory of corrupted whiteness, Newton's was the more original in so far as it produced an entirely new and—like Hooke's—an experimentally-based hypothesis for colour. After 1672, Hooke and Newton had already begun to disagree about the interpretation of each other's experimental results—about the generation of colours between 'thin plates', and within mother of pearl, and about the nature of refraction itself—so that given their respective temperaments, the foundations for a permanently damaged mutual relationship had already been laid.

It is in his studies of light and colour that one glimpses many facets of Robert Hooke's scientific imagination and personality, and is made aware, from the richness of the language with which he describes them, of his love of colours and beauty for their own sake. But most of all, one catches Hooke's sense of sheer wonder when considering the nature of light and what its study could lead to, for as he says, it is through optics that one can 'open not only a cranney, but a large window ... into the Shop of Nature'.[29]

CHAPTER 12

FROM PENDULUMS TO PLANETS: EXPERIMENTS AND THE UNDERSTANDING OF GRAVITY

For most of his active life as a research scientist, Robert Hooke wrestled with the problem of gravitation. He wrestled with it in much the same way as he struggled to comprehend the other great forces of nature, such as light, magnetism, motion and colour, though it is clear from his writings that he ascribed a fundamental importance to light and to gravity in particular, for light and gravity encapsulated the two great moving forces of nature, enjoying as they did a primacy that derived from Scripture itself.[1] For just as God in Genesis had summoned forth light as that wonderful agent which overcame darkness, so was not gravity the secondary cause whereby He separated the great waters and moved the earth?

Yet while light and gravity were as old as the earth itself, and classical philosophers had theorized about them, Hooke was of the opinion, as he told the Royal Society in 1666, that not until 'the Inquisitiveness of this later Age' and the rise of experimental science had the nature of gravity, being 'one of the most Universall Active Principles in the World',[2] come to seriously engage the attentions of learned men. And one might suggest that a major reason for this recent attention had derived from the new astronomy. The discoveries of Tycho Brahe, Galileo and other astronomers between 1580 and 1620 had undermined the credibility of the ancient doctrine of the nine crystalline spheres which were supposed to carry the stars and planets around the earth. Yet if new observational data relegated the crystalline spheres to obsolescence, what made the planets move? Indeed, Hooke had brought up this very problem before the Royal Society as early as May 1666, when he stated that 'I have often wondered why the planets should move about the Sun according to Copernicus's supposition, being not included in any solid orbs', or attached with visible 'strings'.[3]

Likewise, the terrestrial physical experiments of Galileo and Pierre Gassendi into the swinging of pendulums and the orbital paths of projectiles had revealed that earthly objects could be just as mathematically exact in their motions as the planets themselves. Could it be that the planets were kept in their orbits by some all-powerful yet intangible

197

force of nature that radiated throughout space from the sun, and that a similar agent caused a body falling from a tower to increase its velocity by a factor of 32 for every second of fall?

Between the work of Galileo, Kepler and Gassendi in the early decades of the seventeenth century and the presentation of the manuscript of the first part of Newton's *Principia* to the Royal Society in April 1686,[4] the nature and characteristics of this 'gravitating principle' were to occupy the attentions of some of the finest intellects in Europe. In particular, numerous observations of physical phenomena were made by, among others, Hooke's old Oxford mentor, Dr John Wilkins. Indeed, Wilkins's and Hooke's early experiments into the nature of flight were related to their attempts to understand and defy gravity. Long before he had met Hooke, in fact, Wilkins had noticed that while large birds seemed to have difficulty in first becoming airborne, once they were off the ground they seemed to gain altitude with greater and greater ease the farther from the earth's surface they flew. He had concluded by 1640, therefore, building on the earlier magnetic researches of William Gilbert, that the earth's 'Sphere of Magnetick Virtue', which at that date he saw as an aspect of gravity itself, only extended about 20 miles upwards from the earth's surface. So Wilkins came to suggest that if one could ascend 20 miles above the earth's surface, one could escape the terrestrial pull, and navigate a course across space until one entered the moon's own sphere of attraction. Then, using mechanically-powered wings to slow down the descent, one might land on the moon's surface.[5]

In these ideas of Wilkins, which Hooke must have picked up in Oxford by conversation and by reading Wilkins's books, one can discern a set of ideas about the rules which the 'gravitating principle' was believed to obey. For one thing, weight seemed — *contra* Aristotle — to be relative to location in so far as a given object fell faster, and a bird seemed to need to beat its wings harder, the closer they were to the earth's surface. Likewise, those planets which were nearest to the sun in the Copernican system always moved faster than those that were farther out: Mercury orbited the sun in about 88 days as against Saturn's $29\frac{1}{2}$ years, while the moon, being closer to its parent body than any other object in the solar system, spun around the earth in a month. As these planetary orbital velocities had been found by Kepler to conform exactly to a geometrical sequence, and Galileo's falling bodies experiments had discovered a similar sequence acting between terrestrial bodies, it seemed that the 'gravitating principle' acted in accordance with a mathematical or geometrical rule.

Yet in none of these observations was there a coherent concept of *gravitas* as a universal principle. Instead, it seemed to act through local spheres of influence, each planet having its own gravity, such as that

198

which made Jupiter's moons rotate around the parent body; though these lesser, local gravities must somehow operate within greater ones, such as those of the planets *vis-à-vis* that of the sun. In substantiation of the fact that the planets had gravity in their own right, Hooke drew attention to their spherical shapes as revealed in the telescope: was it not clear, he argued in *Micrographia*, that there must be some inherent power of gravity which pulled down and compacted all the matter of which a planet was made around a centre, to make it spherical? The appearance of the moon was a perfect example of this power, for were not its 'pits' or craters suggestive of volcanic formation, where matter expelled from the moon's interior had been pulled back down to its surface?[6] Almost two decades later, Hooke would reiterate that 'The Endeavour of Gravity acts or tends always towards the Center of the Globe of the Earth, as far as any Observation has been made'.[7] He would also have applied this same principle of gravitation to the sun, moon or any other astronomical body he happened to be discussing.

What emerges from Hooke's writings, especially before the early 1670s, is the idea of gravitation being peculiar to given bodies rather than of its being a unitary force that penetrated the length and breadth of the universe. In this respect, he was more or less at one with his contemporaries. It is also clear that Hooke envisaged that gravity might, like light or colour, be pinned down and defined in terms of clear physical principles, and one of the great differences in the approach to gravitation which would emerge between Hooke and Newton after 1686 lay in Hooke's belief that a mathematical definition of gravitational action which failed to say what gravity was in physical terms was somehow ducking the issue. Nature's great forces, in Hooke's mind, needed to be grappled with experimentally and not just mathematically.

This physical approach shines through one of Hooke's first investigations into gravity, when on 24 December 1662 he presented a paper to the Royal Society on the subject. It also acknowledged Sir Francis Bacon's founding thoughts on the subject – 'my lord VERULAM' experiment concerning the decrease of gravity' – as well as the prior experiments of Henry Power.[8] In this 1662 paper, Hooke described a series of experiments which he had conducted high in the roof vaults of Westminster Abbey, which provided a clear drop of 91 feet down to the Church's floor.

A delicate balance was set up in the roof vaults. A 16-ounce weight and 91 feet of fine string were accurately weighed aloft, and after the string had been tied around the balance arms and released to hang just above the Abbey pavement, the apparatus was weighed again. When the suspended 16-ounce weight and string had damped, Hooke found that it was necessary to add 10 grains, apothecaries' weight, to the other pan to bring the arm of the scales back into balance. At first it seemed that an object became heavier the closer it approached the earth, though when the experimentally-scrupulous Hooke re-weighed the weight and string after hauling it back up the vault, he found that the 10-grain increase in weight was still present. He now attributed the 10 grains to moisture absorbed by the string from the air.[9]

Even so, the balance, weight and string method clearly appealed to Hooke's experimental imagination and, as he related to the Royal Society on 21 March 1666, he repeated it from the soaring roof vaults of Old St Paul's Cathedral (soon to be burned down in the Great Fire of September 1666) and also down a well in Surrey. The 90-foot-deep well was in Banstead Downs, and the experiment which Hooke tried therein was during that summer of inventive ingenuity which John Evelyn had commented upon, when Hooke, Wilkins, Petty and others had fled the plague-ridden capital for the sweeter air of Lord Berkeley's Mansion, Durdans, on Epsom Downs. Using a very accurate balance, a weight on an 80-foot line was suspended down the well to see what would happen. Would the suspended piece of metal, as Bacon had suggested, weigh *less* underground than it did upon or above the earth's surface? Or would it be in any way affected by the earth's magnetic field? For at 80 feet below ground, there was a great mass of terrestrial matter — including chalky flint, wood and brass — *above it*, the gravity of which might attract the suspended weight *upwards* by a fraction. When suspended down the well, however, the weight was found to be no heavier than by a single apothecary's grain than it was on the surface. Again, Hooke's experimental acuity told him to be cautious of the result, for atmospheric dampness or air currents could have occasioned the tiny discrepancy.[10]

Over the next few years, Hooke seems to have re-tried these balance and suspended weight experiments in various configurations, including the letting down of a weight on a wire from a balance inside the 202-foot Fire Monument, or 'Piller', in Fish Hill Street, but without ever drawing a reliable conclusion.[11] The reason why all of these experiments failed to demonstrate a weight discrepancy due to a gravity variation, of course, was not that Hooke's physics were wrong, but that the relatively crude balances of the period were unable to detect the tiny gravitational discrepancy that would occur over a mere 80 or 90 feet, especially if the results were being further confused by the presence of air currents or moisture absorption.

Then quite by chance during the 1670s, Jean Richer at Cayenne and Edmond Halley at St Helena discovered that vibrating clock pendulums were sensitive to gravitational changes.[12] Richer had found that his pendulum clock, regulated to beat dead seconds in Paris, lost 2 minutes 28 seconds per day when set up on the equatorial island of Cayenne. Could it be that the earth was an oblate sphere, broader across its equatorial diameter than it was across its polar, as a result of which the clock when in Cayenne and consequently farther away from the centre of the earth than when in Paris, was in a slightly weaker gravitational field, and hence ran slower?

Things really seem to have fallen into place for Hooke when the 22-year-old Edmond Halley, on commencing astronomical observations at St Helena,

> 'told me that his pendulum at the top of the Hill [more correctly, *mountain*] went slower than at the Bottom which he was much Surprised at, and could not imagine a reason. But I presently told him that he had solved me a query I had long Desired to be answered, but wanted opportunity, and that was to know whether gravity did actually Decrease at a height from the Centre.'[13]

Of course, by 1680 there was no reason why Hooke should not have verified this phenomenon for himself, by swinging pendulums for a period of time at the top and bottom of the 202-foot Fire Monument. But quite simply, he did not; and I suspect that it was the complex logistics and long periods of observing time likely to be required for such an experiment that put him off. Hooke, we must not forget, was a virtuosic experimenter, with no research team or back-up staff behind him, and who depended upon obtaining quick and hopefully decisive results from relatively simple pieces of apparatus, used over a relatively brief period of time. Letting a weight attached to a balance down inside a cathedral vault, after all, could be done in an afternoon, but running synchronized and carefully-monitored pendulums at the top and bottom of a 202-foot tower, perhaps for several weeks or months, was beyond his resources. Not until the nineteenth century, indeed, did Sir George Biddell Airy try to quantify gravity by swinging pendulums on an earth radial line that passed through identical observing stations at the top and bottom of a deep coal mine, and in 1854 Airy had a large staff, several weeks of time, and electromagnetic horological relays at his disposal.[14]

Hooke's fundamentally experimental approach to the study of gravitation runs through everything that he wrote. He is constantly trying to measure proportionate relationships whereby he could comprehend gravity physically, as well as substantiate his belief that gravity, like light and other great natural forces, acted through a form of pulsation or vibration. On 9 March 1671, for instance, he was demonstrating a curious experiment where he put fine flour inside a broad flat dish. By percussing the dish in various ways, he was able to generate waves which he argued had a radiating power analogous to that of gravity.[15] Then some years later in 1679, he was trying 'to explain the different Gravitation of the Air, and to show that Vapours press only according to their own Gravity, and not according to the space they take up in the

Atmosphere'.[16] These experiments seem to have been related to his work on air compression and density under natural atmospheric pressure, and one can see how he viewed this compression as also producing the Inflexion of Light, as discussed in *Micrographia* p. IX, and in particular the generation of red light from pulses entering a medium of increasing resistance.

His pulsation model drove Hooke to thrash around for analogies to clarify and make concrete his understanding of gravity, and demonstrate it to his Royal Society colleagues. The analogies he came up with reinforce in our own minds the radical difference of approach to the problem of gravity displayed by Hooke and by Newton.

One problem which dogged Hooke regarding his pulsation-from-a-centre model of gravity was how waves that radiated *outwards* had the physical effect of drawing things *inwards* to the centre. He was to wrestle with this paradox over many years, and around 1680 came up with a model which exemplifies his hands-on, non-abstract approach. Hooke drew his analogy from the way in which a tradesman secures the iron head of an axe or hammer on to its wooden helve (or handle). What do the tradesmen do, asks Hooke? They hit the bottom of the helve upon a hard surface, and as the shock waves travel up the helve, the iron head, instead of shooting off, actually creeps up the helve 'even to their very hand'. In short, a shock wave radiating *out* from a centre seems to be attracting a heavy object *towards* that centre. A homely demonstration it may have been, but it made an analogical point. Hooke hypothesized that gravity might act as a pulsation vibrating at, let us say, 1000 times per second, radiating up from inside bodies, and out into the ether.[17]

Perhaps because he was a classical scholar and a geometer, Hooke was an instinctively symmetrical thinker. In consequence, not only did waves going outwards draw things downwards, but gravitation towards a centre should also be matched by a *Levitation* radiating away from it, as he indicated in his Royal Society paper 'Of Comets and Gravity', written some time after the appearance of the bright comet of 1682.[18] Generally speaking, Hooke did not consider comets to be regular or permanent solar-system bodies, but rather itinerants acted upon by an ethereal *Menstruum* that produced 'a kind of Dissolution' that dissolved them as water dissolves a lump of salt.[19] If comets were insubstantial in this way, however, and seemed to follow peculiar paths that were not, in 1680, generally believed to correspond to any Keplerian orbital criteria, how could they be gravitational bodies in their own right, or else be acted upon by gravity? But what was evident, Hooke argued, was the sensitivity of cometary tails to the sun. Most tails, for instance, pointed *away* from the sun, thus indicating that the sun possessed some kind of *levitating* power that made the tails radiate away from it, as Hooke had suggested was the case with the great comet of 1664.[20] And while some cometary streamers had been seen to be pulled down towards the sun, this, it was believed, was only a temporary phenomenon, for before long they *levitated* away from it again. While by 1680 Hooke still had no idea why some cometary tails apparently turned towards the sun, whereas others did not, he had come to recognize that the shape of the comet's path in space was the result of a gravitational attraction.

Physical and experimental as Hooke's approach to gravitation may have been, he had nonetheless come to draw some profound conclusions on the subject as early as 1674. He presented them as a supplementary section or postscript to his Cutlerian Lecture of that year, *An Attempt to prove the Motion of the Earth*, the greater bulk of which was devoted to detecting a stellar parallax with a zenith-sector-type instrument, as shown in chapter 5. Very clearly, however, Hooke saw gravitation theory as occupying a central role in cosmology, and as leading to the 'true perfection of Astronomy'.[21]

By the 1670s, in fact, researches across Europe were already making significant inroads into the gravity problem, and in *Horologium Oscilatorium* (1673) Christiaan Huygens published his own major study of oscillatory and centrifugal motions. No matter what the absolute physical *cause* of gravity might be, therefore, people were coming to think of stable planetary systems as being in a state of *equilibrium*, wherein one force pulled an object, such as a planet, towards a centre (e.g. the sun), while another force inclined it to spin away into space. But even as early as May 1666, Hooke seems to have come to think that 'all the celestiall bodies, being regular solid bodies, and moved in a fluid, [the ether] and yet moved in circular or elliptical lines, and not strait, must have some other cause, besides the first impressed impulse, that must bend their motion into that curve', or orbit. What he seemed to be groping towards here, in developing his ideas from the curves described by swinging pendulums, was the notion that a body possessed an initial linear motion, or impulse, from which it was drawn or attracted by some other, no doubt central, 'cause'. But by 1674, no doubt with a lot of reading, thinking and experimenting behind him, he formulated three laws or conditions of orbital motion that were necessary before one could hope to explain the nature of gravity. Between them, these three laws or conditions constituted a major breakthrough in the elucidation of gravity, and he set them out under the following headings:

> *Firstly*. All bodies in space possess Gravity and an internal attraction within themselves. All of these bodies, including the moon, the sun, and the planets, moreover, have an inter-mutual attraction 'within their sphere of activity', and 'hath a considerable influence upon every one of their motions also'.
>
> *Secondly*. Bodies acting under their own initial impulse will move in a straight line, 'till they are by some other effectual powers

deflected and bent into a Motion, describing a Circle, Ellipsis, or some other more compounded Curve Line'.

Thirdly. 'That these attractive powers are so much the more powerful in operating, by how much the nearer the body wrought upon is to their own Centers. Now what these several degrees are I have not yet experimentally verified; but it is a notion, which if fully prosecuted as it ought to be, will mightily assist the Astronomer to reduce all the Coelestiall Motions to a certain rule, which I doubt will ever be done true without it. He that understands the nature of the Circular Pendulums and Circular Motion, will easily understand the whole ground of this Principle, and will know where to find direction in Nature for the true stating thereof.'[22]

One notices that Hooke by 1674 is *not* speaking of gravity as a force which acted in equilibrium with a centrifugal force. What he seems to be saying (in spite of his subsequent axe head and helve analogies) is that attraction towards a centre is the key force, although this force acts upon an object's tendency, when otherwise undisturbed, to move in a straight line. Even so, by 1674 he had not, by his own confession, drawn any clear idea from his experiments as to what mathematical properties were acting in such relationship. On the other hand, the straight line of the object's original motion was being made to change *not* to a direct collision course with the disturbing gravitational body, but to a curved orbit which in itself corresponded to a conic section.

By the beginning of 1680, however, Hooke had clearly developed his ideas on planetary motions, and the proportionate relationships of gravity existing between bodies, considerably further. And whether this was the result of further experiments or 'tryalls' with iron balls is not clear, but one suspects that Richer's and Halley's pendulum observations were not without significance. As we saw above, Hooke admitted to Newton on 6 January 1680 that Halley's St Helena pendulum clock observations of 1678 'had solved me a query I had long Desired to be answered'.[23]

Robert Hooke's controversy with Sir Isaac Newton regarding the priority of discovery of certain key concepts in the development of gravitation theory has become one of the legends of the history of science. Since it has become fashionable for certain types of historical popularizers to see the affair not only in stark confrontational terms, but even as a case of a high-handed and powerful 'Establishment' figure stealing ideas without acknowledgement from a hard-working 'underdog', a number of basic home truths must be first established.

In addition to the difference of intellectual style, or approach, displayed by each man, as indicated above, it must not be forgotten that when Hooke initiated the gravity correspondence with Newton on 24 November 1679, Hooke was by far the weightier scientist of the two.[24]

Newton, it is true, had distinguished himself, and had fallen out with Hooke, through his work on light and colours in 1672, and through his invention of the reflecting telescope. In 1679 Hooke, by contrast, was already a Fellow of the Royal Society of formidable international standing, with a 20-year reputation, a major book (*Micrographia*), the universally-admired published versions of his Cutlerian Lectures, a string of lesser publications in half a dozen sciences, and scores of original research communications entered in the Register Books of the Royal Society, to his credit. What is more, Hooke's City Surveyorship made him a figure of clout in the City of London—where the 37-year-old Newton was still a virtual nonentity—not to mention a gentleman worth thousands of pounds when Newton was but a minor Lincolnshire landowner scarcely worth hundreds. To see Robert Hooke as a clever mechanic from whom the lordly Sir Isaac had stolen a pearl beyond price, therefore, is frankly a travesty of the truth considering their respective circumstances in the early 1680s. Of course, Newton would become immensely famous after he published his solution to the gravity problem in *Principia* in 1687, and his behaviour thereafter would become a source of intense chagrin to the elderly Hooke. But Newton's truly formidable power, and his increasingly tyrannical conduct towards those who crossed him in any way, such as Flamsteed, would not really become evident until he had become a knight, a figure of state, and President of the Royal Society after 1705, and by then Robert Hooke would be securely in his grave.

And when Hooke did write to Newton about gravity and other issues in contemporary physical science in the early winter of 1679–80, he did so from a position of authority within the Royal Society, for Hooke wrote as one of the Society's Senior Officers—its Secretary—requesting a reopening of communication with a long-silent yet respected Fellow. It is also very clear in his opening letter that Hooke was aware of the bad feeling which Newton had developed towards him as a result of his *private* criticisms of Newton's work on light and colours of 1672, which had been forwarded on to Cambridge by Oldenburg, and how very keen Hooke was to re-establish amicable relations once again. Indeed, Hooke's 1679 overtures to Newton go well beyond those that a senior officer of a learned society was required to make to an offended Fellow, for they tell us much about Hooke's own approach to the cut and thrust of learned discourse. For one thing, the tone of Hooke's letter is friendly, and even chatty, as when, in that of 24 November 1679, he solicits Newton's views on recent French scientific discoveries: Picard's and De la Hire's latest surveys of France 'have already found that Brest in Britaigne is 18 Leagues nearer Paris than all the mapps make it'. Then to cap it all, and speaking from the heart one suspects, Hooke says 'but Difference of opinion if such there be (especially in Philosophicall matters where

Interest hath little concerne) one thinks should not be the occasion of Enmity—tis not with me I am sure'.[25] In short, we see Hooke as a man who was always willing to apply the most rigorous standards to scientific research, and would always defend his own corner, but who did not see honest scientific criticism as a ground for personal enmity. What we must remember, however, is that Hooke was not only an accomplished practitioner in the art of scientific debate, but that he also thoroughly enjoyed it. Newton, on the other hand, clearly did not, and whether because of temperament or other reasons, he was always likely to translate criticism into 'Enmity'. Here, I believe, is the root of the disastrous relationship which developed between the two men in the wake of Hooke's request for Newton's views upon the 'compounding [of] the celestiall motions of the planetts of a direct motion by the tangent & an attractive motion towards the centrall body'.[26]

Newton's speedy reply, dated 28 November 1679,[27] was cautious. He stated that he had just returned from Lincolnshire, and was 'cumbred with concerns amongst my relations till yesterday when I returned hither', and that he did not wish to get involved in 'Philosophy' or science once again, wishing to devote his time to other studies (these were probably theology and alchemy). But we must also note that 1679 had been an emotionally devastating year for Newton, for the long sojourn which he admitted having recently made in Lincolnshire would have been concerned with his mother's death. While the young Isaac had never been able to come to terms with his mother's remarriage and his stepfather's apparent resentment towards him, Hannah Smith, formerly Newton, was the only person who would ever become at all close to the lonely Isaac, and her death was for a variety of reasons traumatic for him.

In spite of his misgivings about getting involved in 'Philosophy' once again, Newton could not resist communicating his thoughts to Hooke on the *spiral* path which he supposed would be described by a projectile falling from a tower as it descended to the centre of the earth.[28] But the embers of Newton's clearly latent resentment of Hooke were inadvertently rekindled by Hooke's reply of 9 December 1679,[29] in which Hooke corrected Newton's confessedly hasty and incorrect dynamic analysis. For Hooke, no doubt building on Kepler's earlier work on moving bodies, went on to suggest that the path of such a body in free fall to the centre of the earth (the general body of the earth, for sake of argument, being absent) would not be a spiral to the centre, but an *ellipse* about the centre, without actually arriving at it. Newton quickly realized the error of his hasty demonstration 'wherein I explained this I carelesly described the Descent of the falling body in a spirall to the center of the earth' and, as he later made clear to Halley, bitterly resented being corrected by Hooke.[30]

Hooke and Newton exchanged six letters between late November 1679 and 17 January 1680, their main topic being gravitation. (They were to exchange two more letters in December 1680, but these would be upon the less contentious topic of medical recipes.) In the November 1679 to January 1680 correspondence Hooke told Newton about his physical experiments at Westminster Abbey, of his work on the motion of falling iron spheres, and of other researches which displayed the direction of his thinking. Then on 6 January 1679, building upon his ideas about the dynamical processes which produced the elliptical shape of orbiting bodies, and his key suggestion that planets rotated around the sun in these orbits by a combination of tangential and sunward attractive forces, Hooke delivered his revelation:

> 'But my supposition is that the Attraction always is in a duplicate
> proportion to the Distance from the Centre Reciprocall, and
> Consequently that the Velocity will be a subdulplicate [?]
> proportion to the Attraction and Consequently as Kepler Supposes
> Reciprocall to the Distance.'[31]

Hooke does not seem to have delivered this passage as a *conscious* revelation, but as part of the development of an argument, but it appears to have set off an important train of ideas in Newton's mind. For though, as Newton was aware, Hooke's geometrical proofs were imperfect, Hooke's connection of the problem to Kepler's Laws, and especially his working out of gravitation in the light of Kepler's Second and Third Laws (which dealt with the proportionate velocities, distance and orbital characteristics of the planets with regard to the sun) contained momentous possibilities. What Hooke seemed to be saying was that an Inverse Square relationship was in operation. One also wonders how far Hooke had come to envisage this as a *universal* law, moreover: to see that gravity was not just a thing of multiple peculiars, acting between the earth, moon, planets and such, in a range of local systems, but that the same force and the same proportion acted in the *same* Inverse Square relationship between all bodies throughout the length and breadth of the universe. We do not know, for Hooke never developed the idea in any writings that have survived.

On the other hand Newton, having gained a crucial insight, did begin to develop the idea. Newton also broke off correspondence with Hooke on the matter of gravitation after Hooke's letter of 17 January 1680, in which Hooke had reported the success of certain experiments, which amongst other things should 'prove a Demonstration of the Diurnall motion of the earth, as you very happily intimated'.[32]

What seems to have brought gravitation to the intellectual forefront of the early 1680s, though not directly involving Hooke, was the appearance of two bright comets in 1680 and 1682. As indicated earlier, and in chapter 5, the nature and orbits of comets were subjects of much and varied discussion among seventeenth-century astronomers, the general opinion being that they were singular, transient and probably not gravitational bodies. Yet when the comet that had first been sighted on 14 November 1680 vanished into the morning light early in December, only to be replaced on 15 December with an *evening* comet seeming to back away, tail first, from the sun, John Flamsteed suggested that the two comets might indeed be one and the same body. Could this comet, which sported a spectacular 70° tail before shrinking and vanishing from view in mid February 1681, have passed behind the sun in an acutely curved orbit? If so, was it driven by solar gravitational forces?[33]

Then in August 1682, another bright comet made its appearance. Observed by astronomers on both sides of the Atlantic, in South Africa and elsewhere, but especially accurately by Flamsteed at Greenwich, the comet of 1682 was to become of particular interest to Newton — and to Edmond Halley, who in 1705 predicted its return in 1758 within a 76-year cycle, and after whom it is still named. It was the excellent observations of both the 1680 and 1682 comets, taken to an entirely new level of angular accuracy with modern instruments incorporating telescopic sights and precision micrometers — instrumental features actively promoted by Hooke — which made them so significant.[34] It was the precise observations of the paths of the 1680 and 1682 comets across the sky, measured each night with relation to the fixed stars, which subsequently enabled Newton to realize that they behaved like planetary bodies in so far as they were affected by the sun's pull. While at first Newton had been wrong in his belief that the comet of 1680–81 did not pass behind the sun but was turned away as it approached it, the comet of 1682 convinced him that these comets at least moved in an oval path around the sun.[35] And the curved orbital behaviour of comets, and especially that of 1682, was to become the subject of one of the great Gravitation Theory analyses in Newton's *Principia Mathematica* (1687), in the concluding section of Book III entitled 'The System of the World'.[36]

Perhaps following on from the interest which these comets had aroused, and also from Huygens's proportions for centrifugal forces, Edmond Halley, Sir Christopher Wren and, of course, Robert Hooke had started to talk about gravitation in the solar system after a Wednesday meeting of the Royal Society in January 1684, for Halley informed Newton of the fact two years later. Halley, moreover, who possessed acute historical instincts, tried to find out, presumably by subsequent inquiry, who 'had the first notion of the reciprocall duplicate proportion' and what would be the geometrical shape of the path described by an object falling towards the sun or other gravitational source. Had it really been Hooke? Wren declared upon being asked that he himself had quite independently made such a realization 'but that at length he [Wren] gave over, not finding the means of doing it'. Hooke also made a similar claim and had frequently told his friend Wren 'that he had done it, and attempted to make it out to him, but that he never satisfied him, that his demonstrations were cogent'. Then Halley himself maintained that 'I, having from the sesqualter [threefold] proportion of Kepler, concluded the centripetall force decreased in the proportion of the squares of the distances reciprocally, came one Wednesday to town, where I met with Sr Christ. Wrenn and Mr Hook, and falling in discourse about it, Mr Hook affirmed that upon that principle all the Laws of the celestiall motions were to be demonstrated'.[37] Hooke assured his friends that he could indeed provide a demonstration of this phenomenon and, to spur on the efforts of Hooke and Halley, Wren promised to give a book worth 40 shillings to whoever, within two months, came up with a convincing proof. 'Mr Hook then sd that he had it, but that he would conceale it for some time that others triing and failing, might know how to value it, when he should make it publick.'[38]

Yet by March 1684, no one had produced a sufficiently cogent demonstration to enable them to win the 40-shilling book, while Hooke, being of 'the philosophically ambitious temper he is of, he would, had he been master of a like demonstration, no longer have concealed it, the reason he told Sr. Christopher & I now ceasing'.[39] It is well known how Halley later in 1684 visited Newton in Cambridge and, on finding that Newton had already made progress towards a solution, encouraged him to develop his theory into a full treatise. Then the manuscript of that treatise, which would become *Principia Mathematica*, was presented to the Royal Society by Dr Vincent on 28 April 1686, and on 19 May the Society ordered that it should 'be printed forthwith in quarto in a fair letter'.[40] Hooke felt angry not so much on the grounds of Newton's achievement as it stood, but because Newton had failed to pay the intellectual debts that Hooke felt were owing to him: for the ideas freely imparted in the correspondence of 1678–80 which, so Hooke alleged, had alerted Newton to the importance of Keplerian proportions and the Inverse Square Law. Yet whatever anger had erupted, and whatever harsh accusations may have been made, they failed to leave any record in the Register Books of the Royal Society which, in spite of the receipt and ordering of the printing of *Principia* in the spring of 1686, continued in their urbane and matter-of-fact tone. Robert Hooke continues to offer useful pieces of information for record—about the Chinese language, Chinese varnishes, and fossil shells—without a hint of contention. But what might have been said verbally could well have been another story, for as Halley reported to Newton on 22 May 1686, 'He [Hooke] sais you had the notion from him'.[41] One suspects, however, that Hooke's case was more warmly advocated when the Society 'adjourned to the Coffee house, [where] Mr Hook did there endeavour to gain belief that he had some such thing by him'.[42] Indeed, from such a passing remark by Halley, one senses how important the coffee houses were to the amplification of Royal Society business, and how things were probably said in them that never got entered into the Society's official Registers.

From Halley's own statement to Newton, it is hard not to draw the conclusion that what Hooke demanded was no more than fair and due acknowledgement for having contributed towards another man's work, for 'he owns the Demonstration of the Curves generated thereby to be wholly your own'.[43] He was, it seems, in no way trying to steal the glory of Newton's achievement; it was simply that 'Mr Hook seems to expect you should make some mention of him, in the preface, which, it is possible, you may see reason to praefix'.[44] Surprisingly, at this juncture, Halley seemed puzzled as to why Newton was so adamant about not giving any credit to Hooke, arguing 'that nothing but the greatest Candour imaginable, is to be expected from a person, who of all men has the least need to borrow reputation'. Yet Halley's turn of phrase is not without ambiguity, for one could also take him to mean that it was Hooke—whose international scientific reputation stood higher than Newton's in 1686—who had no need to borrow reputation.[45]

At this time, of course, Edmond Halley, as Clerk to the Royal Society and a man who would have worked closely with Hooke, especially in an administrative capacity within the Society, would have been privy to all that was said in the unfolding gravity débâcle. And as a 30-year-old astronomer with an already acknowledged international reputation, moving among men who were 15, 20 or more years his seniors, Halley was stepping with care. For one thing, Halley was a naturally clubbable and congenial individual, who had no especial taste for faction fights and who, as Clerk, was trying to keep some of the Fellows from each other's throats. Secondly, since at least 1684, Halley had been especially impressed by whatever Newton had confided to him regarding a universal solution to the problem of gravitation. It was to be Halley, after all, who not only first recognized the importance of Newton's work, but who persuaded and cajoled him to write those great essays which would when brought together become *Principia*. Halley was, in addition to being an astronomer of brilliance in his own right, an astute scientific diplomat who was honing his skills in the art of dealing with prickly eminent men.

By the early summer of 1686, at a time when, no doubt, Halley was breathing slightly easier under the impression that the ructions following the April deposit of the *Principia* manuscript in the Royal Society were calming down, Newton had worked himself into full eruption.[46] For Newton was now incensed not only at the very idea of giving any acknowledgement whatsoever to Hooke, but even at the suggestion that Hooke had ever said anything of any originality at all, or that he was anything more than a boastful plagiarist.[47] The furious Newton then accused Hooke of plagiarizing the work of Giovanni Alfonso Borelli, to whom Hooke had made a passing and in no way disparaging remark about mathematicians who were 'Drudges and Devotons' in his letter of 9 December 1679, but which, by June 1686, Newton had suddenly construed as a cheap swipe against himself.[48] Hooke, he raged in a massive three-page postscript to his letter, not only stole his ideas on gravity from Borelli, but sees

> 'Mathematicians that find out, settle & do all the business [as simply needing to] content themselves with being nothing but dry calculators & drudges & another that does nothing [Hooke] but pretend to grasp at all things must carry away all the inventions as well of those that were to follow him as of those that went before.'[49]

By this stage in Newton's rancour, therefore, Hooke had become a plagiarist, a dilettante and a lightweight; and, in addition, a trouble-maker, a complainer and 'a man of strange unsociable temper'.[50] An odd thing, indeed, to be said by a Cambridge recluse about a man who was so much at home in social gatherings! But one really senses the irrational nature of Newton's fury when he claimed that 'Mr Hooks letters in several respects abounded too much with that humour wch Hevelius & others complain of',[51] for no matter what Hooke might have said to Hevelius around 1674, the tenor of Hooke's letters to Newton, while every bit as acute and rigorous as one would expect for such a level of discourse, was never insulting, insinuating or gratuitously offensive. Indeed, as shown above, Hooke had been at pains in 1679 to establish a friendly working relationship with Newton and to avoid 'Enmity'. For when one reads through the six letters which Hooke and Newton exchanged between November 1679 and January 1680, what is most striking, in addition to the scientific quality of their content, is a sense of the high mutual regard in which each man held the other. Newton's first letter to Hooke, for instance, concludes 'So wishing you all happines & success in your endeavours I rest ... Is. Newton', and was addressed 'For his ever Honoured Friend Mr Robert Hook at his Lodgings in Gresham College in London'.[52]

Likewise, Hooke addresses his epistle of 9 December 1679 'For his much honoured freind Mr Isaac Newton', and concludes 'Goe on and Prosper and if you succd and by any Freind Let me understand what you think fit to impart, any thing from you will be Extremly Valued by Your very Humble Sarvant Ro: Hooke'.[53] Newton compliments Hooke in turn on his 'acute Letter',[54] and even after writing his last letter 'for my much honoured freind'[55] (Newton already still owing him a reply, for his letter announcing the Reciprocal distances) Hooke returns a generous compliment to Newton on his work on orbital curves under gravity, saying 'I doubt not that by your excellent method you will easily find out what that Curve must be and its Propertys and suggest a physicall Reason of this Proportion'. Hooke also tells Newton that his work on this curve had been passed on to the Royal 'Society (where it has been debated)'.[56]

Even if one were to cynically attribute all of this courtesy to mere epistolary convention, which I do not think was the case, it is hard to see much evidence in these letters of that ill-humour, high-handedness or cunning plagiarism of which Newton was to accuse Hooke in 1686. Yet sadly by the summer of 1686 Newton interpreted every critical point which Hooke had raised in their correspondence of six years previous as an example of 'Magisterial' high-handedness, while every serious thing which Hooke had said became by definition an act of intellectual robbery.

Far from Hooke's getting the acknowledgement he wanted, Richard Westfall has shown how, when perfecting those parts of *Principia* which dealt with attraction and comets, Newton progressively reduced and finally eliminated altogether any recognition of Hooke.[57] One senses that, as a consequence of this dispute, both men suffered. Hooke felt deeply hurt and cheated, as much by the Royal Society for its corporate failure to stand up and see justice done, as by Newton himself. The gravity débâcle certainly pushed Hooke around one of the great corners of his life, making him increasingly suspicious of the intentions of his colleagues, and probably damaged his physical as well as his psychological health. Newton also suffered in so far as the virtually unchecked tantrums of the summer of 1686, followed as they were a year later by the adulations heaped upon *Principia*, fuelled the hitherto latent narcissism and lordliness of the Trinity College Professor. While Hooke had become the first major scientist to fall foul of Newton's high-handedness and suffer accordingly, he would not be the last.

Twenty-five years later, the now elderly Astronomer Royal, John Flamsteed, would feel trapped in a web which the new President of the Royal Society had spun about him, for while Newton's intention of getting the observations of the Royal Observatory, Greenwich, finally printed and made generally available was entirely laudable in itself, the terms of the publication bond which Flamsteed rather trustingly signed in 1705 placed all the power in Newton's hands. Newton, having got possession of Flamsteed's observing books, spanning 30 years of his life's work at Greenwich (a good part of it done at Flamsteed's personal expense and using his own instruments), now let the project lapse. Blaming the hamstrung and exasperated Flamsteed for being dilatory and uncooperative, he allowed Halley, in breach of the original legal agreement, to use Flamsteed's observations to produce the notorious 'pirated' *Historia Coelestis* of 1712. Only when the change of government ministry which followed the death of Queen Anne in 1714 came about was Flamsteed able to get his Observing Books restored to him. And while the editors of Flamsteed's posthumously completed *Historia Coelestis Britannica* (1725) felt it proper to tone down their old master's language about Newton prior to publication, the surviving manuscript draft of its 'Prolegomena', and Flamsteed's letters, pull no punches. As Flamsteed recorded, recalling his dealings with Newton, 'I had formerly tried his temper, and always found him insidious, ambitious, and excessively covetous of praise, and impatient of contradiction', and when he had pointed out some errors in *Principia* 'instead of thanking me for, he resented ill'.[58] Newton, in Flamsteed's view, was too fond of 'criers-up' and fawning 'creatures' to be content with impartial treatment from those around him;[59] and while it is true that Flamsteed and Hooke were by no means the best of friends, it is nonetheless clear that their views on Sir Isaac Newton were very similar.

While it would be absurd to attribute the mature development of gravitation theory to Hooke—a claim which Hooke himself never seems to have made—it is very likely that he did produce a number of crucial breakthroughs that made that development possible. Through his experiments with weights and balances and falling balls, and his simulation of orbits by rolling balls inside shallow upturned cones,[60] and through his recognition of the gravitational importance of the pendulum, his wide reading of and familiarity with the contributions of Kepler, Galileo and Huygens, combined with his acute awareness of geometrical relationships, Robert Hooke beat the first trail through the darkly thicketed woods of gravitational confusion. What is more, he was probably the first scientist to make the imaginative leap of connecting the ratios displayed in terrestrial experiments with the ratios implicit within Kepler's laws of planetary motion. But whether that imaginative leap took him—as his letter of 6 January 1680 in which he speaks of reciprocal distances could imply—to a full concept of *universal* gravitation is a matter of scholarly opinion.

It is certainly the case that John Aubrey—who may have been partial in so far as he was a friend of Hooke, and not perhaps the best judge on technical matters because he made no claims to being a mathematician—firmly believed that Hooke had a distinct prior claim as far as the early elucidation of gravitation theory was concerned. But while not a mathematician or astronomer, Aubrey was a great gossip and collector of gossip, a Fellow of the Royal Society in his own right, a clubman, and a man who was acutely attuned to whatever stories were currently circulating in the London coffee houses and Oxford Common Rooms. In Aubrey's frank opinion, Hooke's work on gravity was 'the greatest Discovery in Nature that ever was since the World's Creation'. In an undated note, but probably written around 1689, Aubrey recorded that

'About 9 or 10 years ago, Mr. Hooke writt to Mr. Isaac Newton, of
Trinity College, Cambridge, to make a demonstration of his theory,
not telling him, at first, the proportion of the gravity to the distance,
nor what was the curv'd line that was thereby made. Mr. Newton,
in his answer to the letter, did expresse that he had not thought of
it; and in his first attempt about it, he calculated the Curve by
supposing the attraction to be the same at all distances: upon which
Mr Hooke sent, in his next letter, the whole of his hypothesis, *scil.*
[*icet*] that the gravitation was reciprocal to the square of the
distance: which is the whole coelastiall theory, concerning which
Mr. Newton haz made a demonstration, not at all owning he
receiv'd the first Intimations of it from Mr. Hooke. Likewise Mr.
Newton haz in the same Booke printed some other Theories and
Experiments of Mr. Hooke's, without acknowledging from whom
he had them.'[61]

While Aubrey's account undoubtedly contains errors of fact and also
confusions, at least the general drift of its chronology is right. And with
his historian's instinct and fascination with the motives of his fellow-
men, one suspects that John Aubrey's general conclusion regarding
Hooke, Newton, and gravitation was not too far from the truth.

CHAPTER 13

FRIENDS, MISTRESSES, RELIGION AND POLITICS: HOOKE'S INNER WORLD

It is unfortunate that Robert Hooke's controversy with Newton, which undoubtedly embittered the last 16 years of his life, has somehow been allowed to become so formative in history's wider judgement of the man. It was especially tragic for him, moreover, because the gravitation controversy synchronized exactly with what was probably the single greatest tragedy in his personal life, the sudden death by fever of his beloved niece, Grace Hooke, who at the time of her passing at the age of 27 had lived with her uncle Robert in Gresham College for about 15 years. Hooke's biographer Richard Waller emphasized the damaging effects of Grace's death upon him, and soon after his health began to deteriorate as well, so that all these factors contrived to produce the melancholic, secretive and increasingly unkempt personage of popular mythology.[1] Yet up until his late fifties, Robert Hooke was a very different figure, and the man who meets us in the pages of the Diary, between 1672 and 1680, is a smart dresser, with a remarkable taste as a collector and bibliophile, proud of his fine wineglasses and tableware, intensely sociable and warmly attached to his friends, and relentlessly and perhaps even manically active. In the days before Newton's own very different genius began to make its mark on the Royal Society, Robert Hooke rightly basked in the regard and the fame in which he was held as the most ingenious man of the age. One might suggest that, like so many latter-day celebrities in whatever walk of life, Robert Hooke did not necessarily warm to rival celebrities in the same field, especially if the newcomer were undisputedly brilliant, had a difficult personality and a very different intellectual style, and would not defer to the reigning celebrity.

We should not forget that Robert Hooke was a celebrity scientist, with a reputation that reverberated across Europe by the time that he was 35. Indeed, the very concept of celebrity was still a new one in European society; for while one might defer to the illustrious dead, such as Aristotle or Julius Caesar, the hero-worship of the living or the recently deceased was a novel thing. Michelangelo, Sir Walter Ralegh and Galileo were amongst the first modern members of that remarkable breed, and Galileo was the first celebrity scientist. Galileo was a celebrity, moreover, not only because he fundamentally changed the nature of science in his day, but because he also had a driving ambition for fame; and when that ambition led to conflict with the Pope Urban VIII (though by no means with the rest of the Roman Catholic Church), first Protestant Europe and then, in particular, anti-clerical eighteenth-century Europe reinvented him in the guise of a martyr to intellectual freedom.[2]

Robert Hooke held Galileo in the highest veneration, and one senses that, after the lad from the Isle of Wight had acquired a taste for the regard in which his Oxford and early Royal Society encouragers clearly held him, he began to glimpse where his scientific ingenuity might take him, especially within the intellectually freer climate of Anglican England. In *Micrographia*, however, he expressed trepidation at the thought of walking in the footsteps of Wilkins and Wren; and indeed, one never ceases to be struck by the genuine humility, personal simplicity, honesty and even gratitude that were never far below the surface of the self-conscious genius. One insight into this more modest Hooke comes from his reflection upon his circumstances at the end of 1676, when he was at the height of his powers and already rich, when he records 'Much Love to all my friends I owe'.[3]

ROBERT HOOKE'S FRIENDS

It is friendship which is the key to understanding the social Hooke. We have already seen that there is no evidence to suggest that Hooke ever seriously contemplated marriage, and while he was certainly drawn to women sexually and, as we shall see, had several short-term mistresses, it was to men—to very clever men—that he always turned for his real social life; for Hooke was overwhelmingly an intellectual, to whom the life of the mind was pre-eminent. Yes, he may have enjoyed good clothes, jaunts to the playhouse and sociable meals but, unlike his friend Samuel Pepys, Hooke was not especially drawn to sensual pleasure. His senses, rather, gave him the maximum delight when they were revealing hitherto concealed wonders of nature, rather than exciting his palate or arousing more conventional bodily sensations.

One of the first social glimpses that we have of Hooke, from the hand of a third party, and which certainly captures his modesty, is in the Diary of Samuel Pepys. In mid-February 1665 Pepys attended a meeting of the Royal Society, after which the company adjourned from Gresham College to the Crown Tavern to continue the discussion and where, among these luminaries, Pepys recorded 'Above all, Mr Boyle was at the meeting, and above him Mr Hooke, who is the most, and promises the least, of any man in the world that I ever saw. Here excellent discourses till ten at night, and then home.'[4] Was Pepys's ambiguous 'promises the least' a comment on Hooke's unprepossessing appearance, which stood in stark contrast to his transparent power of intellect? Mid-February 1665, we must not forget, was only five or six weeks after the publication of *Micrographia*, the book of wonders that was on everybody's lips at the time, which had kept Pepys out of his bed until 2 a.m. on 21 January, and which he described as 'the most ingenious book that ever I read in my life'.[5] Quite simply, by early 1665, Hooke had become the talk of the town.

Pepys's Diary, moreover, for the period immediately following his own election to Fellowship of the Royal Society in early 1665, makes several references to Hooke, which not only show the very high esteem in which Hooke was held by Pepys personally, but also indicate Hooke's presence at the very heart of the Royal Society's inner circle of Fellows. For example, on May Day 1665, Pepys 'met my Lord Brouncker, Sir Robert Murrey [Moray], Dean Wilkins and Mr Hooke, going by coach to Colonel Blunt's to dinner. So they stopped and took me with them.'[6] Here is Pepys, Clerk of the Acts to His Majesty's navy and a senior civil servant, feeling flattered to be offered a lift in a crowded coach and a presumably free dinner by the inner ring of the Society, which included Mr Hooke. What is more, Pepys was to record future social meetings and private conversations with Hooke over the years, which only hammers home the point that Robert was both a part of and at ease in such company, and to see him as a tolerated inferior, or else as a man with some kind of long-standing grudge against 'the Establishment', is nonsense.

Robert Hooke had the capacity to make and to retain long-standing male friendships. His friendship with Pepys is a case in point, and in 1676 Robert's Diary contains several references to Pepys, including one to a not very good speech which Pepys made to the Elder Brethren of the Navy's Trinity House,[7] and even to passing on a recipe to make white varnish to 'Mrs Pepys'.[8] Exactly who this lady was, however, is uncertain, for the real Mrs Elizabeth Pepys had died of smallpox back in 1669. Most likely she was Miss Mary Skinner, Pepys's long-standing girlfriend and mistress, whom the 40-odd-year-old widower never married or regularly lived with in spite of her being of good family with a country estate. One wonders how many people, including Samuel's Royal Society friends, actually knew Mary's real status.

What even the most cursory glance at Hooke's Diary brings out is the sheer range of his social circle. Samuel Pepys, because of the pressures upon his own time, being a hard-pressed civil servant, was not a regular part of what might be called Hooke's 'circle'. The names which most regularly crop up in the Diary include Robert Boyle, Sir Christopher Wren, Sir John Hoskins, Abraham Hill, Theodore Haak, John Evelyn,

John Aubrey (when in London), Mr Wild and Francis Lodowick. To these Royal Society 'regulars' one can add a good many more, who, because they lived or worked outside London, appear more spasmodically, such as John Flamsteed, the Astronomer Royal, who lived at Greenwich, Thomas Tompion who earned his living making fine clocks and watches in the City and, of course, Pepys.

The almost invariable venues for these gatherings were coffee houses rather than taverns. Coffee houses were brand new social institutions in England at the time, the first commercial coffee house having been opened in 1650 in Oxford, on the corner between St Edmund Hall and the High Street, by one Jacob the Jew. The fervently Royalist apothecary-grocer Arthur Tillyard began to sell the boiled liquor of the toasted Turkish bean by All Souls College in 1655, and found that there was a massive market waiting to be tapped for the rich, brown, aromatic fluid.[9] Also during the 1650s, Chinese tea first came to be sold commercially, along with 'chocolat' which was derived from the New World cocoa bean. All of these new substances were the fruit of English and European trade with distant places, and resulted in beverages that, with their active ingredients of caffeine and theobromine, were able to stimulate without inebriating. Hooke's Diary records visits to no less than 65 coffee houses that were trading in London between 1672 and 1680 while his later Diary, kept between 1688 and 1693, mentions more.[10] These beverages were not as cheap as tavern ale but, as one could drink coffee, tea and 'chocolat' all day long without becoming drunk, they were ideal for talking marathons where the drinkers, instead of falling into progressive alcoholic stupor, seemed to get ever more quick-witted and alert as time went on. However, they often paid the price in sleepless nights. Garaways, Childs and Jonathans were the favourite 'intellectual' coffee houses beloved of the Royal Society, and Robert Hooke was a very regular customer.

Such gentlemanly venues became ideal meeting-places for lawyers, clergy, medical men, City merchants and scientists and, as Hooke's Diary shows, they often fulfilled a social function which from the late eighteenth to early nineteenth century onwards would be increasingly catered for by the gentlemen's clubs of St James and Pall Mall. Indeed, informal 'clubs' or groups of like-minded gentlemen frequently met in such venues and, as Hooke was an inveterate 'club' man himself, the records of several of these clubs can be traced in his Diary.

One such club first assembled on 1 January 1676, being the famous 'Saturday Club' which has been considered the forefather of what later became the Royal Society Dining Club. While the Club met, and was to continue to meet, in St Christopher Wren's house, Hooke's day began and ended in coffee houses. For instance, Hooke spent the morning of New Year's Day 1676 in the 'Green Roome' of Garaways Coffee House in Change Alley in the heart of the City, some 300 yards down Bishops-gate Street from Gresham College, before meeting up with his chums Mr Abraham Hill and Dr Whistler at Mans coffee house near Lincoln's Inn, just over a mile west, and then going to Wren's house together. What followed turned out to be the longest and most intellectually scintillating single day entry in the entire Diary.[11]

'We began our first Discourse about light' recorded Hooke, in which Robert argued for his own wave theory as against that of Newton (who was not present). Then they seem to have moved on to discussing the nature of the moon, with Hooke and Wren disagreeing as to whether or not there was water on it. Hooke believed there was, and gave an account of the 'Selenoscope' which he had devised and with which he had been making lunar observations over the past six months. Then one Mr Wind came in and told the company how one might use floats, attached to the feet, to walk across water, after which the talk switched to the nature of combustive material, and 'Mr Wild told of the fellow that kindled Tow [tinder] in his mouth [and] mention [was] made of the fellow that held the Red hot iron in his teeth seen by the Royal Society'.[12] The nature of explosions and the practicable ranges of cannon next came up, with a report of experiments made from the Dover cliffs to see how far out to sea a gun could fire. On and on the discussions went, from the nature of phosphorescence to botanical experiments performed on plants sealed inside glasses, with Wren relating one of the late Dr Wilkins's (the 'Bishop of Chester's') experiments, to new aspects of musical notation. The meeting at Wren's house eventually broke up at 9 p.m.—whether or not food had been served in the interim, Hooke does not say—but the day was not yet over for Hooke. He left Wren's house by coach, and at Ludgate met John Aubrey, and now went off with him to Childs coffee house where they talked till 11 p.m. When he eventually arrived back home, Hooke found the 'colledge gate lockd', and no doubt had to rouse the porter to get back into his rooms, where he did 'eat meat and drank chocolat that Rectifyed stomack and made me sleep well'.

Next morning he got up at 9.30 a.m., was visited again by Aubrey, drank warm ale, smoked pipes, negotiated prices for telescopes and microscopes with Christopher Cox [Cocks] the optician, visited a coffee house with a friend, then drank 'chocolat'. But when he went to bed on 2 January, he could not sleep![13]

Without any doubt, New Year's Day 1676 was a marathon—sustained high-level conversation extending over many hours, coffee drinking, the chance meeting of friends, and keeping late hours—but it was only an especially well-documented occasion in a regular pattern of behaviour. 28 February 1674, for instance, must have been equally felicitous, or gruelling, depending upon one's viewpoint: first at Lord

Brounker's, then with Wren, followed by the usual coffee-house crawl talking about everything under the sun ('At Gar[a]ways till 10 with Mr Wild. Mr Aubery [*sic*] Hoskins, Hill, Lodowick', and so) on before going home. On this day, moreover, Hooke and his friends, along with the mass of London citizenry, also celebrated the newly-proclaimed peace with Holland which ended the last cycle of Dutch wars, a celebration which led to the lighting of 'Bonfires' in the streets of the City.[14]

Yet not all the social meetings were exclusively scientific or, perhaps, so intense in their content as the above. On 16 February 1667, for instance, a group of Fellows of the Royal Society, and even some ladies, it would seem, were invited to the house of Lord Brouncker, President of the Royal Society, for a musical entertainment. As Pepys, a near neighbour of Brouncker, recorded, 'Here came Mr Hooke, Sir George Ent, Dr Wren and many others' to hear the 'master composer' Signior Vincentio and six other musicians perform, one of whom was a woman. There were two harpsichords as well as the voices and other instruments, and it was clearly a grand entertainment. Even so, the urge to bring up aspects of science was never far away, for 'Sir Robert Murray [Moray]', the FRS who had first proposed Hooke for the Curatorship back in 1662, discoursed with Pepys on the 'doctrine of musique', whilst a medical conversation about eunuchs also got under way. Two of the Italian singers, it seems, were castrati, or eunuchs, and were unusually tall, which led to a discussion about whether the gelding of men, as was the case with oxen, led to exceptional growth. Hooke is not mentioned with relation to these discussions, but as he had learned to play the organ at Westminster School, and was also fascinated by medical curiosities, one suspects that he was not silent.[15]

Hooke had claimed the patrician John Evelyn as a 'friend' since at least 1664, when, as narrated in *Micrographia*, Evelyn had given him some fossil material.[16] And it is clear that Evelyn held Hooke's scientific brilliance in the highest esteem, and had spoken of him thus in his own Diary when seeing Hooke with Wilkins and Petty at Lord Berkeley's Surrey mansion Durdans in 1665.[17] On the other hand, their friendship never seems to have been especially close, and one can perhaps understand why, as the forthright and self-confident Hooke was a man of very different temperament from the rather reserved, urbane and deeply religious Evelyn. Indeed, in the Diary, their contacts often seem to have been about architecture in one way or another, for Evelyn was an acknowledged connoisseur whose opinions and tastes in such matters were highly valued.

Three of Hooke's closest friends, whose names appear regularly throughout the Diary and elsewhere in his writings, are men with whom (like Evelyn) he shared a direct Oxford connection: Sir Christopher Wren, Robert Boyle and John Aubrey. In each case, the friendship was ended only by death. Hooke probably first met Wren at Dr Wilkins's Wadham meetings, by which time Wren, his slightly older Westminster and Oxford contemporary, had left Wadham to become a Fellow of All Souls, and then a Gresham Professor, before returning to Oxford to take up the Savilian Chair of Astronomy. Wren's precocity and brilliance were recognized right from the start, and when Hooke himself grew to prominence in the early Royal Society, he felt 'affright' at the prospect of following in the footsteps of this English Archimedes.[18] Their backgrounds were very different, for while the curate's son from Freshwater was a minor gentleman in his origins, Dr (later Sir) Christopher had good Wiltshire connections, being descended from bishops, royal chaplains and Deans. Even so, their friendship seems to have formed quickly, and to have been broken only by Robert's death in 1703. Wren, on the other hand, lived on to 1723, dying suddenly at the age of 91.

Hooke and Wren shared a great deal, including not only school and university, and the exhilarating patronage of Dr Wilkins in their youth, but also a commonality of genius. Both men were 'mechanical philosophers', both were intellectually inspired by the experimental method and the 'artificial organs' of scientific instruments, both had a partiality for physiological experimentation in the wake of Harvey's theory of blood circulation, both possessed an outstanding talent for draughtsmanship and design, and both made fortunes as architects and planners of post-Fire London. They also seem to have been distantly related, as one of Sir Chistopher's sisters had married a Hooke cousin of Robert, and in February 1677 Hooke laconically noted 'Cozen Wren Hooke buried.'[19]

One has only to glance through the Diary to find constant reference to 'Sir Chr. Wren'. For one thing, it is clear that they worked very closely with each other on architectural commissions, and their mutual involvement with the Royal Observatory at Greenwich, where Wren designed the buildings and Hooke devised the instruments, has been mentioned above. An astronomical theme also conjoined their interest in designing the Fire Monument or 'Piller' in Fish Hill Street in 1678. Both men hoped that the 202-foot hollow tube might be used to measure a stellar parallax — and hence prove the earth's motion in space — if only a lens of sufficiently long focal length could be found. And there were many other architectural projects in which it is clear that both men participated.[20] Indeed, it is not easy in many cases to know exactly who was doing what on a new London church — a class of Royal building which fell primarily under Wren's architectural jurisdiction — as when Hooke recorded in March 1676 'to Sir Chr. Wren's. Agreed with Cartwright for Bow Tower for £2550'.[21] For instance, what part had Wren played in the new Physician's building in September 1675, when Hooke 'Dind at Sir Ch. Wrens... With Sir Christopher to Physicians Colledge view 1 sh[illing]'?[22] The College building in Warwick Lane, surmounted with its fine dome and 3-foot-diameter 'gilded pill', was a 'City' building and arguably Hooke's masterpiece. Conversely, what part did Hooke – who was fascinated by the strength of materials and the geometry of how buildings stayed up – play in the development of those 'caternary arches' which Wren was later to use as his solution to the dome of St Paul's Cathedral, and the properties of which Hooke had both recorded in Latin in his Diary on 26 September 1675 and would state in an anagram the following year in his *Helioscopes* Cutlerian Discourse?[23]

The social and professional lives of Hooke and Wren were so intertwined that it is sometimes impossible to separate the one activity from the other: '...to Sir Ch. Wrens. Dind with him passed Wises accomt';[24] 'To Sir Chr. Wrens, back with him to the chamber, thence again to Scotland yard', where Wren had his official residence;[25] 'With Sir Ch. Wren... at Childs Coffee House';[26] 'At St Paules with Sir Ch. Wren' on architectural matters; and the obvious social *coup* 'Walkd in to the Park with Sir Chr. Wren. The King calld me to him, bid me show him experiment. Followd him through tennis court garden & c. into closet. Shewd him the Experiment of Springs.'[27]

Equally enduring in the social fabric of his life was Hooke's friendship with the Hon. Robert Boyle, which extended across some 33 or 34 years, and progressed in stages from a relationship of employer to employee, or of teacher to pupil, to one of regular friendship; and in 1691 Hooke became the beneficiary of his 'best Microscope, and my best Load-Stone' following Boyle's death.[28] And by that time, the wealthy Hooke had long since ceased to need money, so that valuable and useful scientific keepsakes were more appropriate mementoes to receive from an old friend than cash.

As with Wren, Hooke's friendship with Boyle spanned a wide range of social situations, including private dinners, Royal Society meetings and get-togethers in coffee houses. As a rich landed proprietor, with properties at Stalbridge, Dorset, and in Ireland – for he was a younger son of the very wealthy Earl of Cork – Boyle was often away from London, especially in the summer, as one can gauge from the seasonal pattern of Hooke's letters to him, whereby the absent Boyle was kept abreast of the London scientific gossip.[29] Until 1668, moreover, Boyle still maintained his rented premises and laboratory at Beam Hall, Oxford, where the young Hooke had first acted as his Assistant in the ground-breaking experiments on air and combustion. Boyle's London residence was at the Pall Mall house of his sister, Lady Ranalaugh, and in his early days in London, when down from Oxford, Hooke seems to have lived there as well, and was using 'Pall Mall' as his address on letters in July 1663 before moving into Gresham College at the beginning of September 1664.[30] Boyle continued as Hooke's employer and the principal provider of his income until Hooke received his Royal Society and Cutlerian

Lecturership monies and, by 1665, his Gresham professorial stipend. It is clear that, in spite of Hooke's early client or employee relationship with Boyle, a warm and enduring friendship sprang up between the two men; and while Hooke was not born a '*great* gentleman' in the way that the landowing Boyle, Wren and Evelyn were, he was undisputedly a *gentleman*, and was so by birth as the son of an Anglican clergyman, by education, by his cultured tastes and by his friendships. With his City Surveyorship after 1666, of course, he soon became rich in his own right.

'Dind with Mr Boyle'[31] and its variants became a common entry in Hooke's Diary. And it is plain from the Diary, and from the surviving correspondence that passed between Hooke and Boyle, that all manner of scientific discourse passed between them. At a time, for instance, when the young Isaac Newton's reflecting telescope was exciting admiration in the Royal Society in the autumn of 1672, and Hooke was busy trying to develop his own telescopic configuration using mirrors, he was 'With Mr Boyle, he told me of the improving of Mr Newton's way of telescopes in Italy'.[32] Boyle's role as an international correspondent, indeed, clearly provided Hooke with all sorts of interesting pieces of information, as when in 1677 he was 'At Mr Boyles, he told me of the Parisian flying & C'; and on the same occasion Boyle, the great connoisseur of natural and artificial curiosities, showed Hooke 'his Leather cap that had held water 6 days without being moyste on outside'.[33] And never very far away was Boyle's and Hooke's common interest in medicines and practical therapeutics, though as we have seen, many of these, in their folklore-like origins, seem to us to be strikingly at odds with both men's international reputations as scientists. While one can perhaps see some rhyme and reason behind Hooke's entry 'to Mr Boyles for Spirit of Harts Horn' (ammonia tincture),[34] especially considering Hooke's frequent references to his own need for head-purges, one is taken aback by 'Mr Boyle told me that *stercus humanum* [human excrement] dryd ... is good for eye films'.[35] Yet whilst both Hooke and Boyle lived to good ages for their day, being 67 and 64 respectively when they died, they were not robust, or at least did not think of themselves as robust. Both men were constantly concerned with finding or keeping their 'health', as part of what they perceived to be a game of hide and seek with death. In this preoccupation they formed something of a contrast to contemporaries such as Wren or Edmond Halley, both of whom enjoyed more vigorous constitutions, seemed less morbidly absorbed in being physicked, and who lived to 91 and 86 respectively.

The most colourful of Hooke's friends was undoubtedly John Aubrey, or 'Aubery' as Hooke often wrote his name. A decade older than Hooke and the inheritor of both broad acres and crushing debts, Aubrey had been up to Trinity College, Oxford, at the beginning of the Civil Wars, though the two men may not have met until 1663, by which time Hooke was an already famous man on the way up, while Aubrey was still living grandly but beginning his long slide into serious and insoluble debt.[36] On the face of it, Hooke and Aubrey were poles apart: one was financially shrewd, an accomplished professional 'networker', overworked yet utterly focused and determined; the other was a child with money, a born charmer with an enormous circle of friends but with no real idea about how to make life's choices, and with a mind so enchanted by the curious and the strange that he always ended up getting sidetracked and bringing nothing to proper completion. Both Hooke and Aubrey were extremely clever, Fellows of the Royal Society, and equally fascinated by the realm of ideas. Aubrey was not an experimental scientist like Hooke, but as an antiquary who excavated and recorded Stonehenge, Avebury and other ancient monuments, and whose life's passion was rescuing the past from oblivion, he has been rightly considered the father of archaeology and folklore studies.

Both men, moreover, were great conversationalists, coffee-drinkers and bachelors. They were in their element in the all-male ambience of clubs, coffee houses and groups of like-minded chums, where 'Discourses on the Realm of Learning' could go on all day and deep into the night, seemingly without the need for intermission. All the evidence suggests that Aubrey could be wonderfully entertaining, and sometimes downright good fun. One also suspects that Hooke and Aubrey struck sparks off each other, made one another laugh—that is, when Aubrey was not bemoaning his chaotic finances—and enjoyed an easy informality that would not have been present in the same degree in the much more formal company of Evelyn, Wren or Boyle.

While Aubrey's 'Monumenta Britannica', 'Brief Lives', and studies of his native Wiltshire and adjacent counties' folklore were all unpublished at his death in 1697 (New Style), he was to be subsequently immortalized as the greatest English biographer of the seventeenth century, as well as an incisive observer of the social scene. In the same way that he had observed—somewhat exaggeratedly—that John Wilkins's group at Wadham College after 1649 had introduced scientific curiosity and research into England,[37] so likewise Aubrey identified the coffee house as the age's fundamental institution of social exchange: 'the modern advantage of coffee houses, in this great citie, before which men knew not how to be acquainted, but with their own relations or societies'.[38] As Hooke's Diary makes clear, he and Aubrey were regular frequenters of these institutions whenever Aubrey was up in London from the west—and sometimes trying to dodge bailiffs. One wonders, moreover, how many times it was Hooke who ended up paying the bill for both of them, as the permanently financially embarrassed Aubrey may have relied on his friend to ease him out of tight corners with a cash loan, as when in October 1673, Hooke 'Sent Aubery 20 sh[illings] he promised to repay it'.[39]

Coffee houses, and the clubs that met in them, were also places where gentlemen could write and receive letters and transact business, as when in 1679 Aubrey informed Anthony Wood in Oxford of a letter 'I writt . . . on Thursday night late, at our Club. [Then] Mr Hooke delivered me a chiding letter from Mr M[ariet], that I was slow of my pen and desired to write him some newes' about the Royal Society.[40] This 'Club', of which Hooke and Aubrey were part, was probably the one made up of Royal Society Fellows which adjourned after the meeting to a coffee house to keep the discussion going 'late'.

Even if Aubrey did not always lodge with Hooke in Gresham College when he was in town (for Aubrey's letters indicate that he did rent rooms in London), he certainly left important belongings in Gresham for safe keeping. In 1688, by which time Aubrey had become so indigent that he was living with friends around the country and no longer had a fixed abode, he wrote to Wood to say 'All my bookes are now at Mr Hookes, or Mr Kent's; so that if I had leisure, I cannot enjoy them'.[41] He had already appointed Hooke as his executor in October 1687,[42] and by May 1692, when John Evelyn returned a loaned manuscript, Aubrey was writing 'I desire you to leave it with Dr Hooke at Gresham College to be put into my chest marked *Idea*'.[43] Similar letters went to other friends by the 1690s, specifying Hooke as the custodian of his literary treasures.

It is abundantly clear that Aubrey held Robert Hooke in the highest esteem. Hooke's *Attempt to Prove the Motion of the Earth* (1674) was held in especial awe by Aubrey, and he both recorded its principal points in his 'Brief Life' of Hooke in 1689, and defended Hooke's claims against an unnamed Newton in the gravitation dispute which was then raging. Indeed, Aubrey had been singing the praises of Hooke's Cutlerian Discourses to Wood as early as March 1674.[44] Then in January 1691, when Anthony Wood was completing his famous *Athenae Oxonienses* — the great biographical gazetteer of Oxford writers — Aubrey hammered home the intellectual priorities regarding gravitation, saying 'Whatever you doe, pray take special care to doe Mr Hooke right against Mr Isaac Newton; it [gravitation] is the greatest discovery ever yet was made, and Mr Newton runnes away with the glory of it. When I sent you my box last, I enclosed an exact account of Mr Hooke's case, writt with his own hand: I perceiv'd you had not taken notice of it.'[45]

Aubrey, in his letters and jottings, enables us to glimpse aspects of Hooke's mind and ideas which do not seem to be recorded elsewhere. Among them, and no doubt deriving from his youthful contacts with Dr Wilkins, were Hooke's views on language and education. Instead of the traditional 'written-out grammars' used in schools, for instance, Hooke argued that a 'rational or natural grammar' should be developed for education that would stimulate the pupil's imagination rather than

225

using tedious rote methods.[46] Also on the subject of education and the rhetorical declamatory exercises of the schools 'Mr R. Hooke says that declaiming is of no use, but only in the universities. But Sir Chr. Wren is of another opinion: he says that it makes them speak well in Parliament.'[47] Aubrey also makes us privy to the interesting snippet that 'neither are Sir Chr. W[ren] or Mr R. H[ooke] great readers'. This did not, of course, mean that Wren and Hooke were not thorough scholars, but rather that, instead of following the traditional reading practices of the age and concentrating on classical poetry or history, they preferred to read 'what kind of learning their geniuses do lead 'em' to.[48] Thomas Hobbes was of a similar frame of mind.

How far did this attitude to learning and to education relate to the Royal Society ethos that true originality of mind came from the study of *things*, rather than from following in the literary footprints of other men? I would suggest that Aubrey himself was also of this persuasion and, as what one might call a social and historical empiricist, he put on record aspects of Hooke's mind and attitudes which were disclosed in innumerable private conversations and discussions conducted in Hooke's Gresham College rooms and in coffee houses, probably under the aromatic spell of hot coffee and pipes of tobacco. It is also interesting to see how often Aubrey conjoins Hooke's and Wren's views on a given topic, as is sometimes confirmed in the Diary, not to mention Hooke's personal habits, such as his preference for late retiring to bed and late rising in the mornings, such as when in 1690 Aubrey comments 'He lies abed all the morning, sitting up at study all night; and the afternoons he goes about business'.[49] Very clearly, Hooke and Wren were often together, and it is interesting to note the distinct sense of reverence in which Aubrey held both men. One suspects, however, that as a family man, Sir Christopher was not able to enjoy Hooke's fondness for late nights and late rising.

The Robert Hooke who emerges from the Diary, be it the Diary kept in the 1670s, or the less detailed one which ran from 1688 to 1693, is an immensely social figure. The Diary abounds with such entries as 'At Gar[a]ways, Flamsteed, Hawley [Halley], Tompion',[50] 'Told Sir Chr. Wren my Invention of flying, airpump and my anagram. He approved it',[51] 'Drank brandy with Captain fudled, drank coffe with Odell',[52] all of which impress upon us the energy which he put into his social activities. On the other hand, one does find that in addition to endless conversation, Hooke took pleasure in more contemplative pastimes, and one entry running through much of the Diary during the 1670s was 'Haak chesse' for he met up regularly with Theodore Haak FRS, the German scholar who had been living in London for nearly 40 years, with whom he enjoyed playing chess. The Diary also brings out Hooke's passion for book collecting, and one senses the pleasure which he took in striking a good bargain to obtain a desired volume, as when, in January 1674, he recorded 'At Mr Pitz. Viewd books. Bought mechanicks, 2 sh[illings]. Viewed old books. At Gar[a]ways. Treated with Pits for *Hortus Estitensis* 10 sh'.[53] His recorded book purchases over the years cover all aspects of science, architecture, mechanics and classical literature and even include odd items of modern English literature, such as when he recorded 'to Maitlands Auction, pd 2½ for *Tempest*'.[54] He had seen Shakespeare's play on stage some years before, and one wonders if Prospero's great storm caught Hooke's imagination, at a time when, in the late 1680s, he was at work on his Earthquake Discourses for the Royal Society.

Indeed, we should not forget that Hooke possessed a vivid visual imagination, and though he rarely describes their content, he does make occasional references to his dreams—'wild, frightful dreams'[55] after taking syrup of poppy, 'Str[ange] Dreams etc',[56] and '... dremt of riding and eating cream with Capt. Grant'[57]—while one wonders what levels of imagination were touched when 'Mr Wild told me... a story of witches'.[58] But when it came to stories of ghosts, spirits, and spine-chilling prophecies, then the folklorist John Aubrey surely had no equal.[59]

One also suspects that, while he loved the conviviality of meals eaten in the company of the friends with whom he dined regularly, Robert Hooke was not a gourmet. Indeed, for a man whose curiosity covered innumerable useful arts including bricklaying, tailoring and oil painting, his lack of interest in cooking is noteworthy; and while his Diary abounds with references to congenial meals, eaten in the company of Lord Mayors, Peers of the realm, Church dignitaries, scientists, or simply 'at home', he only occasionally describes menus.[60] One wonders if that dietary abstemiousness to which Waller referred was simply an aspect of a man who valued the company seated round a table more highly than the delicate viands that the chef placed upon it. Indeed, I would suggest that this culinary lukewarmness—which contrasts so sharply with the avid interest of his friend Pepys—is in keeping with a man for whom subtle bodily sensations had a fairly low appeal.

On the other hand, Hooke clearly took pleasure in plain, straightforward public shows and spectacles, such as London's August St Bartholomew's Fair, where in 1680 he saw a man walk on 12-foot stilts,[61] he was 'at Scaramuches [a play] at York House' (at whch the King, Duke of York and several other great dignitaries were also present),[62] and there was the memorable occasion when he 'Saw Elephant wave colours, shoot a gun, bend and kneel, carry a castle and a man etc.'[63] One even wonders whether this was the same elephant which, four years previously, he recorded had been sold for £1600.[64]

Yet in all of this social round, except for occasional jaunts out to St Bartholomew's Fair, the theatre or other entertainments with young Tom Giles, his niece Grace Hooke, or members of his household, Hooke was in the company of men, and one cannot help but wonder what part women played in his life.

ROBERT HOOKE, WOMEN, AND THE HOOKE FAMILY

Robert Hooke's relations with women are interesting. It has already been mentioned that he never appears to have had any clearly-recorded intention to marry, and while his Gresham Professorship required him to remain unmarried, nonetheless, by the time he was keeping a Diary in the early 1670s, he was earning substantial sums of money from his City Surveyorship and architectural commissions, so that he could easily have resigned his £50 professorial stipend to marry had he so wished, especially if it were to a young woman or widow with money of her own. But quite simply, as we have seen, he seems to have preferred the life of a bachelor Gresham College professor to that of a substantial London householder with a family.

On the other hand, while the terms of his Professorship required him to be a bachelor, there appears to have been no effort on the part of the College authorities to enforce celibacy. He was clearly allowed to have female servants, whom he paid at the rate of £3 or £4 a year, and even his live-in niece, Grace Hooke, the daughter of his elder Isle of Wight brother John. Grace seems to have lived with her uncle from the age of 12 down to her death at 27.

One suspects, however, that Robert Hooke may not always have found it easy to relate to women of his own social status. As a scientist first and foremost, he may have felt somewhat shy in the company of ladies, and lacked the conversational and other skills to relate to them comfortably. While Hooke's intellectual range may have been staggering in its breadth and imaginative power, his ability to engage with non-scientists, or people who were not ingenious artisans such as Tompion, may have been limited. It may well be that high-status women were alien territory to him, unless, as with Elizabeth Tillotson the Archbishop's wife, a relationship had grown from long and gradual acquaintance, in this instance going back to his days with Dr Wilkins, and from a common interest which possessed a scientific dimension. In the case of Mrs Tillotson, this common interest seems to have been home-made medical preparations. On the other hand, the Diary does record social gatherings at which ladies were present, such as that attended by several eminent senior Churchmen in December 1678, before Tillotson had risen to the Arch-episcopate when '... at Dean Tillotsons where was Stillingfleet, Lloyd, Fowler and their wives'.[65]

Hooke's relations with Lady Ranalaugh, Boyle's sister Katharine, are interesting. He had no doubt known her for a decade by the time that he began his Diary in 1672, as he had been in Boyle's employ in

London and would have been familiar with, and even perhaps have lived in, the Pall Mall house where Boyle and the Ranalaughs resided. She is, indeed, one of the few ladies to appear relatively frequently in Hooke's Diary, where phrases such as 'Dind at Lady Ranalaughs' occur.[66] It is clear that Hooke was invited to social gatherings at the Pall Mall house at which other ladies were present, as when in November 1674 he recorded 'At Mr Boyles. Dind with Lady Ranalaugh, Lady Slane, Lady Inchiquin' the two latter of whom were probably Irish friends of the Boyles.[67] On the other hand, at least one encounter with Lady Ranalaugh led to Hooke's feelings being hurt, as when in 1678 he recorded 'At Lady Ranalaughs, she scolded & C, I will never goe neer her againe nor Boyle'.[68] There is no clue as to the cause of the scolding: was it occasioned by some social *faux pas* committed at table, or did it relate to the architectural work which Hooke undertook for Lord and Lady Ranalaugh?[69] But it is clear that Hooke's friendship with Boyle and even her Ladyship was sufficiently strong as to overcome a passing tiff, for five days later he 'Calld at Mr Boyles and Discoursd with him' while dropping off 12 copies of the French *Journal de Scavans* at a cost of six shillings.[70] It is true that when he next dined with Boyle and Lady Ranalaugh in early July 'She [was] still finding fault',[71] though by September things seem to have smoothed over at last, as 'Dind with Boyle and Lady Ranalaugh' Diary entries resume without further comment about scoldings or fault-finding.[72]

Katharine, Lady Ranalaugh is undoubtedly the most often mentioned high-status woman in Hooke's Diary, yet even in her case no actual conversational topics are recorded. Only her presence—and references to scolding! The wives of his married friends and colleagues are much more shadowy figures in the Diary. Lady Wren, Sir Christopher's first wife Faith, for instance, scarcely makes an appearance before Hooke's record of her death from smallpox in 1675,[73] while even Jane, her successor as Sir Christopher's second lady, fares only slightly better. It is true that we do, in her case, have a brief flash of a conversational topic—'Lady Wren spoke of fair'[74] (which may have been the recent St Bartholomew's)—though other 'Dind with Sir Chr. Wren and Lady'[75] entries record no more details of her. Indeed, the most detailed mention of Jane Wren in the Diary is in connection with one of those bizarre folk therapies beloved by seventeenth-century scientists and peasants alike which Sir Christopher tried out on her, when he attempted 'curing his Lady of a thrush by hanging a bag of live boglice about her neck'.[76] We are not told of the outcome of the treatment. The somewhat exotic and risqué 'Mrs' Abigail Williams, the live-in girlfriend of Lord Brouncker, President of the Royal Society, also makes her sole appearance in Hooke's record in a medical context. In August 1673, ear-syringing was being tried out, and he mentions 'Mrs Williams syringed' with no further comment. But when it came to Hooke's own turn, he was more fulsome:

'Dr Carew syrrenged my ears, fetch out a core but removed not the noyse'.[77] This social ear-syringing took place at Lord Brouncker's house.

What one suspects from the above is that Robert Hooke, unlike his friend Samuel Pepys, probably did not enjoy a social ease with women—unless he had got to know individuals very well. While we should not be surprised to find that he did not form intellectual companionships with women—for the seventeenth-century intellectual world was a man's world—he never seems to have gone in for the mixed theatre parties, picnics, jolly dinners and evening entertainments which could contain cultivated middle-class wives and daughters, actresses, courtesans and other independently prosperous 'working' women, in whose company Pepys so delighted.

On the other hand, there is no question of the fact that Hooke was physically attracted to women, though the ones who played the biggest part in his life in this respect were working-class girls and his niece Grace. It may strike us as incongruous, moreover, that Hooke's biographer Aubrey could refer to him as 'a person of great vertue [*sic*] and goodness'[78] when his Diary records a series of distinctly seedy encounters with his own maidservants and the commission of incest with his 16-year-old niece, to whom he was standing *in loco parentis* in London. Perhaps, because Hooke seems to have done all his sinning at home, at least between 1672 and 1680, he was able to keep it under wraps, though as Aubrey was a regular visitor to Gresham College when he was in London, it is surprising that this curious antiquary never sensed anything and mentioned it in his own voluminous manuscript remains. But there again, even if he did sense misbehaviour, perhaps he regarded it as irrelevant for, discounting the well-concealed incest, it seems that the ability to 'play' with young working-class women was simply one of the privileges of being a gentleman. One is amazed, for instance, at how Samuel Pepys pawed and 'played' with a succession of shop and tavern girls, female stallholders in Westminster Hall, and his own maids—some of them no more than 14 or 15 years old[79]—and one surmises that Hooke, and no doubt Aubrey and other men of their class, saw it as completely natural to do so when opportunity and inclination came together. How such conduct was squared with personal friendship with bishops is not clear, but human beings, after all, are capable of compartmentalizing their lives in an extraordinary way, especially if their awareness of Christian teachings is not always uppermost in their minds.

On a more innocent level, however, it does seem to have been customary, in that draughty and poorly-heated age, for old gentlemen to use young women as human hot water bottles, in accordance with the impeccable Biblical precedent of King David of Israel who, when aged, enjoyed the non-carnal company of a young woman in his bed to keep him warm. So likewise, John Aubrey recalled, the septuagenarian

Dr William Harvey 'kept a pretty young wench to wayte on him, which I guesse he made use of for warmeth-sake as King David did, and tooke care of her in his Will'.[80]

Robert Hooke always noted a sexual encounter in his Diary with the astrological symbol for Pisces, the Fishes. The symbol is to be found with relation to his various maid-servants, his niece Grace, and sometimes himself alone. It is never found with relation to a man's or a boy's name.

Nell Young, an Isle of Wight girl, was Hooke's servant around the time of his commencing his Diary, though whether she was the same 'Isle of Wight mayd [who] lodged here' in mid September 1672 is not clear.[81] Most of his early references to her are noted with the Gemini sign, but they are sometimes more unequivocal such as 'Nell lay with me ♓. Slept ill',[82] or the rather sadly humorous 'Played with Nell — ♓ hurt small of back'.[83] Even so, Hooke felt an undeniable affection for her, and when she absented herself to get married, in August 1673,[84] he was upset. After her wedding night, however, she returned to work at Gresham College where 'Nell came home at noon',[85] and while this continuity of employment could no doubt be acceptable for a newly-married servant, what does strike us as unseemly is that the following morning she was back in Hooke's bed: '*mane* ♓ Nell'.[86] Waiting, no doubt, until she and her new husband could find a place to live, Nell continued at Gresham College until she formally moved out shortly before 22 November 1673, when Hooke recorded 'Nell this week setled at Fleet Ditch. Bridget [Taylor] fetched her trunk this night'.[87] Even a month later, however, Nell is still buying cloth and doing other errands for Hooke.[88] It is likely, however, that Nell was the sister of Jane and Mary Young — both of whose names appear in Newport, Isle of Wight records as Hooke family servants — and it is quite possible that the Young women could have been, amongst other things, a useful line of contact with the Isle.[89] One also wonders if the woman invariably referred to by her Christian name, Mary, in London, was the same Mary Young, and whether Mary Robinson could have been the same person after marriage. Nell Young's successor as Robert Hooke's servant, Bridget Taylor, was also an Islander, and seems to have been a cousin of the Youngs.[90]

During her last month with Hooke, Nell Young was no doubt helping to train up her successor, Doll Lord, whom he took on on 16 October 1673.[91] He eventually got Doll into bed just after Christmas 1673 and on several further occasions prior to her leaving his service at the end of March 1674, when he paid her outstanding wages.[92] On that same day, however, he was 'At Nells 9^{d}', though as there was no reference to sexual services, one presumes that he was merely continuing to employ Nell for her skills as a seamstress.

One cannot help but wonder, however, about the paternity of the child which Nell Young brought into the world on 20 April 1674: 'Nell brought to bed of a daughter and Xned [Christened]'.[93] Very likely it was a child conceived out of wedlock with the man whom Nell married in mid August 1673; but it is not beyond the bounds of possibility that it could have been Hooke's. Yet while there are several ♓ Pisces signs in the Diary for July 1673, they are not given in conjunction with any woman's name.

The Diary would record many future non-sexual meetings between Hooke and Nell, the purpose of which seems either to have been social or else the purchase of her services as a seamstress, and a clear friendship came to develop between them.[94] They were still in close touch in the early 1690s and, in spite of her married status, Hooke always referred to her by her maiden name of Young. Almost certainly the 'N. Young' of the 1690s Diary is the Nell of the 1670s, for not only is this individual always referred to in the feminine, but she had at least two daughters and was paid for skilled services as a needlewoman. There are for instance quite a number of straightforward seamstressing references, such as 'N. Young pd. her $2\frac{1}{2}$ sh[illings] for making sleeves [*sic*] etc.'.[95] But the social side is also plain: 'N. Young and daughter', who could well have been the nineteen-year-old born in 1674;[96] 'N. Young & 2 children here: trouble', possibly a reference to Nell's younger and clearly mischievous children;[97] and '1 paire of sleeves from N. Young. Gave her girle 2'.[98] Gave her girl two what, one wonders? Two shillings to the nineteen-year-old for stitching done on her mother's behalf? Or two pennies to a playful little girl to buy some cakes for a treat? Then on 21 May 1693 we have 'N. Young, that Giles left widdow well, her cozen coming up: other chat'.[99] This sounds like Isle of Wight Hooke and Giles family gossip being relayed, perhaps via Nell's cousin coming up from the Island to London, for Hooke's mother had been a Giles (or 'Gyles'). By the age of 58, in 1693, therefore, it seems that Hooke had become a long-established friend of Nell, not only employing her, but receiving social visits from her and her playful younger children, usually on a Sunday, and enjoying their 'chats' and exchanges of gossip. Had the great scientist, by middle age, even become a popular surrogate uncle to Nell's younger children (who no doubt enjoyed causing 'trouble' in the Aladdin's Cave of his College rooms), and was giving them treats in their own right?

Back in the 1670s, Doll was succeeded by Bette Orchard, and on 3 April 1674 an annual salary of £3 was agreed for her services.[100] But poor Bette had her shortcomings as a domestic servant, and 'brok one of my white glasses',[101] while she does not seem to have regarded the provision of sexual services to have been included in her £3 a year, for on 20 June Hooke 'Wrastled with Bette' but seems to have got no further. Then in early July Hooke 'Lost my Cock by Bette's carelessness',[102] though exactly what was lost is unclear. By 17 July 1674 Bette had become 'intollerable Lazy', and Hooke seems to have had no more sexual success with her, until on the night of 25 July he 'Dreamt of [sex]', which seems to have spurred him on to approach Bette again. The next day, which was a Sunday, he 'First saw Betty ♓ SI: Menstr', and wrote down his Pisces symbol for a second time in the day's entry.[103] Had he made an advance at a time when Bette was having her period and still carried on regardless?

Bette Orchard was eventually sacked and paid off at the end of September 1674, and on the very same day a new girl named Mary, perhaps the same individual as the above, was engaged at £4 a year. She seems to have remained unmolested, for whatever reason.[104]

By 1676, however, Robert Hooke's sexual attentions were being drawn to his niece Grace. Born on the Isle of Wight, the daughter of Hooke's elder brother John and his wife Elizabeth, and baptized at Newport, 2 May 1660,[105] Grace had been destined for betrothal to the son of Sir Thomas Bloodworth, a former Lord Mayor of London and a rich vintner, and one can only presume that the betrothal had gone some way along to a settlement, for in September 1672 Hooke recorded 'Bloodworth here, when he resolved to continue to have Grace and to send me his dymands next day'.[106] By the summer of 1673, however, the betrothal was clearly off. Perhaps the mighty Bloodworths regarded the Newport Hookes as insufficiently grand in their dowry provision, for though John Hooke was a burgess and had even served as Mayor of Newport, his finances were in a shaky condition. Grace, therefore, remained unbetrothed, and by the time she was 16 was clearly a very attractive young woman—judging from the attentions young men seem to have paid her.

We do not know what factors led Hooke to make sexual advances to his niece, for incest was anathema to the whole religious and cultural ethos of that society. Even so, incest *did* happen in that society, for Hooke noted a few years later that he 'Heard of Doll. child by her unkle': this was Doll Lord, his own ex-maidservant and mistress, now got pregnant by her own uncle.[107]

On the night of Wednesday 14 June 1676, Hooke 'Slept with Grace', though as no Gemini sign appears in the text, he may simply have cuddled her, though by 16 October of the same year he plainly states ' ♓ Grace in Bed', after which there was to be a succession of similar encounters.[108] During this time, he developed a distinctly peculiar relationship with the girl to whom he was acting as guardian. It is true that buying her a looking-glass for 8/6d,[109] items of jewellery, and fine material for dresses need not mean any more than an affectionate generosity, but when they are combined with an active sexual relationship, and a resentment of any young man who showed any interest in her, then things do look rather unsavoury. Hooke tried to protect Grace from young men by doing his best to keep her inside the College, but this strategy was doomed, as Grace seemed as willing to secretly entertain young men as they were to be in her company. There was, for instance, one Pettis who kept hanging about and whom Hooke caught with Grace in the cellar, though the Diary does not record what they were up to.[110] And Hooke even resented Grace's friendship with other young women, and feared their gallivanting around town and receiving the attentions of young men. Her friend Mary Kerry was one such person, leading him to complain: 'Jade Kerry here. Denyd her Grace, saw her afterwards with a blade [young man] in Morefields, she raild and scolded'.[111] Was this 'jade' the same woman as the 'Mrs Kerry here Wheedling' of the year before?[112] The title 'Mrs', meaning 'Mistress' in seventeenth-century usage, was sometimes applied to young single women as well as married ones. And it would be interesting to know what her 'Wheedling' was about.

Continuing research into Robert Hooke confirms that he was never out of touch with Isle of Wight affairs, especially if they pertained to the Hooke and Giles families, and Grace herself made several recorded visits back to the Isle to see friends and family. On the night before her departure from London for one of these visits, on 10 August 1677, Hooke slept with her. But once on the Isle the 17-year-old Grace also began to receive the attentions of the Isle's 55-year-old Governor, Sir Robert Holmes (whom Hooke had probably known since the sea-clock trials in 1662), a notorious womanizer, for as he recorded, he 'heard of Sir R. Holmes courting Grace'.[113] Exactly how Holmes was courting her, and with what intentions, is never stated, though there is no evidence that they were ever matrimonial. Then in the early part of 1678, while still on the Isle, Grace contracted measles, from which she seemed to make a good recovery. During the winter of 1677–8, however, Hooke seems to have tried, unsuccessfully, to make contact with Holmes when he was visiting London, though the Diary does not say on what business. But by May 1678 the two men were in correspondence, for Hooke 'wrot to Grace. Received letter from Sir R. Holmes... sent to Grace by Hewet' (or Huet), the Portsmouth to London carrier.[114]

The ten months that Grace was on the Isle of Wight (she arrived back in London on 7 June 1678)[115] were overshadowed by traumatic events. Quite apart from Holmes's worrying interest in Grace, and her own bout of measles, Hooke's 'Brother John', Grace's father, was found dead, having committed suicide by hanging himself on 27 February 1678. This was an awful thing to happen to any respectable family, and when Hooke first 'heard of Fatall news of Brother John Hooke's death from Newland Hayles', while visiting Nell Young a couple of days later, he was clearly shaken.[116]

John Hooke, eldest son of the Revd John Hooke, was five years older than his brother Robert. He stayed on the Isle of Wight, and after completing his apprenticeship as a grocer, went on to become a leading citizen of Newport, a burgess, and on two occasions, in 1669 and 1677, Mayor of the town.[117] He had married Elizabeth, the daughter of William Maynard, another grocer, and a leading member of a powerful Newport family which had produced Chief Burgesses, Mayors and other civic dignitaries. So John Hooke had clearly been a leading member of Newport and Isle of Wight mercantile society, and should have been rich and respected. 'Brother John', as Hooke called him, sent hampers of food up to Gresham,[118] especially around Christmas time, and there are numerous references in the Diary to correspondence passing between the Hooke brothers, not to mention money for Grace, or to cover her London living costs.[119] Brother John, who no doubt wished to go beyond his Newport grocery business and become a member of the Isle of Wight's landowning gentry, had his eye on the purchase of land at Alvington ('Avington' in the Diary) to the west of Carisbrooke Castle in 1675,[120] and Robert in London seemed keen initially about putting up a hefty £4000 for the purchase of the estate, though in the end everything fell through.[121]

What was the concatenation of circumstances that led to John Hooke's suicide, however, is hard to be certain. But one suspects that financial and social failure may have played their part. At the time of his suicide, not only had he failed to become Lord of the Manor of Alvington (even though this would have been on the strength of Robert's money), but he was also badly in debt. He owed his brother at least £250 (a sum five times the size of Hooke's annual professorial stipend), not to mention other monies to Newport Corporation. Very clearly, something had gone seriously wrong with John Hooke's business finances, and he may well have been staring ruin in the face, though there is no evidence that Robert, who was rolling in money by 1678, was putting any pressure on him.

Following the suicide of his Brother John, however, three things appear in Hooke's Diary and Isle of Wight records, all of which have a bearing on Grace. Firstly, Hooke 'Spake to the King for Brother J. Hooke's estate'.[122] Secondly, when approaching His Majesty, Hooke discovered that Holmes had already put in a prior plea, for ' [King Charles] said Sir R. Holmes had begged it for wife and child ... Sent Crawley upon Mr. Davys horse to Isle of Wight. Gave him 40 sh[illings]'.[123] Thirdly, Hooke began a correspondence with, and also probably met in London, Dr Edward Harrison, a Newport, Isle of Wight physician. Indeed, not many provincial suicides could have had two men as powerful and well-connected as Hooke and Holmes personally entreating the King on their family's behalf within a week of death.

The reason for such entreaty lay in the legal understanding of suicide in 1678. Quite apart from its religious implications, as an affront to God's mercy, suicide was seen as a felony in law, being the wilful murder of one of His Majesty's subjects—even if that subject was killing himself. One of the automatic consequences of such a proven felony was the forfeiture of the victim's estates to the Crown, and the consequent ruin and disgrace of the surviving family. It was because of this humiliating outcome that Coroners' Juries invariably tried, wherever possible, to decide against suicide if the evidence would in any way allow. But when a man was found hanging on his own rope, it was hard to return a verdict of accident or natural causes.

While John Hooke appears to have been in financial difficulties at the time of his death, he could not have been without some property—the 'Brother J. Hooke's estate' to which Robert referred when speaking to the King—against the forfeiture of which both Robert Hooke and Sir Robert Holmes pleaded with His Majesty. But no matter how personally sympathetic or otherwise Charles II might have been, the forfeiture clearly went ahead, for by May 1678, three months or so after John Hooke's suicide, Dr Harrison was putting in a claim for 'diet and bord' for Mrs Hooke, Grace and Jane Young: the widow, daughter and maid-servant of the late John Hooke.[124] Had Mrs Elizabeth Hooke even lost her home and was now lodging with the Harrisons? Grace seems, moreover, to have been a friend of Harrison's daughter. Yet though the surviving Newport Hookes had crashed spectacularly, from John's second Mayoralty in 1677 to having no home of their own a year later, one presumes that their social standing was deemed such that they were allowed to keep a maidservant at the Corporation's expense. Elizabeth Hooke thereafter became a pensioner of Newport Corporation, and received £10 per annum from the Civic Purse down to her death in 1684.[125] There appear to be no references to monies being paid to Elizabeth Hooke in Robert Hooke's Diary, however.

It would also be interesting to know the subject of the correspondence which Hooke was exchanging with Dr Harrison over April and May 1678, not to mention the purport of the conversation Hooke had with the Isle of Wight physician when he was in London. But at least one of their recorded exchanges related in whole or in part to Grace.[126] It seems, moreover, that on some occasions Nell Young in London was acting as Hooke's correspondence go-between with the Isle of Wight—no doubt she knew Hewet, the Portsmouth to London carrier—for at the end of April 1678 Hooke 'from Nell Young Receivd Letter from Harrison'.[127]

One can understand the flurry of correspondence which Robert Hooke had with the Isle of Wight following his brother John's tragic death. Yet one wonders too how much his sexual possessiveness of the absent Grace also played a major part in his concerns, for she was clearly an attractive young woman, being pursued by a womanizing Governor and by goodness knows who else on the Isle. If the Jane Young who served the Hooke ladies in Newport was the sister or close relative of Nell Young in London, and gossip did pass by letter or by word of mouth via Hewet or other carriers, then any flighty conduct on Grace's part could well have reached the burning ears of her uncle in London, and helped to drive his worried pen.

It has also been suggested that Grace Hooke was the mother of Mary, the illegitimate daughter of Sir Robert Holmes: a child who seems to have been born in 1678.[128] Very surprisingly, no mother's name was ever mentioned, and as she seems to have been his only offspring, Sir Robert laid down conditions in his will whereby she, through marriage to a young male Holmes, would inherit his name and property. After all, the chronology would fit: Grace was on the Isle of Wight for ten months, between August 1677 and June 1678, though we do not know how soon after her arrival she began her relationship with Holmes.

On the other hand, as was mentioned above, she spent the night immediately prior to leaving London for the Isle, 9 August 1677, in bed with her uncle Robert. This has led the present-day writer, Monica Mears, to suggest that if Grace was indeed the mother of Mary Holmes, then Robert Hooke himself could very well have been her father.[129] Holmes, in spite of a lifetime of relentless philandering, never seems, in that pre-birth-control age, to have had any bastard children brought to his door, and may well have been sterile. On the other hand, the same could be said about Robert Hooke.

I am cautious, however, about accepting the idea of Grace becoming pregnant by either of the two Roberts, and then staying on the Isle to give birth, to the spectacularly public disgrace of herself and of her family. But it is true that any sixth-month announcement of her pregnancy, in the days when even the slimmest women wore 'tyres' to give them extra girth around the hips in accordance with prevailing fashions, may have helped tip her father John into suicide, and that references to her 'measles' and Hooke's correspondence with Dr Harrison about Grace may all have been oblique allusions to her condition.[130]

On the other hand, I find three historical objections to the idea of Grace being pregnant. The first and most obvious of these is the total absence of any reference to the event in the record; and why should Robert Hooke, who kept an intimate diary of sometimes breathtaking frankness, be totally silent about his beloved Grace going through the very great dangers of childbirth in a place that was a two or three days' journey away from London? After all, fornication, adultery, incest, quasi-rape and masturbation are all dutifully set out on his daily pages — so why not the slightest hint of an obviously worrying pregnancy, if such a pregnancy existed? Hooke was clearly not a prude.

Secondly, if Grace had found herself to be pregnant by Christmas 1677, the last place she is likely to have chosen for her lying-in was the Newport house of Dr Harrison. Even working-class girls who found themselves pregnant usually headed to the relative anonymity of a city to give birth. Westminster, Southwark, Kensington, Chelsea and the other settlements around London were famous for providing such services to girls in difficulties.[131] Also, if a girl still entertained good marriage prospects, was the daughter of a Mayor, and had two rich lovers, she is likely to have been whisked away before her condition became obvious, to give birth in a secret location, before handing the child over to adoptive parents and making her seemingly innocent return to her usual haunts. While secrecy was naturally the *sine qua non* of such a situation, one might at least have expected a record in Hooke's Diary, if not necessarily in Newport Corporation's records.

Thirdly, following Grace's return to London on the Portsmouth coach in early June, things seem to have gone on as before, though the absence of Pisces signs in Hooke's Diary would suggest that the relationship was no longer sexual.[132] Hooke even seemed to be presenting her in respectable society, as when in July 'by water with Grace to Mrs. Mayors to Jonathans with Mayor, met Lodowick'.[133] As there is no-one in the Diary named Mayor, one might surmise that Hooke was taking Grace to visit the Lord Mayor's Lady, while he himself went off to Jonathan's Coffee House with his Lordship. Yet one looks in vain in the Diary for any hint of concern for a child farmed out to a wet nurse or foster parents, or of Grace wishing to see such a child. Either the infant Mary Holmes was some other woman's daughter, or else Grace Hooke, and her kind-hearted albeit morally irregular uncle who was now her guardian, were entirely devoid of natural affection.

Yet no matter what might have happened on or off the Isle of Wight between August 1677 and June 1678, Grace had not lost her wayward charms, for she was still being pursued by men, much to the chagrin of her uncle, for over a year later, in November 1679, he felt obliged to 'Chid[e] Grace about Edwards'.[134] One wonders why.

How troubled was Hooke's conscience regarding Grace? It is hard to be sure, although what cannot be denied is that she disturbed him deeply. His recorded Diary references to Grace indicate the development of a very complex, compulsive and anxious love affair. An almost certainly one-sided love affair, moreover, between a middle-aged and far from handsome man, and a flowering, beautiful, coquettish and sexually exciting young woman living under his roof. Six months after the beginning of their sexual relationship, towards the end of 1676, he recorded 'Grace out. I resolved to rid myself of her',[135] and though he was to express similar sentiments over the next four years, affection, lust, family duty and the pleasure of her company all conspired to keep uncle and niece together down to her death in 1687.

What is very clear from the Diary and other documents is that, in addition to matters relating to John Hooke's suicide and Grace's visits, Robert Hooke maintained very close links with the Isle of Wight. His sister Katherine occasionally visited London,[136] and the Diary contains various references to messengers being sent to and from the Isle, especially in the custody of Hewet, whose waggon would have conveyed gifts of poultry, letters, and even servants such as the Young girls on their way up to London, though those with money or with claims to genteel status (such as Grace) came by the more comfortable stage coach.[137]

Indeed, it would be interesting to know more about the Young family, for they were clearly Islanders, and some of them seemed to be in regular service to the Hookes. There was, of course, Nell Young, Robert's maidservant, sleeping partner, and even, after her marriage, his seamstress, hair-cutter, servant-trainer, and as we have seen by the 1690s, Hooke's drop-in friend and source of Island gossip, who even brought her children around to see him. Was she, for instance, the same Mrs Young—now married but still referred to by Robert by her more familiar maiden name—who saw 'all the Livani and Pewter here'[138] in Gresham College, and to whom he paid ten shillings for stitching clothes? And back on the Isle, of course, there was Jane Young, and possibly Mary. One wonders whether the latter was the Island girl whose father seems to have performed some service for Grace in November 1676: 'Mary's father with Grace'.[139] Hooke's Diary further contains references to one Nicholas Young, a stonemason. Was he a relative of the Young women? Did a working-class Island family in the employ of the Newport and possibly Freshwater Hookes see Robert in London as the provider of economic and even matrimonial opportunities in the big, wide world of London?[140]

On the other hand, the disease-ridden city of London could often destroy those who came to live in it, and what must surely be one of the most touchingly described death scenes to come from any seventeenth-century pen related to the passing of young Tom Giles or Gyles, the son of Robin Giles, a cousin of Hooke on his mother's side. Young Tom, 'a pretty boy, good at Reading Arithmetic & C. his mind for sea' came to London in the care of Hewet the Portsmouth carrier in July 1675[141] and seems to have lived in Gresham College with Grace, the maidservants, Harry Hunt, Hooke's assistant and the rest of his uncle's entourage. The clever boy's presence brought out some of Hooke's kindest and most generous traits and makes it clear, as did Grace and even Nell, that Hooke undeniably had a gentle and affectionate side — though in relation to Grace and Nell, this came to be clouded by a sexual jealousy which was not of course present with regard to Tom.

But then, in September 1677, 'Tom complained of crook in his back'.[142] It grew rapidly worse, so that he 'Raved',[143] and when Mr Guidly, Hooke's surgeon friend, refused to let blood from the choking boy who was now 'pissing blood' anyway, things began to look alarming, and Hooke wrote to the lad's father Robin on the Isle of Wight. Dr Diodati diagnosed the dreaded smallpox, and declared that in spite of his own efforts, and those of Dr King (later Sir Edmund King, a Royal Physician), Mr Guidly, Mr Whitchurch and a brace of Hooke's medical friends, the case was hopeless.[144] Very significantly, Hooke, the heroic self-medicator, never seems to have tried out any of his own favourite recipes on the sick boy, and left his care entirely to the professionals, as he also did when Grace fell 'exceeding dangerously ill' in July 1679.[145] Then, after sinking during the morning of 12 September 1677,

> 'Tom spoke very piously, began to grow cold, to want covering, to
> have little convulsive motions and after falling into a slumber
> seemed a little refresht and spoke very sensibly, and heartily, but
> composing himself for slumber he ratled in the throat and presently
> Dyed. It was 14 [minutes] after 12 at noon, he seemed to goe away
> in a slumber without convulsions.'[146]

Could Charles Dickens have written a more poignant child's death scene, and at the same time have been so clinically observant?[147]

By the time that Grace Hooke died at the beginning of 1687, Hooke had abandoned his daily Diary, so we have no first-hand account of the event. But perhaps it is as well, for this was probably the most emotionally fraught experience, and certainly the blackest bereavement, of his life. According to his biographer Richard Waller, Grace's death from fever at the beginning of her 27th year was one of the decisive events of Hooke's life, 'the concern for whose Death he hardly ever wore off, being observ'd from that time to grow less active, more Melancholly and Cynical'.[148] One wonders how Grace's death also impacted upon Hooke's wider moral and religious awareness.

RELIGIOUS AND POLITICAL OPINIONS

It is impossible to understand the life of any seventeenth-century man, especially a famous and prominent one, without addressing his spiritual and political allegiances. The seventeenth century, after all, was riven with spiritual and ideological cleavage planes, and they in turn related to political loyalties. Robert Hooke seems to have been born into a Royalist, perhaps Laudian, Anglican or High Church family, for these political and religious beliefs were not only part and parcel of 1640 Isle of Wight culture, but the Goodman, Oglander and Fell families with whom the Revd John Hooke seems to have been closely associated, and who probably patronized the young Robert, were of those persuasions. His Oxford years, moreover, would have seen him being recognized and assisted by men of conspicuous piety. Seth Ward was to become Bishop of Exeter and then Salisbury, and Wilkins, who was a Broad Church, or Latitudinarian, Anglican went to the See of Chester. John Tillotson, who became Primate of England and Archbishop of Canterbury, was a long-standing friend, a famed preacher and theologian, and anti-Catholic, while several of Hooke's lay colleagues and friends were notably devout in their manner of life and loyal to the monarchy. Dr Thomas Willis, with whom Hooke had probably lodged and who had certainly employed him around 1658, was a leading member of the High Church, anti-Puritan party in Oxford, and married into the distinguished High Church Royalist Fell family. Sir Christopher Wren was descended from a similar background, while Hooke's early employer and lifelong friend, the Hon. Robert Boyle, was so distinguished for both his theological learning and his personal devoutness that Aubrey was inclined to style him a 'Lay-Bishop'.[149]

Yet in that deeply theological age, and notwithstanding friends like the above, Robert Hooke's religious life has left so weak a trace upon the historical record as to be invisible to the casual glance. But three things certainly come over. Firstly, he seems to have had no truck with Puritans or Puritanism. Secondly, his private Diary is entirely lacking in any serious religious content. He makes no reference to private prayers, Bible study or religious conversations. His book-buying, moreover, rarely included theological works. Only on odd occasions does he mention God or Providence, and that usually in a purely conventional way, such as *miserere mei deus*—'Lord have Mercy on Me'—following the recent news of and problems surrounding his brother John Hooke's death at the beginning of March 1678.[150] Indeed, one might wonder why he did not call for God's mercy on the soul of poor Brother John himself who, as a suicide, could well have been burning in Hell! Thirdly, his church attendance for purposes of worship—as opposed to architectural contracting—seems to have been virtually non-existent, at least before 1693. His Diary entries for the 1670s show that Sundays were passed pretty much like other days—in the coffee houses, talking with friends on scientific subjects, and if necessary even transacting business, for instance 'Mr Tompion here from 10 till 10. He brought clockwork to shew',[151] or even 'Examined Accounts'.[152] On some Sundays he even committed acts of incest with Grace.

One outstanding religious reference, however, comes from a Sunday in early May 1674, when Hooke recorded 'I Received the Sacrament at St Peters Poor with Dr Pope. Dr Croon Sir James Oxendine' and several other Royal Society and learned friends.[153] Dr Walter Pope was the now late John Wilkins's half-brother, and a long-standing friend of Hooke. Then on a Tuesday five years later, Robert 'walked with Dean of Canterbury [Tillotson, as he then was] to St Laurence Church and with

Dr Whitchcot ... Missed Sir Chr. Wren',[154] but it is not clear whether this expedition had anything to do with partaking in an act of worship.

One wonders what conversations passed between Hooke and friends such as Wilkins, Tillotson, Boyle and the others mentioned above, each of whom, be he priest or layman, took his Christian faith in earnest. It is true that they all saw God as the Great Designer: the rational supreme being Who had set His universe in motion and which universe, as Hooke had pointed out in *Micrographia*, we men of this latter age could observe, measure and study as a way of understanding Him more deeply. But this facile delight in nature would not have been enough for men like Seth Ward, Robert Boyle, Thomas Willis and the rest. They were all too aware of the problems that needed resolution. Why, for instance, did God's Providence, to an unthinking observer, seem so unfair? Why did little Tom Gyles, who was a lad of such promise, die of smallpox, while the buccaneering, philandering Sir Robert Holmes was still seducing girls well into his mid fifties, and would live on to reach his Biblical span of 70 years? In addition to the contemplation of the beauties of Creation, the disturbing and theologically meaty issues of sin, repentance, redemption and an individual soul's relationship with Christ lay, and still lie, at the heart of the Christian religion; and it is impossible to imagine regular social gatherings at which devout scientists, medical men and Archbishops dined and drank coffee together without these issues arising in conversation — and if they did, what did Hooke say? Indeed, we have only the vaguest of hints, such as when in September 1678 he was in the company of Dean Tillotson, Bishop Lloyd and Dr Gale, Head Master of St Paul's School, and they all 'Discoursd much of Criticall Learning of French Bible',[155] and when on another occasion Hooke, Gale and Wren 'talked about the Alexandrian Bible of Tecla. Sir Chr. Wren would have it printed from copper plates'.[156] And in that age of cheek-by-jowl living, with the inevitable gossip that must circulate within a relatively closed world in which servants abounded, did any of his friends know what went on with Nell, Grace and the other women in Hooke's life?

There were, after all, figures who bridged Hooke's domestic and professional worlds, and who must have known and seen a good deal. The most obvious of these was Harry Hunt, who started to work as a sort of boy laboratory assistant to Hooke in the early days, and who took over the Curatorship of Experiments to the Royal Society when Hooke became too over-burdened, and then became Secretary to the Society, and was to help in the laying out of Hooke's body after he drew his last breath in 1703. Hunt seems to have lived in Gresham College, and to have been part of Hooke's 'household'; on the day after Christmas 1674, he 'drew Grace her picture'.[157] And by the early 1690s, Hunt had also gone on to become a companion and friend of Hooke, and in Hooke's later Diary, kept between 1688 and 1693, was frequently and cryptically alluded to in 'H.H. tea', for the now early-middle-aged Harry Hunt regularly drank tea and conversed with his boss. Yet Hunt seems to have been both loyal and tight-lipped, for as far as the historical record goes, no hint of sexual indiscretions with the girls at home ever got into the wider world — although by the 1690s poor Grace was long in her grave in St Helen's Church, Bishopgate Street, close by Gresham College.

Though neither the 1670s nor the 1688–93 Diary records any explicit religious exercises, Richard Waller said in his 1705 'Life' of Hooke 'He always exprest a great Veneration for the eternal and immense Cause of all Beings, as may be seen in many Passages in his Writings, and seldom receiv'd any remarkable Benefit from God without thankfully acknowledging the Mercy'.[158] The Diary may not contain pointers to a deep personal faith or to revealed religion, but nonetheless it does make many acknowledgements to God's 'Omnipotent Providence'. Waller also mentions that Hooke was 'a frequent studier of the Holy Scriptures in the originals',[159] which would have meant in Latin, Greek and Hebrew which were, no doubt, languages which he had learned at Westminster School under that formidable linguist Head Master, Dr Richard Busby.

In particular, when on his 61st birthday in 1696 the Court of Chancery eventually found in Hooke's favour for the payment of his outstanding monies due under the terms of the Cutlerian Lecturership, he thanked God most heartily for this Providence, even if those thanks were prefaced by a pagan Roman prayer to the Great Being: 'D.O.M.S.H.L.G.I.S.S. *Deo Opt.[imo] Max.[imo] Summus Honor [Laus] Gloria in Secula Secularum Amen*' [To the best and most high God be the greatest honour, praise, and glory for ever and ever, Amen]. 'I was Born on this day of July 1635, and God has given me a new Birth, may I never forget his Mercies to me: whilst he gives me Breath may I praise him.'[160] It is possible that by the 1690s, following the death of Grace and in the wake of his own close brush with death in what seems to have been a dangerous illness in 1691, not to mention his becoming increasingly troubled by loneliness and infirmity, Hooke's ideas did move towards a more personal and deeper religion, but the evidences for this are very sketchy. I cannot accept the suggestion, made by some, that Robert's statement after his 1696 lawsuit victory that 'God has given me a new Birth' be read as indicating that he had become a born-again Christian. While being born again in the Holy Spirit is an ancient and enduring part of Christian experience, going back to Christ's own conversation with Nicodemus in *John* III.3–5, with Nicodemus's honest question about how an old man might re-enter his mother's womb, such religious experience, by the late seventeenth century, was seen as a part of the ecstatic Christianity of the Puritans. Indeed, Hooke would probably have found ecstatic Christianity quite uncongenial, belonging as it did to the opposite end of the ecclesiastical spectrum from the High and Broad Church Anglicanism in which he had been brought up and educated, and a highly emotional form of Christian expression which was at odds with his rational, experimental and cool-headed temperament. It is my suspicion that Hooke's reference to his 'new Birth' did no more than indicate the profound relief which he experienced upon winning a long and stressful lawsuit, and had nothing to do with any immediate religious experience.

Yet irrespective of Hooke's private religious beliefs, one must not forget that his standing within the Church of England must have remained unsullied, for soon after his friend John Tillotson had been consecrated Archbishop of Canterbury in 1691, he conferred a Lambeth Doctorate of Medicine degree on Hooke. Such an honour is not likely to have been bestowed upon a man whose personal and religious reputation and loyalty to the Church of England were seen as being in any doubt. When after a 21-month gap Hooke's Diary narrative resumes at the beginning of December 1692, however—a gap of maddening silence for the year 1691—one suddenly notices a fairly regular series of cryptic entries for Sundays which look like church attendance: 'M. St Helens', 'M. St Peter Poor', 'M. Westminster'.[161] As Hooke sometimes used the letter 'M' to signify the Latin word *mane*, or 'in the morning', could it be that he had started attending Sunday Morning Prayer at St Helen's—his parish church, just across the road from Gresham College, and where both Grace Hooke and Tom Gyles already lay in their graves—the nearby St Peter's, or the less handy but musically more glorious Westminster Abbey, where Henry Purcell was organist? Was the 58-year-old Hooke sensing 'Times winged Charriot hurrying near'?[162] But even so, there is an impenetrability about the nature of Robert Hooke's inner religious beliefs and, as Waller himself summed them up, 'If he was particular in some Matters, let us leave him to the searcher of Hearts.'[163]

On the other hand, in an age when religious and political beliefs were often so intimately connected, it is important to note that neither of the two most damaging and socially destructive appellations of the age were ever—as far as I am aware—applied to Hooke: 'atheist' and 'papist'. Being regarded as an atheist could lead to one being anathematized in society or, as in the case of Thomas Hobbes, viewed a species of bogeyman. Both Puritans and High Church Royalists came together in their mutual detestation of Hobbes, though his learning, wit and, one suspects, his mischievous sense of humour won him friends and admirers as diverse as the soberly Royalist Dr William Harvey (who left Hobbes £10 in his will as 'a token of his love'),[164] King Charles II, John Aubrey and Robert Hooke. But to be called a *papist* was far more dangerous, for it implied that one was in league with the Pope, the King of France, and no doubt the Antichrist, and against Protestantism, liberty and the

Church of England. Popery, after all, was seen as striking at the very heart of the political constitution. A good barometer of how a public man was regarded in this respect were the accusations made against him during those mad months in the autumn of 1678, when the notorious Titus Oates was making his revelations of a 'Popish Plot' and denouncing some of the leading men of the day for not only being closet Catholics, but also secret Jesuit agents bent upon killing the King, burning London and handing England over to the despotic Catholic regimes of Spain and France. Hooke's friend Samuel Pepys suffered brief imprisonment in the Tower simply because of his professional association with the declared Catholic Prince James, Duke of York, and the absurd accusation that he possessed a picture of the Crucifixion. But not all the accused, even if they were prominent gentlemen, got off as lightly as Pepys, for several innocent men were executed.[165]

When one goes through Hooke's Diary for the months of the Popish Plot, one finds no record of accusations or of serious fears on his own part. It is clear, however, that Hooke was all too aware of the madness that was exploding in society. In October 1678, for instance (soon after Sir Edmondsbury Godfrey, the London Magistrate who had received Oates's depositions, had been found mysteriously murdered), Hooke noted 'News that Dr Mazarine in Plot',[166] and earlier 'St Peters minister interrogated at All Hallows', and in December of that year '5 Lords found guilty of High Treason by inquest, Bell, Arun, Pow, Pet. Staff. [Bellasis, Arundell, Powis, Petre, Stafford]. Scaffolds erecting in Westminster Hall',[167] while just before Christmas he noted 'Mr Montacue impeached'.[168] But the nearest that Hooke got to feeling personally threatened — and whether by Catholics or Protestants he does not specify — was when he 'Received from Weeks a paper found in the custome house which seemed to threaten my life, which I burnt the next night supposing it sent by some rogue to fright me'.[169] Even so, in early December he purchased a pair of French pistols for £3 from one Mr Davys, who was probably one of the several tradesmen of that name with whom we know he had dealings.[170]

What shines through all of this swirling madness of the autumn of 1678 is that Hooke stood in no personal danger from the Catholic witch-hunting authorities, and the reason is clear: his Protestantism and loyalty to the powers that be were beyond question. He obviously knew King Charles II well enough to be invited over to partake in conversations when he casually chanced upon His Majesty walking with his courtiers in the Park, not to mention being able to secure speedy, if financially unsuccessful, access to the Royal Person when his brother John's estate was forfeit in early March 1678. Yet very importantly, Hooke never seems to have been especially close to the King's younger brother James, Duke of York, whose conversion to Catholicism in 1672 lit the powder train which would come to detonate both the Popish Plot of 1678 and the Glorious Revolution of a decade later. No one ever called Hooke a 'Jacobite' as they did his friend Samuel Pepys, and while Hooke says little about his own political beliefs in either the main Diary covering the 1670s or the more cryptic one which ran intermittently for five years after 1688, the very things which he chose to record, and the way in which he used language, make it clear where he stood. On the evening of Guy Fawkes' Night, 1673, for instance, he laconically recorded 'Returnd to Gar[a]ways Coffe House and heard of the burning of the Pope' (in effigy no doubt),[171] a favourite activity of loyal Londoners during their annual celebration of the foiling of the Gunpowder Treason of 1605. Hooke clearly was not disapproving of the symbolic burning. Then in the Diary which began on 1 November 1688, he referred to those Catholic churches that the unloved King James II (formerly James, Duke of York) had permitted to open, and which were now being destroyed by the Protestant mob, as 'mass houses':[172] a term, indeed, which in its Protestant zeal could have fallen from the lips of Oliver Cromwell himself!

Robert Hooke may not have been a churchgoer, or—considering his indulgence in fornication and incest—a man who conducted deep spiritual self-examinations, but his political tenor was overwhelmingly Protestant, English and libertarian. And so, one might argue, was that of poor imprisoned Samuel Pepys, though Hooke, unlike the Protestant Pepys, had no *personal* loyalty to the Catholic King James, nor any jealous enemies who were willing to commit perjury in an attempt to destroy him.

It is interesting that, after a gap of eight years, Hooke chose to recommence his Diary at the beginning of November 1688, for this was a time when the fate of the English nation seemed to hang in the balance. Catholic King James was on the verge of abdication, and his ultra-Protestant nephew and son-in-law, Willem, Prince of Orange, Stadholder of Holland, who was married to the English Princess Mary (daughter of King James, but herself an unequivocal Protestant), was bringing the Dutch fleet and an invasion force over from the Netherlands.[173] Nerves were on edge as people pondered what would happen, and one can only assume that Hooke was sufficiently stirred by the times through which he was living to recommence his daily journal. The London coffee houses were clearly buzzing with news of the rapidly shifting political currents, always, inevitably, a day or two old because of the slowness of transport. On another Guy Fawkes' Day, in November 1688, Hooke recorded: 'Dutch seen off Isle of Wight'.[174] But by the time that this stirring intelligence was mixing with the aroma of coffee and clouds of tobacco smoke in the talking clubs of Cornhill and Change Alley, Stadholder Willem, King William III—as he was soon to be crowned—had already made his decisive landing a few score miles

246

down Channel from the Isle of Wight at Brixham, in Torbay. Hooke recorded this landing two days later when the news reached London,[175] along with many of the key events in what later came to be called the 'Whig' history of Britain, such as the British fleet and key regiments in the army deserting the King and declaring for Protestant William,[176] the final abdication of King James and his flight to France, and the public acclamation of William of Orange in London. Yet long before the dramatic events of 1688, and even back to the 1670s and 1660s, Robert Hooke's 'Whig' political credentials can be gauged from the company he kept. His numerous friends in the upper echelons of the government of the City of London and in the Anglican ecclesiastical hierarchy give testimony to this fact, and in particular Sir John Laurence FRS, a former Lord Mayor of London, a prominent Whig with whom Hooke shared many a good dinner, a dish of coffee, and a conversation during the first Diary years of the 1670s. But on matters of politics and religion Hooke does not seem to have been loud or especially expressive of his views, for only science seemed to be capable of drawing declamatory passions out of him. However, one senses that the men who mattered knew that Hooke was both loyal and sound.

The historian cannot help but be struck by Hooke's lack of conventional worldly ambition. As we have already seen, he never appears to have valued the social advancement that a good marriage could have brought. On the other hand, he was very financially astute when it came to striking bargains or chasing debts, and he made a large fortune. Yet he was content just to hoard his thousands of guineas, and never used them as a more conventionally ambitious man might have done. Why, for instance, did he never choose to purchase a country estate, such as that at Alvington in the Isle of Wight, a transaction which his brother John had been so keen to encourage, in 1675 and 1676?[177] After all, Hooke was rich enough, on the strength of his architectural earnings alone, to have become the Lord of a Manor, giving him the property qualification to enter Parliament if he wished, and maybe even negotiate through his royal contacts for a knighthood. Indeed, even if he had chosen not to reside on such an estate, he could easily have generated an income from its rental, which would at least have been making his gold do some work, rather than remaining locked up in chests under his bed in Gresham College. But Hooke never wanted to do any of these things, probably because the kudos bestowed upon him by scientific research was the only kind of power that he ever desired, added to which that by the time he was 45 he already had far greater savings, and a far bigger income, than he ever knew what to do with, given his obvious priorities.[178]

Ultimately, one of the most fascinating things about Robert Hooke's life is its very narrowness of perspective. It is true that his creative genius

247

forces from us a gasp of amazement at its sheer fertility and diversity, for he was, after all, England's Leonardo. On the other hand, the very driven-ness of his scientific, technological and architectural activity left no space for anything else. The allurements of power and estate seem to have passed him by, his spiritual life was without intensity, his sexual needs were satisfied by perfunctory encounters, and even wealth, once the money had been made over to him, was simply a thing to record, lock up in a chest and ignore. Hooke lived intensely through his friendships, and while he undoubtedly possessed the ability to make and to retain long-lasting friends, each friend had the advantage, in his eyes, of being an autonomous being with his own creative agenda, and was in no way a dependant. In this way, Boyle, Wren, Aubrey and all the rest could come together to enjoy each other's company, at set times and in communal venues, such as Royal Society meetings or coffee houses, and then leave each other in peace to pursue whatever was occupying them. It was an essentially bachelor world, even if some of the figures moving within it, such as Wren, were married men; and one suspects that ordinary family life, with its long-term dependencies and emotional demands, would have driven Hooke mad. And when his publicly-expressed emotions did break through and enter historical record, they were invariably rancorous and directed against colleagues such as Henry Oldenburg or Sir Isaac Newton. On the other hand, he showed a genuine kindness and sympathy to his close friends, Hooke and Giles relatives, Isle of Wight dependants both on the Isle and in London, and to his servants, for Hooke was no snob. One surmises, however, that these were the sort of relationships which he liked best, entailing as they generally did a clear emotional distance: he preferred to help people without the need for too much personal involvement. When Grace Hooke died in 1687, there ended the only recorded love affair of his life, and one senses that it broke through the essentially intellectual orderliness of his existence, and the shock wave which it set in motion overwhelmed Hooke and led to his gradual decline.

One is constantly struck by the parallels that existed between Robert Hooke's life and that of his arch-rival, Newton, for not only had Hooke and Newton endured sickly yet inventive childhoods, and were brilliant, driven and obsessive bachelors, but both had passed through what might be called disturbing irregular love-affairs—Hooke's with Grace, and Newton's celibate passion for the young Swiss mathematician Fatio de Duillier which came to grief in 1693. The ending of both men's respective relationships marked each for life, making Hooke more withdrawn, melancholic and less creative, and pitching Newton into the bizarre 'black year' of nervous breakdown.[179] Over the last 30 years of his life, Newton was to re-invent himself as a Great Man of State—Knight, President of the Royal Society, Member of Parliament, Master of the

Mint and self-promoting Apollo of intellectual Europe; and part of that mythology required the marginalization of the work and memory of Robert Hooke among learned men.

So finally, how did Robert Hooke see himself, his life, and his career? While, as has been said above, Hooke's documentary remains are teasingly thin when it comes to his own inner life, one cannot help but be struck by that sudden shaft of self-revelation which appears in the manuscript of a lecture on '... The Penetration of two Liquors' delivered to the Royal Society in 1689, a couple of years after the death of Grace and while the gravitation controversy was still raging, where he comments on the usual career patterns of successful professional men, thus:

> 'We therefore find that the Greatest part of Learned men Respecting the Reward, Soon List... themselves in the Societys of Divines, Lawyers, or physicians, where their way to Canaan is already chalked out[?]. And if some Straglers chance to be left behind by the Caravan, they aim at diverting their private... [unclear word]... to Some parts where they think there may be somewhat [?] more than comen advantages reaped. We generally Observe therefore that of the few that Remain for Experimentall philosophy, and inquiries into the Knowledge of Nature or Art, the Greatest part have been seekers after the philosophers stone or the Perpetuall motion which every one at first thinks he has a Prospect of the prove but a... [unclear word]... sight & that he never living long enough to arrive at it. In true such Seekers doe oft discover many pleasant prospects in their way and find many curious experiments, but they are only Regarded as *in transitu* and not further sought into for other uses.'[180]

Was this the mature Hooke's distinctly disillusioned assessment of science as a career, not to mention a comment upon many of his fellow natural philosophers? Could his words about seekers after the Philosopher's Stone have been intended as a dig at Newton, whose activities in his Trinity College alchemical laboratory must have been common knowledge in the world of the *virtuosi*? In that use of Biblical analogy which was so commonplace in the discourse of the time, yet which is otherwise almost absent from Hooke's writings, did he feel regret that, unlike the ancient Jews, who had been led by God from their Egyptian bondage through the wilderness of Sinai to Canaan, the land that flowed with milk and honey, he as a young man had not followed the caravan of learned men into the Church, the law or medicine, to reap the earthly glories that could accrue thereby. So did Robert Hooke, in spite of his fame and undoubted wealth by 1689, see himself as a 'Stragler... left behind by the Caravan...' of life?

CHAPTER 14

DEATH AND HISTORICAL LEGACY

It is hard not to feel both sorrow and sympathy for Robert Hooke when looking at the last years of his life. For unlike Wren, Halley and Newton, Hooke's life did not decline into a comfortable and respected old age. Nor did it end suddenly at the height of his powers, as had been the case with Thomas Willis, whose irritating winter cough suddenly lurched into a fatal pneumonia, which carried him off to a peaceful death at the age of 57.[1] In many ways, it resembled the death of his close friend, John Aubrey, whose life was gradually worn away by a surfeit of problems, and who succumbed to a sudden death when visiting his 'beloved Oxford' in 1697.[2] Yet unlike Aubrey's, Hooke's problems were not in any way financial (though it is true, he did develop a miser's obsessive parsimony in the late 1690s), but rather the product of decades of overwork, gruelling self-medication and progressive loneliness. Death crept upon Robert Hooke at a very leisurely pace and, so his contemporaries believed, could have struck at any time over the last couple of years of his life, as the ingenious and now retired Curator of Experiments became progressively infirm.

On the other hand, Hooke's obsessive pursuit of experimental ingenuity never ceased. It is true that the sheer brilliance of his creativity was in a way diminished, as he tried to extract extra research mileage from a corpus of optical, mechanical, technological and conceptual creations which had gone as far as the resources of the age would allow, although during the 1690s he was still producing original insights in geology and the earth sciences. He attempted to find further evidence to back up his Polar Wandering theory in order to explain fossil distribution and the present-day positions of the continents, and seemed increasingly willing to countenance the possibility that the earth had already been through an ancient geological history by the time that God cleared away the Old Chaos prior to commencing the final Genesis Creation and the crowning of that Creation with the fashioning of Adam and Eve. His last Earthquake Discourse was delivered on 10 January 1700.[3] During the 1690s, and even after 1700, moreover, Hooke still presented papers to the Royal Society on a variety of topics that spanned geology, telescopes and instrumentation, along with reports of noteworthy creations by other scientists, whenever his health permitted. On 27 June 1698, for instance, according to Waller, he delivered a lecture discussing the ideas put forward in the recently-deceased Christiaan Huygens's *Cosmotheoros*, though Hooke's speed of response was remarkable, for the Latin text of Huygen's book was fresh off the presses at the time.[4] One wonders whether a manuscript version, or perhaps proof sheets, had already crossed the North Sea from Holland and found their way to the Royal Society by 1698. Huygens's book had not only discussed the possibility of life on other worlds, but even the possibility of intelligent and technologically-advanced life upon them, and while over the years Hooke had not always seen eye to eye with 'Monsieur Zulichem', he certainly believed that living creatures could exist on the planets. As much of Huygens's research had focused upon Saturn, Hooke accompanied his 1697 lecture with 'a Module of *Saturn* and his Ring'.[5] It was not, however, until 10 June 1702 that Robert Hooke delivered his last recorded utterance at a Royal Society meeting. Yet even when sunk into infirmity and within weeks of his death, on 17 December 1702, Hooke could not give up his passion for invention, and 'sets down a Memorandum about an Instrument to take the Horizontal Diameter of the Sun to a Tenth of a second Minute, but discovers not the way'.[6] Quite simply, his scientific intellect never stopped working.

The 1690s and early 1700s were a time of increasing bleakness for Hooke. Yes, it is true that his Lambeth medical doctorate and his successful Chancery lawsuit against the Cutler Trustees were high points during these years, providing him with academic recognition and a moral and financial victory respectively, but in fact the £550 payment[7] which he received from the Cutler Trustees does not seem to have affected his life-style in any way, and once won, the money appears to have been consigned to the heap of several thousand pounds of unused hard cash that he kept locked up in his Gresham College rooms. As Waller points out, however, it was the death of Grace Hooke which signalled a turning-point in Robert's life, making him more 'Melancholy, Mistrustful and Jealous'.[8] It is, indeed, likely that there was a deep strain of melancholy running through the male members of the Hooke family (though we know nothing about the women), as witnessed by the melancholic father, the unsuccessful depressive and suicidal elder brother and in Robert's case, by a temperament which inclined him to manic activity, relentless socializing, headaches, floundering 'searches for his health' and sustained coffee addiction. When he began to fail physically in the 1690s, no doubt those demons which he had struggled to keep at bay during his years of comparative vigour now began to close in upon him.

Hooke's first close brush with death seems to have been that illness in 1691 which had led Aubrey and others to despair of his life. Then in June 1696, Hooke—perhaps in tacit recognition of the fact that his truly creative days were over—signalled to the Royal Society his intention of drawing up a full account of all his instruments and inventions, 'but by reason of his increasing Weakness and general Decay, he was absolutely unable to perform it'.[9]

A year later, in July 1697, his general physical condition began to deteriorate alarmingly, for as his friend Waller recorded:

> 'he began to complain of the swelling and soreness of his Legs, and was much over-run with the Scurvy, and about the same time being taken with a giddiness he fell down Stairs and cut his Head, bruis'd his Shoulder, and hurt his Ribbs, of which he complain'd often to the last. About *September* he thought himself (as indeed all others did that saw him) that he could not last out a Month. About which time his Legs swell'd more and more, and not long after broke, and for want of due care Mortify'd a little before his Death. From this time he grew blinder and blinder, that at last he could neither see to Read nor Write.
>
> Thus he liv'd a dying Life for a considerable time, being more than a Year very infirm, and such as might be call'd Bed-rid for the greatest part, tho' indeed he seldom all the time went to Bed but kept in his Cloaths, and when over tir'd, lay down upon his Bed in them, which doubtless brought several Inconveniences upon him, so that at last his Distempers of Shortness of Breath, Swelling, partly of his Body, but mostly of his Legs, increasing, and at last Mortifying, as was observ'd after his Death by their looking very black, being emaciated to the utmost, his Strength wholly worn out, he dy'd on the third of *March* 1702/3, being 67 Years, 7 Months, and 13 Days Old.'[10]

So what did Robert Hooke die of when these symptoms are considered from a modern clinical point of view? As Waller's narrative is so rich in clinical and behavioural detail, it occurred to me that a tentative diagnosis might be possible, and from my own knowledge of medicine, picked up over the years as part and parcel of my researches as a medical historian, I thought that certain underlying syndromes could be recognized.

In consequence, I showed Waller's account of Hooke's last years and death to several experienced physicians whose admittedly speculative and cautious diagnoses agreed remarkably well on the main points.[11] Of course, as they all emphasized, any attempt to diagnose the causes of death of a person who died three centuries ago is fraught with pitfalls, for not only did seventeenth-century doctors think of the disease process in fundamentally different terms from their twenty-first-century colleagues, but a medical man of three centuries ago would leave— from our point of view—maddening gaps in his narrative, deriving from his different clinical perspective, which a modern clinician would need to fill in before attempting to diagnose with confidence. Whilst Hooke himself was a Doctor of Medicine and a brilliant experimental physiologist, and many of his Royal Society friends were physicians of eminence, their own diagnoses were predominantly qualitative: based on a meticulous record of the gross pathological changes and accidents through which the patient passed, but with no measurements, tests or quantification. The clinical recognition of body temperature, blood pressure and blood chemistry, and the techniques of retinal inspection and urine analysis, lay two centuries into the future. While Thomas Willis had written a treatise on urine (and had even recognized the sugar content of diabetic urine), the examination of body products in 1700 went little beyond what one could learn from colour, smell and taste. A modern doctor, on the other hand, would have wanted to know what Hooke's blood pressure was doing, whether there were changes in his blood chemistry, and whether his blindness could be put down to serious retinal haemorrhages as observed with an ophthalmo-scope, and to determine, by modern urine and blood tests, if his kidneys were failing.

Yet there is one very conspicuous set of symptoms which Hooke started to display after 1697: severely swollen legs on an otherwise 'skin and bone' body, becoming 'Mortify'd', or probably gangrenous, shortly before his death; shortness of breath; and giddiness and a tendency to bad falls, such as that which he suffered down the stairs in 1697.[12] Chronic cardiac failure, in which the muscular structure of the heart has become so degenerated and impaired that it can no longer pump the blood properly, can produce the above symptoms. Such a failure is not sudden, as in the case of a heart attack, but tends to develop slowly over a period of time; and the overall vascular damage that results can lead to an accumulation of fluid in body tissue—in the thoracic cavity where it impairs respiration and can lead to breathless-ness, around the middle of the body and, most notably, in the legs. In the lower legs, the skin can become drum-tight, cold to the touch, and very thin, so that a slight bruise can produce a large running ulcerous sore which even with modern methods of treatment can take months to heal, and which in the unhygienic seventeenth century probably 'wept' and irritated the patient to the end of his or her days. Waller tells us that Hooke's legs 'broke', most likely into such running ulcers.[13] As the blood circulation degenerates under weakening heart pressure, the extremities, usually the feet and toes, can become black, gangrenous or 'Mortify'd'.

Such symptoms would form part of a clinical syndrome known as atherosclerosis. This condition could then become the cause of other physiological failures within the body, such as renal failure or ischaemic heart disease. Hooke's decades of self-medication with mineral drugs, especially mercury compounds, could have done untold damage to his kidneys; and while it is true that his old-age blindness may have been due to cataracts, glaucoma or diabetic complications, we have no way of knowing for certain. Hooke's sustained partiality for caffeine and theobromine drinks could have aggravated any ischaemic heart problems and essential or malignant hypertension, which could have led to cerebral and retinal haemorrhages. Cerebral haemorrhages can cause strokes of varying intensities, and retinal haemorrhage can result in blindness. But circumstantial evidence indicates that the 'blinder and blinder' condition to which Waller drew attention may not have led to total blindness until the end, if at all, for as late as 10 June 1702 Hooke presented a paper to the Royal Society and in December of that year left a Memorandum about measuring the solar diameter.

It has also been suggested that Hooke's serious illness of 1691, which as we know from Aubrey's letter was a winter disease, may have been acute pneumonia.[14] Now if Hooke's spinal deformity which caused him to go 'awry' at 16 and become 'very crooked'[15] was, as indicated in chapter 1, a kyphosis, scoliosis or a compounded kyphoscoliosis, then it is likely that the interior of his thoracic cavity would already have been deformed and its major organs somewhat displaced. People with such internal thoracic deformities, indeed, are often more susceptible to pneumonia, and if Hooke's lungs, heart and major thoracic airways and blood vessels were not in their normal positions, then bouts of severe coughing could well have weakened him. It is not unlikely, therefore, that his 1691 illness could have set the scene for many of those syndromes which would develop, especially after 1697.

But what about Hooke's 'scurvy'? While it is true that seventeenth-century doctors sometimes used this term in a portmanteau fashion to describe all manner of skin conditions, and perhaps even leg ulcers, one wonders how far, in Hooke's case, it may even have been the product of chronic malnutrition. Since latterly he became 'nothing but Skin and Bone, with a meagre Aspect',[16] and even in middle age had had 'popping' eyes, it could be that these features of his appearance were the result of a 'temperate' man not eating enough at the best of times and in old age probably starving himself, for chronic malnutrition can cause serious skin damage.

Yet why, one might ask, should so rich a man starve himself? Well, as we have already seen in previous chapters, Hooke's lack of interest in the techniques of cooking on the one hand, and his omission of the menus for most of the dinners recorded in his Diary on the other, suggest that he was not a gourmet and was possibly not especially interested in food as such. It is true that on at least one occasion when in his prime, Hooke did dream of eating cream, and quite often mentions eating seasonal fruits with friends.[17] But from the culinary sparseness of the Diary one senses that, for Hooke, food was largely fuel. As he became less well, it was a part of life which he might well have come to ignore. It is further possible, considering his Diary concerns with purges and bowel movements, that Hooke may have suffered from long-standing digestive problems which inclined him to eat sparingly at the best of times. As his general health deteriorated, he left his hair 'very long and neglected over his fact uncut and lank', rarely went to bed, slept in his clothes and was 'scarcely affording himself Necessaries',[18] with insufficient 'foode and Rayment';[19] and it is very likely that, living a 'dying Life for a considerable time', he became chronically undernourished.[20]

All of the above circumstances, further, have a bearing on Robert Hooke's mental health, for in addition to any general underlying Hooke family melancholy, his last years were clearly a time of withdrawal and psychological retrenchment. Here we find him 'to a Crime close and reserv'd', and 'living like an Hermit or Cynick too penuriously'.[21] For he had now become a miser and obsessed with the fear that he would outlive his money and become a pauper, while apparently oblivious of the fact that he was worth many thousands of pounds. It was from this last part of his life, moreover, that people such as Sir Godfrey Copley subsequently reported that Hooke not only starved himself, but starved his servant too; and one wonders how far, by the age of 67, he was even becoming a little deranged.[22] Hooke had always been generous to his dependants, and this obsessive parsimony and fear of penury contrasted sharply with his expressed wish to amply endow the Royal Society and use his wealth to give it a sure institutional and academic future.

But in addition to a deepening physical infirmity and melancholy, Hooke's old age was lonely. Not only did he have, as far as we are aware, no family to hand (though there were certainly maternally-related Giles and probably some distant Hooke relatives, who received part of his estate), but most of his old friends were either dead or increasingly ailing. Boyle had died in 1691, Archbishop Tillotson in 1694, and Aubrey in 1697, and while he and Sir Christopher Wren (who was to live on to 1723) remained on excellent terms, they no longer saw as much of each other as they once had. One also senses that Pepys and Hooke must have drifted apart after the former President of the Royal Society fell increasingly under the intellectual spell of Newton, besides which, by 1700, Pepys's own health was failing, and he would follow Hooke to the grave in May 1703.

Those who seem to have been closest to Hooke by 1700 included Richard Waller FRS, who would be entrusted with trying to secure a fitting memorial for his friend, and would act both as biographer and editor of his *Posthumous Works* (1705). Then there was his loyal protégé,

Henry or Harry Hunt, who had come up from the country as a lad in January 1673 to act as Hooke's assistant,[23] did 'draw Grace, her picture' at Christmas 1674,[24] succeeded to the post of Curator of Experiments at the Royal Society, became the lonely old man's companion and confidant, and would finally help to prepare his corpse for burial. But it is clear that by the first decade of the eighteenth century, if not before, Harry Hunt had become a man of very considerable means in his own right, for in 1711, when the hard-up Royal Society moved from Gresham College to Crane Court in Fleet Street, he made it loans totalling £900.[25] Another companion in old age was Captain Robert Knox, the sailor, traveller, and author of a book on Ceylon, who left a manuscript account of Hooke's death which was eventually published in 1911.[26]

By the early weeks of 1703, Hooke must have become convinced that the shades of death were enfolding him, for on 25 February 1703 he drew up a will.[27] This document, however, was incomplete and unsigned, and when Waller said 'he dy'd at last without any Will and Testament that could be found', he may well have meant that, though he knew of the existence of this draft document, he knew of no complete will which would stand up in law.[28] Whoever wrote this will is not clear. The hand does not resemble Hooke's own, besides which he was probably blind by 25 February. It could have been drawn up by Harry Hunt or another friend, for the blots would suggest that it was not the work of a professional scrivener.

This will, which only recently came to light in the Public Record Office, clearly intended to divide Hooke's estate between four unnamed people, who are simply referred to as A, B, C and D. Perhaps he could not make up his mind who they should be, but very clearly, there was no specification of the Royal Society as a corporate body. Six days after Hooke's death, Thomas Kirke in London wrote to his father in Yorkshire to say that 'Dr. Hook hath left this World', but mentioning that Hooke had intended to appoint four men 'Sr. Christopher Wren, Sr. John Hoskins & one Reeve Williams & Captain Knox Execōrs and Trustees but some accident or other Hindred his Executing his Will, for it seems there was one made'.[29] Had it been Hooke's wish to leave his money to Wren, Hoskins, Williams and Knox, and in the flurry of gossip that clearly surrounded Hooke's death, beneficiary and executorship names were being confused?

As far as we can tell from Robert Knox's narrative, Robert Hooke died, perhaps alone, on 3 March,[30] in the same set of rooms in Gresham College that he had occupied since 1665. Knox tells us that when Hooke sensed that death was finally upon him, he sent his maidservant around to Knox's lodgings to summon him to the bedside. But whether Hooke slipped away untended before the two—and any other friends— could get to him is not clear, for there is no record of any last words, prayers, death-rattles or any other of those memorials of passing which seventeenth-century people recorded so assiduously. But there again, it is not uncommon for people with advanced degenerative cardiac disorders to hang on for months and then suddenly take flight 'like a thief in the night'. There was certainly no reference to a clergyman being summoned, the Eucharist being administered, or the anointing of the dying man with oils. After Knox had been summoned by the 'Girle', his narrative switches to the laying out of the body for burial. This appears to have been done by Knox and Harry Hunt soon after the death. Robert Hooke does not seem to have been ritually washed or laid in a winding sheet. Instead, he was buried in his ordinary 'Cloaths, Goune [gown] & Shooes as … as he Died', and which, according to Waller, he had slept in.[31] The reason, probably, for this peremptory dealing with the corpse was that Hooke was 'soe lowsey when he dyed that there was no comeing near him & his Cloathes or rags wrapt him like Searcloth',[32] or like a shroud. It is interesting to note, however, that whether lousy or otherwise, Hooke died and was wrapped in his academic gown, to be sent to his Maker as he had lived: a scholar.

After laying him out, Captain Knox and Hunt sealed up Hooke's Gresham College rooms—an intended legal security precaution made plain by Knox attaching his seal to the door—for beyond the corpse and its squalor, Hooke's chambers contained items of very great value.

Then someone must have been employed to coffin the body, for

> 'His Corps was decently and handsomely interr'd in the Church of *St Hellen* in *London*, all the Members of the Royal Society then in Town attending his Body to the Grave, paying the Respect due to his extraordinary Merit.'[33]

Mourning rings and other formal funerary gifts were also made.

It was not until 23 March, three weeks after Hooke's death, that a J. Exton, a Public Notary, made and eventually finalized and signed on 22 April 1704 the Inventory that itemized, valued and reckoned up the contents of his Gresham College rooms.[34] While one wonders what might have been pilfered from this Aladdin's Cave of wonders in the interim—the Inventory, for instance, says nothing about any microscopes, though we know for certain that Boyle had bequeathed his 'best microscope' to Hooke in 1691—Exton's Inventory gives us the only surviving glimpse of Hooke's domestic surroundings. Perhaps the most striking thing about these surroundings was the grandeur of the Gresham professorial apartments, for while an Oxford or Cambridge bachelor College Fellow would have had but one spacious chamber, and a bedroom and perhaps a couple of small closets leading from it, Hooke enjoyed a 'Parlour or Committee roome', a 'Library', a 'next roome to the stairs on the same Floore with the Library', a 'little Roome within [or off] the Committee roome', three large cellars, a garret, and probably some smaller rooms as well.

The rooms had once been elegantly and richly decorated, and from references to wall hangings, bolts of silk and other fine cloths, rugs and various furnishings, one sense that, when Hooke had been in his prime, these would have been chambers in which he could have entertained his gentlemanly friends with style. No-one would have seen anything to look down upon.

The 1703 Inventory also specifies a private library of around 3000 books, some 500 of which were in folio, as well as prints, pictures, silver, china coffee dishes and other luxury items. On the other hand, there were some unexpected omissions: while there was a model of a house in wood, which no doubt related to Hooke's architectural work, very few scientific instruments were mentioned. It is true that there were in the 'parlour or Committee roome... 3 Loadstones, two large Globes in frames, 3 tellescopes and some few odd Instruments in Brass', along with some scales and weights, but where was everything else? Where were the missing microscopes, Hooke's famous barometers, weather-clock—of which drawings were published—airpump, sea-bed sampling device and portable camera obscura (drawings of both of which were published by William Derham in 1726) and a whole range of horological devices? It seems strange that from the rooms of the greatest horological mechanician of the age, whose researches into pendulums, escapements and springs had been the subjects of such controversy, not a single clock is listed. Indeed, the only horological devices that Exton itemized were 'two watches valued at—£1:10:0'. And if these watches were worth only 30 shillings jointly, then they were hardly likely to have been good-quality timepieces, or mounted in precious metal cases. On the other hand, the Inventory did list 'two old pistolls', which one suspects were the guns Hooke bought from Mr Davys for £3 during the Popish Plot scare of 1678.[35]

While Hooke did own 'A picture of a Naked woman', it was lacking a frame and part of a general collection of discarded items which included old clothes, tapestry hangings and 'A paire of old harpsichords out of Repaire' that had found their way into one of the cellars. One would have liked very much to have known the subjects of the other pictures and prints in his possession—in particular, it would have been interesting to know if one of these pictures was a portrait of Hooke himself: perhaps one done by Mary Beale in 1674.[36] Whether such a portrait of Hooke ever existed, or whether it had already been spirited away by the end of March when Exton drew up his Inventory, it is impossible to say. What is clear, however, is that Richard Waller makes no mention of a portrait in his 'Life' of Hooke, nor did he have such a painting engraved to serve as the frontispiece to the otherwise sumptuously produced and illustrated

Posthumous Works of Robert Hooke, which he edited in 1705. It was, after all, the usual practice to preface such a volume with a finely engraved portrait of the subject, and the absence of this probably indicates the absence of an oil painting or other reliable likeness on which such a frontispiece could be based.

The precise value of Hooke's estate was itself a subject of conjecture and gossip, even immediately following his death. Exton's Inventory records the presence in Hooke's rooms of cash and jewellery valued at £8264:5:0, and when the value of his library and of the contents of his rooms was added to it, a 'Summa totalis Inven[tionis] of £9580:04:8' resulted.[37] To this amount there had to be added a sum of £20 due to Hooke in the form of rental from some land which he owned on the Isle of Wight, though unfortunately the Inventory fails to specify the capital freehold or leasehold value of this land. But only six days after Hooke's death, and before Exton had made his Inventory, Thomas Kirke had written to his father that he had been 'creditably Informed by Mr Lewis... one of his neighbours' that Hooke had left a total of £12 000 'all Lockt up in Chests in his House'.[38] Then in 1706, when Isle of Wight relatives began a Chancery lawsuit to lay claim to their intestate relative's estate, a figure of £18 000 was being mentioned.[39] Now whether this was simply a big figure snatched out of the air by the relatively poor claimants, or whether they knew things that do not survive in the historical record—such as Hooke's land holdings on the Isle of Wight and elsewhere—we have no idea.

According to Waller, however, it had been Hooke's intention to endow the Royal Society:

'I indeed, as well as others, have heard him declare sometimes that he had a great Project in his Head as to the disposal of the most part of his Estate for the advancement of Natural Knowledge, and to promote the Ends and Designs for which the ROYAL SOCIETY was instituted: To build an handsome Fabrick for the *Societies* use, with a Library, Respositary, Laboratory, and other Conveniences for making Experiments, and to found and endow a perpetual *Physico-Mechanick Lecture* of the Nature of what himself read. But tho' he was often solicited by his Friends to put his Designs down in Writing, and make his Will as to the disposal of his Estate to his own liking in the time of his Health; and after when himself, and all thought, his End drew near, yet he could never be prevail'd with to perfect it, still procrastinating it, till at last this great Design prov'd an airy Phantom and vanish'd into nothing. Thus he dy'd at last without any Will and Testament that could be found. It is indeed but a melancholy Reflexion, that while so many rich and great Men leave considerable Sums for founding Hospitals, and the like pious Uses, few since Sir *Thomas Gresham* should do any thing of this kind for the promoting of Learning, which no doubt would be as much

for the Good of the Nation, and Glory of God, as the other of
releiving the Poor.'[40]

In the money of 1703, even £9580 was an enormous sum to be owned by a
private gentleman, let alone the larger amounts of £12 000 and £18 000 put
forth by gossips and claimants. One should also bear in mind that in the
1660s Oxford's Sheldonian Theatre, which at the time had been one of the
most spectacular and daring architectural creations in Europe, had cost
£14 470, reckoned to be a fabulous sum for a building paid for by a
private benefaction. So the Royal Society could indeed have been
endowed and transformed by Hooke's money, had it ever received it,
for cash values were relatively stable in the late seventeenth century.

As Robert Hooke died intestate, his executors were obliged to seek
out and deal with beneficiaries, who turned out to be Isle of Wight and
possibly London relatives on the maternal Giles and Hooke sides of
Robert's family. As far as we can tell, most of his fortune went to a
cousin, Elizabeth Stevens or Stephens: 'The said Elizabeth Stephens as
alsoe Joseph Dillon, who transacted and managed for and by direction
of the said Elizabeth were sworne upon the truth of the above-mentioned
Inventory before me.'[41] Elizabeth Stevens, who was illiterate and signed
Exton's Inventory with 'her marke', and other Hooke relatives, described
by Thomas Kirke as 'a Caine chair maker of Charing Crosse and a Horse
Courser', then, 'being his nearest Relaçons have it all'.[42] One wonders if
Robert Hooke had ever met them, or even knew of their existence. So
while various papers, including the invaluable Diary, entered the
possession of Waller, then of William Derham, and of the Royal Society,
the Hooke fortune seems to have vanished into the Isle of Wight and
Charing Cross. It is likely, moreover, that any oil portrait that may have
been in Hooke's rooms on 3 March 1703 either found its way to the Isle
of Wight or was auctioned off and subsequently lost, along with his
library and other possessions.

In conclusion, one might be tempted to speculate about what could
have happened had Hooke, who was very concerned about his historical
reputation and posthumous regard, acceded to his friends' wishes and
produced and signed a proper will in favour of the Royal Society. But
what can be said for certain is that, had he done so, future generations
would have viewed him very differently. Had Hooke's fortune been
used to provide the Royal Society with the buildings, library and research
resources which he intended, and had he endowed a Robert Hooke
Memorial Professorship or Lectureship, not even Newton's *Principia*, or
whatever personal spite Newton harboured against him, could have
detracted from such munificence. Instead of coming down to us in the
distorted guise of an irascible, unkempt melancholic who was somehow
foolish enough to cross Newton, Robert Hooke could have entered the

historical canon not only as an experimental scientist of genius, but also as one of its greatest material benefactors—for had Hooke gone on to embed his scientific reputation within a magnificent public endowment, then it is hard to imagine that reputation ever being denigrated or allowed to slide into neglect. Indeed, one might have regarded him in a similar way to which future generations would regard Sir Hans Sloane or Sir Thomas Gresham.

Yet wills and endowments aside, let us not forget that Robert Hooke was one of the greatest experimental scientists of all time. While modern historical scholarship can now place that genius within a wider intellectual and social context, and enable us to develop a balanced understanding of Robert Hooke as a man, what cannot be denied is that he, more than anyone else, showed that the experimental method actually worked, and could transform mankind's understanding of nature. And this he achieved through decades of ingenious and painstaking application, and through the communication of his findings by writing, by demonstration, and by the spoken word—to the wider world, wherein they could inspire scientists, inventors, and poets.

APPENDIX

PORTRAITS OF ROBERT HOOKE

Human beings are fascinated by faces, for we learn so much about people from what they look like. And when we encounter a famous person for whom there is no known portrait, we instinctively feel a form of frustration: as though we could understand and relate to him or her better, if only we had a face!

Robert Hooke falls into this category, for there is no authentic historical record of his being painted. And as we saw in Chapter 1, the 'Hoock' portrait, which the visiting von Uffenbach reported having seen at the Royal Society in 1710, was more likely to have been that of Theodore Haak than of Robert Hooke.

The contemporary absence of a Hooke portrait, moreover, seems to be borne out by the fact that when Hooke's friend Richard Waller, Secretary of the Royal Society, prepared some of Hooke's surviving manuscript lectures — on earthquakes, gravity and light, along with his own 'Life of Hooke' — for publication in 1705, no portrait was engraved to form a frontispiece to the published *Posthumous Works*. This sumptuous folio volume, moreover, was clearly intended to be a major commemorative work, being elegantly printed with numerous high-quality art engravings on fine paper. Indeed, it is exactly the sort of volume which would have carried a portrait of its subject — no doubt wearing his Lambeth MD gown, a fine lace cravat, and high-status periwig, and looking out from an oval frame replete with scientific and allegorical devices — had such a likeness existed for the engraver to copy. But *The Posthumous Works* has no frontispiece and no portrait, nor is there any written mention of a lost or missing portrait; and one suspects the reason for this conspicuous omission was the simple non-existence of an authentic picture in 1704-5, when the work was being prepared. Nor was any portrait included in *The Philosophical Experiments and Observations of the late Eminent Dr Robert Hooke* (1726), which the Revd Dr William Derham edited from other surviving Hooke manuscripts, although once again, this was another well-produced book with engravings.

Three hundred years later, as part of the celebrations for the anniversary of Hooke's death in 2003, the Royal Society invited art students to try their hand at a portrait or realization of Robert Hooke, and the best efforts were put on display in the Royal Society's apartments at Carlton House Terrace, as part of the Society's summer soirée on 3 July 2003. Six paintings and a bust were displayed. One exquisite painting, done by Guy Heydon in the style of Vermeer and the Dutch School, showed a periwigged, frock-coated Hooke, with an anatomically accurate face in profile, standing in a fine seventeenth-century interior, his rooms in Gresham College, perhaps, with various instruments and aspects of his life around him. The other pictures, however, were in the seemingly orthodox tradition of modern art: facially unrecognizable expressions of something or other!

Also, in 2003, Professor Lisa Jardine suggested that a portrait in the collections of the Natural History Museum, London, and anecdotally attested, in the 1787 benefaction of Dr William Watson, as a likeness of the botanist John Ray, was, in fact, Robert Hooke. I have, however, been unable to find any convincing historical link between this portrait and Hooke, and find Professor Jardine's suggestion that it was removed from the Royal Society during Newton's Presidency, and passed through the guardianship of Sir Hans Sloane, who gave it to Watson, who eventually bequeathed it to the British Museum, from whence it passed to the Natural History Museum, just too speculative. Indeed, even Dr Watson believed it to be Ray — 'I, Ray' having been added to the picture at some stage — and no one seems to have produced any sound documentary evidence to connect the portrait with Robert Hooke.

Although certain newspapers had carried the story of Professor Jardine's claim that the Natural History Museum 'John Ray' portrait was Robert Hooke, it was in her book *The Curious Life of Robert Hooke* (2003), that she most clearly advanced the candidacy of the 'John Ray' portrait as an authentic likeness of Hooke, and related the above details of its supposed provenance.[1] What is more, she even suggested that there was evidence on the darkened canvas that the sitter had some kind of spinal deformity which might be congruent with the 'awry' appearance specified by Waller in his 'Life of Hooke'. I have, however, examined photographs of this 'Ray' portrait against a strong light source, thereby making the sitter's dark blue shoulder-line stand out against the neutral black and brown ground, only to find that the sitter had a perfectly *normal* shoulder-line, without any sign of deformity. I even compared the portrait's shoulder-line with an image of my own, when looked at from the same angle, and still found no significant evidence of any abnormality.

Central to Professor Jardine's claim for the 'John Ray' portrait is an entry in Hooke's Diary for 20 April 1674, to which I referred on p. 1 of this book:

> 'At Boyles. He promised eye water and to sit at Mrs Beales. At Mrs Beales. Shavd and Cut hair at Youngs.'

Professor Jardine interprets this to mean that Robert Hooke sat for Mary Beale, the well-known portrait painter, who had also immortalized Hooke's friend and mentor, Bishop John Wilkins, formerly Warden of Wadham College, Oxford. Yet with the best will in the world, I cannot construe this passage as to mean that Hooke, rather than Boyle, sat for Mary Beale, for Boyle seems to be the subject of the sentence. Nor am I happy with Professor Jardine's suggestion that Hooke's Diary reference to having a shave and haircut was part of the business of being 'tidied up [in] his appearance'[2] prior to being painted, chiming in as this interpretation does with the popular image of Hooke the scruffy mechanical, although it is true that Professor Jardine reminds us that Hooke was a man of standing.

Indeed, let us not forget that in 1674 Robert Hooke was a prominent London personage (not to mention an internationally famous scientist), had already amassed sufficient capital to consider buying land on the Isle of Wight, and moved in the highest social circles. And as we saw in Chapter 13, he liked expensive clothes,[3] showed off 'my Dancing shooes' to his friend Sir John Hoskins,[4] and paid hefty sums to 'the periwig woman',[5] no doubt to keep his periwig in good order. The Youngs, who shaved him and cut his hair on 20 April 1674, moreover, seem to have been, as will be recalled from Chapter 13, an Isle of Wight family often in the employ of the Hookes, the London branch of which performed all manner of services for Robert—including acting as mistresses, tailors, barbers, and stonemasons.

Yet let us assume, for the moment, that Hooke actually was painted by Mrs Beale on 20 April 1674. How much time can he have spent at the sitting on that day? Well, even by Hooke's hectic standards, 20 April was a busy day, the Diary entry containing, amongst other things, visits to Lord Brouncker and Boyle, the transaction of business as City Surveyor, along with the costing of building work for the Bishop of Salisbury. Hooke also performed a series of microscopical observations to verify some of Leeuwenhoek's latest discoveries, took part in discussions on wine fermentation and planetary astronomy, went to see Tomkins the quadrant maker, and made two separate visits to Garaway's Coffee House and conversed with a string of friends. And on top of all that, he had his portrait painted!

Hooke's very detailed daily Diary makes no reference to any subsequent visits to Mrs Beale to provide her with the additional sittings that would have been necessary to complete the portrait to the artistic standard of the 'John Ray' picture. Indeed, Hooke did not return to Mrs Beale's until 8 June 1675—'With Mr Boyle at Mrs Beales'—an entry which said nothing of the purpose of their visit, nor anything about any portraits. It is also hard to imagine a man as astute and as thrifty as Hooke sitting for a portrait but making no subsequent reference to his own personal judgement of the finished picture's quality, or to what it cost in cash terms, even if he were not footing the bill.

But when we remember that seventeenth-century oil-painting techniques required a picture to be slowly built up from layers of whites, pigments and varnishes, each of which needed to dry before the next one was applied, it becomes clear that to develop the skin tones and the translucency of the 'John Ray' portrait, more than one quick visit was necessary. For while certain details might suggest that 'John Ray' still awaited a few finishing touches, it is nonetheless a polished work of art and not an oil sketch. Yet where is the record of subsequent visits to Mrs Beale, as she worked on his portrait? After all, his Diary records the comings and goings of servants, visits to coffee houses and a massive amount of historically valuable ephemera, so why did he not mention any more portrait sittings, if he had in fact done them?

Quite apart from the documentary sparseness and ambiguity regarding the supposed painting of Hooke by Mary Beale, I believe that there is a fundamental discrepancy in the anatomy of the face and neck between the individual depicted in the 'John Ray' portrait, and the two detailed pen portraits of Hooke by two friends who knew him well. I have cited these descriptions at the beginning of Chapter 1, p. 2, but in the present context of Hooke's face, I repeat them here:

> 'He is but of midling stature, something crooked, pale faced, and his face but little below, but his head is lardge, his eie full and popping, and not quick; a grey eie. He haz a delicate head of haire, browne, and of an excellent moist curle. He is and ever was temperate and moderate in dyet, etc.'
>
> Aubrey, 1689

> 'As to his Person he was but despicable, being very crooked, tho' I have heard from himself, and others, that he was strait till about 16 Years of Age when he first grew awry, by frequent practicing, with a Turn-Lath ... He was always very pale and lean, and laterly nothing but Skin and Bone, with a Meagre Aspect, his Eyes grey and full, with a sharp ingenious Look whilst younger; his nose but thin, of a moderate height and length; his Mouth meanly wide, and upper lip thin; his Chin sharp, and Forehead large; his Head of a middle size. He wore his own Hair of a dark Brown colour, very long and hanging neglected over his Face uncut and lank ...'
>
> Waller, 1705

In consequence, I would like to draw attention to the following points of incongruence:

(1)   All those who knew Hooke and described him drew attention to his thinness. Yet this is not the face of a thin man.

(2)   Indeed, the neck on the portrait is thick; he is even 'bull-necked', suggesting a stocky physique. There is, moreover, what looks like

a jowl below the right cheek. But either way, this is not the face of a man who was 'lean, and laterly nothing but Skin and Bone'.

(3)   'Face but little below'. This seems to indicate that Hooke had a chin that appeared small with relation to the rest of his face — perhaps a receding chin. Yet on the 'Ray' portrait the sitter has a rounded, rather jutting, ball-chin — a chin, indeed, which the portrait painter Rita Greer has even likened to that of Mr Punch!

(4)   Waller says that Hooke's upper lip was thin. The lips on the 'Ray' picture are full and rather sensuous.

(5)   Hooke seems to have had prominent *grey* eyes. The eyes of the sitter in the 'Ray' portrait are very dark, perhaps brown, but certainly not grey. Nor are they especially 'popping', as Hooke's were said to be.

(6)   Both Aubrey and Waller speak of Hooke as having brown hair, and since Waller is probably describing him in the last, sad, decade of his life, it is likely that he kept his natural hair colour well into his fifties, or even his sixties. This would suggest that, while two months short of his 39th birthday, as he was on 20 April 1674 when Mary Beale is said to have painted him, he would probably have had little or no greying. Yet the sitter in the 'John Ray' portrait already has what look like silver temples, and something of a 'pepper and salt' appearance in his brown or ginger hair. It is my supposition, there-fore, that if this man lived another 20 years, to 1694, he would by then have been thoroughly grey or silver, and no longer brown. And if Richard Waller (1646–1715?) had met Hooke for the first time around 1680 (Waller was elected FRS in 1681), then I think he would have remembered Hooke not as a *dark-brown-haired*, but as a *greying* man, if the face in the 'Ray' portrait was really Robert Hooke's.

In addition to the anatomy, I believe that the social context of the 'Ray' portrait is also wrong. In 1674, when his Diary records visiting the studio of Mary Beale, Robert Hooke was at the very zenith of his powers, and moreover, as we have seen over the previous pages, he was undisputedly a gentleman, who moved confidently in the world of other high-ranking gentlemen on a daily basis. His Diary attests, as cited above, that he owned periwigs and enjoyed good clothes, as befitted his station as a friend of Bishops, Lord Mayors, Knights of the Shires, and a leading FRS. Yet the sitter in the 'Ray' portrait looks more like an upper servant, such as a loyal steward or Clerk of the Kitchen in a great house-hold, whose visage his generous noble master wished to record. The sitter's humble status is, I would suggest, indicated in three ways: by the absence of fine clothes; by his being bare-headed and not periwigged; and by his being set against a totally neutral background. There are no pillars, Grecian urns, or fine background landscapes, and no rich, plush clothes. And while some gentlemen preferred to wear their own hair, generally long and well curled if they did not go bald—portraits of John Evelyn and Thomas Hobbes (who was bald), amongst others, show them with their own hair—the 'short back and sides' haircut on the 'John Ray' portrait is unusual for the period, to say the least. And knowing what we do about Hooke's intense pride in his own achievement, and of his place in history, I would suggest that if ever he had consented to sit for Mary Beale or for any other artist, he would not have chosen to present himself to posterity in the guise of an upper servant.

I find it hard to accept, therefore, that the 'John Ray' portrait in the National History Museum is the face of Robert Hooke.

Over the years, I have asked various artists to take the Aubrey and Waller anatomical and character descriptions and attempt to build up a credible face. The first was done by Sandra Bates (née Kingsworth) in 1980, and depicts Hooke as he might have appeared as a young man in the late 1650s. This is before the fashion for French periwigs caught on after the Restoration in 1660. He is shown wearing his own hair long, as was the fashion of the day. The anatomical structure of the face fits the Aubrey and Waller descriptions.

In 2002, my wife Rachel did her own independent construction from the same two descriptions, as a pencil drawing. The basic anatomy of the face is the same, but here we have a mature Hooke, in his early fifties, as he was when John Aubrey described him in the mid to late 1680s. He is now periwigged, and wearing a fashionable cravat and a heavily-buttoned coat of the period. And on the assumption that the trauma of Grace Hooke's death still lay in the future, along with his embittered response to the Royal Society's willingness to praise Newton and neglect his own contributions to gravitation theory, he is given an acute, smiling face. This is Hooke in coffee house mode: engaged in lively discussion with his friends and displaying the 'ingenious' look for which he was so famous. True, he is no matinée idol, but by the same token, nor is he the twisted misanthrope of popular fantasy.

Then, while the main text of the book was in press, Rita Greer, a professional artist and skilled portraitist, wrote to me saying that she had finished an oil portrait of Robert Hooke. Like so many people, Rita Greer had become fascinated by Hooke, and intrigued by the lack of a likeness. After she had read some of my previously published work on Hooke, and written to me, I sent her a copy of Rachel Chapman's pencil reconstruction, along with the Aubrey and Waller descriptions. And what I received from Rita Greer was quite breathtaking. For not only had she performed a remarkable anatomical reconstruction of the face, but she had set it in a detailed overall portrait of the man and his surroundings. Rita Greer's Hooke looks thoughtful and very ingenious.

He is periwigged, cravatted, plush-coated and sat at a table with drawing instruments, an architectural plan, a *Cornu Ammonis* fossil, a spring, and a (spring balance) pocket watch. And behind him, in a deepening blue sky, are the stars and other glorious products of the 'Inflection' of light, as it passes through the air. Hooke is now painted in the mannered style — the man and his intellectual world — that was so popular in the seventeenth century.

And, of course, the facial structure of Rita Greer's Hooke follows the Aubrey and Waller accounts with great accuracy. His nose is thin, his mouth is thin-lipped yet in proportion, and his chin relatively small, sharp and a little receding. Hooke's eyes are also grey.

Having internalized the structural geometry of Hooke's face, in the way that a professional portrait painter can, Rita Greer then went on to produce several variants of her main oil painting, including a small watercolour in which Hooke is wearing an elaborate lace cravat, in the style of John Riley's portrait of Hooke's friend Samuel Pepys.

And, moreover, television's love of faces led to the commissioning of another artist's reconstruction of Hooke's face, for *Robert Hooke — Victim of Genius*, BBC 4, March 2003. This picture, however, concentrated on Hooke the emaciated elderly recluse, and I have to say that the white-haired goblin-like countenance which appeared on the screen struck me more as the realization of a minor character from *The Lord of the Rings*, than it did of either Waller's or Aubrey's verbal descriptions of their friend.

Of course, we must never lose sight of the fact that all of these artists' realizations from verbal descriptions are *not* authentic likenesses of Robert Hooke. After all, the human face is a profoundly idiosyncratic thing, and even the most acute pen portraits must omit those tiny and often elusive details of a face that make a person unique. Yet in the absence of an authentic and documented likeness of Robert Hooke suddenly presenting itself to the world of learning, maybe from a private collection (I have already searched without success at Westminster School, Christ Church, Oxford and the Royal Society), I would humbly suggest that the Sandra Bates/Kingsworth, Rachel Chapman and Rita Greer pictures will be as close as we are ever likely to get to the face of Dr Robert Hooke, MA, MD, FRS.

ENDNOTES

Chapter 1

1. Z. C. von Uffenbach, *Merkwürdige Reisen durch...* (Ulm, 1753), translated by W. H. Quarrell and M. Mare as *London in 1710* (London, 1934), 102.
2. Hooke, *Diary, 1672–80*, 20 April 1674.
3. Aubrey, *Brief Lives*, 'Robert Hooke', 165.
4. *Posthumous Works*, 'Life of ... Hooke', xxvi.
5. Rachel E. W. Chapman's drawing was based entirely on the details contained in the above two descriptions, and designed to capture the 'ingenious' expression of countenance that was attributed to him. Similarly, since the completion of the main body of the present text, Rita Greer, a professional artist and portrait painter, has done an oil portrait of Hooke, which has been reproduced on the inside jacket and as a plate in this book. See, also, Appendix, p. 1, 'Portraits of Robert Hooke'.
6. Hooke, *Diary, 1672–80*, 27 Feb. 1677.
7. Hooke, *Posthumous Works*, 'Life of ... Hooke', i–ii.
8. John Aubrey to Anthony Wood, 15 Sept. 1674, cited from the original manuscript in the Bodleian Library, Oxford, in Maurice Balme, *Two Antiquaries. A Selection from the Correspondence of John Aubrey and Anthony Wood* (Durham Academic Press, 2001), 64. Though not a primary source, in so far as it was written 37 years after his death and drew on the material contained in the published Royal Society manuscript record, the first non-contemporary biographical assessment of Hooke, one might say, appears in John Ward, *Lives of the Professors of Gresham College* (London, 1740), 'Robert Hooke', 169–93.
9. Anthony Wood, *Athenae Oxonienes*, **II** (London, 1721), 1039, for 19 July. *Posthumous Works*, 'Life of ... Hooke', ii, for 26 July.
10. Aubrey, *Brief Lives*, 'Hooke', 164.
11. Joseph Foster, *Alumni Oxonienses. The Members of the University of Oxford, 1500–1714*, **II** (Oxford, 1891), 740: 'John Hooke', of Warwickshire, 'cler. fil.', Matric. 14, March 1599–1600, aged 16; 'John Hooke', Bramshot, Southants., Armigeri (son of Esquire), 1622–3, aged 17. Also John Venn and J. A. Venn, *Alumni Cantabrigienses*, **I** (CUP, 1922), 402: John Hooke, 'Matric. sizar from Emmanuel 1602'.
12. I am indebted to Rob Martin of the Isle of Wight History Centre, hookeWEB HOMEpage, who along with his IOW colleagues has put a great deal of Hooke family data, drawn from IOW archives, on to the internet: see 'The Hooke Family Tree (including the Giles family tree)', 1–5, file :// A:\ hookeweb\ tree.htm.
13. 'Hooke Family Tree' (n. 12), 2. I am also indebted to Professor Lisa Jardine for her published research into the early-seventeenth-century Isle of Wight history: Jardine, *The Curious Life of Robert Hooke, the Man who Measured London* (Harper Collins, London, 2003), 22–6, 329–31.
14. Aubrey, *Brief Lives*, 'Hooke', 164.
15. 'Life of ... Hooke', *Posthumous Works*, ii.

16. Hideto Nakajima, 'Robert Hooke's family and his youth: some new evidence from the Will of the Rev. John Hooke', *Notes and Records* **48**, 1 (1994), 11–16: 14–15.

17. Aubrey, *Brief Lives*, 'Hooke', 164. 'Life of ... Hooke', *Posthumous Works*, xxvii, for melancholy.

18. Thomas G. Jackson, *Wadham College, Oxford* (Clarendon Press, 1893), 58.

19. Ellen Tan Drake, *Restless Genius. Robert Hooke and his Earthly Thoughts* (OUP, New York and Oxford, 1996), 217. Also, *Posthumous Works*, 'A Discourse of Earthquakes', 327.

20. Aubrey, *Brief Lives*, 164.

21. Nakajima, ' ... Will ... of Revd. John Hooke' (n. 16), 14–15. Jardine, *Curious Life of Robert Hooke* (n. 13), 330, where she cites *A Royalist's Notebook. The Commonplace Book of Sir John Oglander Kt. of Nunwell. Transcribed and edited by Francis Bamford* (Constable & Co., London, 1936).

22. *Posthumous Works*, 'Life ... of Hooke', iii.

23. Wood, *Athenae Oxonienses*, **II** (n. 9), 1039.

24. Hooke, *Diary, 1672–80*, 13 March 1679, 14 April 1679, for meetings relating to architecture. 22 March 1679: 'Dind with Dr. Busby.' Also, Hooke, 'Diary, 1688–93': many brief references to meeting, dining, etc., with Busby. Hooke also seems to have entertained members of his old school to dinner: 31 Oct. 1689, 'Westminster Scollers dind at the [Gresham] Colledge.'

25. Waller, *Posthumous Works*, 'Life of ... Hooke', iii, says Hooke went to Christ Church in 1653; Aubrey, *Brief Lives*, 165, in 1658. The Registers of Oxford University state that Hooke matriculated at Christ Church on 31 July 1658, with the rank 'cler. fil.', or son of clergy. Foster, *Alumni Oxonienses* (n. 11), 740. Christ Church's own database of old members also records his matriculation date as 1658. I am indebted to Judith Curthoys, Archivist of Christ Church, for this information.

26. Cardell Goodman, *senior* (Revd), a native of Ware, Hertfordshire, Rector of Freshwater, and John Hooke's patron and erstwhile Executor, matriculated at both Oxford (Christ Church) and Cambridge (Emmanuel) in 1625–6: Foster, *Alumni Oxonienses* (n. 11), 582; Venn, *Alumni Cantabrigienses* (n. 11), 235. Cardell Goodman, *junior*, son of the above, became a Pensioner of St. John's College, Cambridge, at the age of 13, 30 November 1666: Venn, *Alumni Cantabrigienses*, 235.

27. *Posthumous Works*, 'Life of ... Hooke', iii, for supposed Goodman servitorship. Wood, *Athenae Oxonienses*, **II** (n. 11), 118, records Samuel Fell's Freshwater Rectory, his eviction from the Christ Church Deanery, and that he 'suffered much for his loyalty' to the King and to the Church of England.

28. Christ Church old members' database (n. 25) for Thomas Newnham: he matriculated in 1656. Foster, *Alumni Oxonienses* (n. 11), 1063.

29. Aubrey, *Brief Lives*, 'Hooke', 165.

30. I am indebted to Judith Curthoys and Dr Paul Kent of Christ Church for their suggestions regarding these possible title changes during the Commonwealth period.

31. Wood, *Athenae Oxonienses*, **II** (n. 11), 1040, seems to be the only known source recording Hooke's receipt of his Oxford MA. Even Foster, *Alumni Oxonienses* (n. 11), 740, cites Wood as the sole source in the absence of any University Register.

32. Aubrey, *Brief Lives*, 'Hooke', 165.

33. Declaiming in fluent spoken Latin remains one of the exercises of Westminster Scholars today, while Oxford's high-summer Encaenia festivities still see the recipients of honorary degrees extolled in rolling Latin cadences by the Public Orator.

34. Hooke came up to Oxford after Petty had left, in the early 1650s, but Petty's initial impulse towards medical and scientific studies at Oxford was significant. See Wood, *Athenae Oxonienses*, **II** (n. 11), 807–11, and Aubrey, *Brief Lives*, 'Petty', 237–41, for early lives.

35. *Posthumous Works*, 'Life of . . . Hooke', iv. There is no mention of Hooke in Walter Pope, *The Life of Seth [Ward] Lord Bishop of Salisbury* (London, 1697), ed. J. B. Bamborough, Lutterell Society Reprints, No. 21 (Blackwell's, Oxford, 1961). Walter Pope was John Wilkins' half-brother. For a fascinating modern study of Wilkins and his influence, especially on Hooke, Wren and others, see Professor Lisa Jardine's Royal Society Wilkins Lecture, 2003: 'Dr Wilkins's Boy Wonders', *Notes and Records*, **58**, 1 (2004), 107–29.
36. *Memoirs . . . of John Evelyn Esq., F.R.S.*, ed. William Bray (London, 1871), 13 July 1654, 231–2.
37. Hooke, *Diary, 1672–80*, 29 Sept. 1672, 8 Oct. 1672, 14 Jan. 1673, 8 July 1673, and many more besides. C. S. L. Davies, 'The Family and Connections of John Wilkins, 1614–1672, *Oxoniensia*, **LXIX** (2004, forthcoming). I am indebted to Mr Davies for a pre-publication copy of his typescript.
38. Aubrey, *Brief Lives*, 'Wilkins', 320, described Wilkins as a 'lustie, strong growne, well sett, broad shoulderd person, cheerfull, and hospitable'. Wood, who did not especially like Wilkins, acknowledged the breadth of Wilkins' intellect, his toleration, and his popularity: *Athenae Oxonienses*, **II** (n. 11), 505–7. John Evelyn, *Memoirs* (n. 36), recorded, 10 February 1656, that Wilkins 'tooke great pains to preserve the Universities from ignorant sacriligious Commanders and Souldiers, who would faine have demolish'd all places and persons that pretended to learning.' A. Chapman, 'John Wilkins and Experimental Philosophy at Wadham', *Wadham College*, ed. C. S. L. Davies and Jane Garnett (Wadham College, Oxford, 1994), 24–35. See also Barbara J. Shapiro, *John Wilkins 1614–1672: an Intellectual Biography* (Berkeley and Los Angeles, 1969); Hans Aarsleff, 'John Wilkins', in *Dictionary of Scientific Biography*, 15, ed. Charles Coulton Gillespie (Scribner's, New York, 1970–80), 361–81.
39. [Thomas] *Willis's Oxford Casebook (1650–52)*, ed. Kenneth Dewhirst (Sandford, Oxford, 1981), 23–40.
40. *Posthumous Works*, 'Life of . . . Hooke', iii.

Chapter 2

1. *Posthumous Works*, 'Life of . . . Hooke', iii.
2. 'Bursar's Accounts 1648–1660', Wadham College, Oxford, archives. Pages not numbered, but in date order: Accounts, Christmas to Midsummer 1658 and to Christmas 1659.
3. Aubrey, *Brief Lives*, 'Life & Times of John Aubrey', xxxviii.
4. John Wilkins, *Discovery of a New World . . . in the Moon* (London, 1638, 1640), 'Proposition XIV'. This 'Proposition XIV' is reprinted in F. K. Pizor and T. A. Camp, *The Man in the Moon* (Sidgwick and Jackson, London, 1971), 3–40. The *Discovery* is also printed in full, including 'Proposition XIV', in *The Mathematical and Philosophical Works of the Right Reverend John Wilkins* (London, 1708). Also A. Chapman, ' "A World in the Moon": John Wilkins and his Lunar Voyage of 1640', *QJRAS*, **32** (1991), 121–32.
5. Francis Bacon, *The New Organon, or True Directions concerning the Interpretation of Nature* [*Novum Organon*], in *The Works of Francis Bacon*, ed. James Spedding, Robert L. Ellis and Douglas D. Heath, **IV** (London, 1860), 115, Aphorism CXXX.
6. Bacon, *New Organon* (n. 5), 110, Aphorism CXXIV.
7. Bacon, *New Organon* (n. 5), 114, Aphorism XXVIII. For the best and most detailed scholarly study of the intellectual origins of the scientific movement in England, and the influence of Bacon's thought upon it, see Charles Webster, *The Great Instauration: Science, Medicine, and Reform, 1626–1660* (Duckworth, London, 1975).
8. A. Wood, *Athenae Oxonienses*, **II** (London, 1721): Wood's *Fasti*, p. 163, says Boyle came to reside in Oxford 'in the time of *Oliver*, about 1657', though a surviving Boyle letter pre-dates his residence: see R. E. W. Maddison, *The Life of the Honourable Robert Boyle, F.R.S.* (Taylor and Francis, London, 1969), 90.

9. Pepys, *Diary*, **VII** (22 Jan. 1665–6), 19–20.
10. Wood, *Athenae Oxonienses*, **II** (n. 8), 1039.
11. *The Diary of John Evelyn III, Kalendarium, 1650–1672*, ed. E. S. de Beer (Clarendon Press, Oxford, 1955), 110–11.
12. Wood, *Athenae Oxonienses*, **II** (n. 8), 1039.
13. Wood, *Athenae Oxonienses*, **II** (n. 8), 1039; also Aubrey, *Brief Lives*, 'Hooke', 165.
14. Thomas Willis, *Anatomy of the Brain* (*Cerebri Anatome*, 1664), transl. Samuel Pordage, in *The Remaining Medical Works of that famous and Renowned Physician Dr Thomas Willis* (London, 1681): see Pordage's unpaginated 'Postscript'.
15. Willis, *Anatomy* (n. 14), Pordage's 'Postscript'.
16. Willis, *Anatomy* (n. 14), Willis's 'Epistle Dedicatory'. Willis reminded his readers of the sheer volume of comparative anatomical evidence that underpinned his neurological researches, having 'slain so many Victims, whole Hecatombs almost of all Animals in the Animal Court' like sacrifices upon the altar of pious scientific learning.
17. Willis, *Anatomy* (n. 14), 'Epistle Dedicatory'. Also Willis, *Two Discourses concerning the Soul of Brutes...* (*De Anima Brutorum*, 1672), transl. S. Pordage (London, 1683). See Dedication to Archbishop Sheldon and 'Preface to the Reader'.
18. Willis's brilliance and range, both as a clinician and as an experimentalist, come over in the texts published in *Thomas Willis's Oxford Lectures*, ed. Kenneth Dewhirst (Sandford, Oxford, 1980), and *Willis's Oxford Casebook (1650–52)*, ed. Kenneth Dewhirst (Sandford, Oxford, 1981). See also K. Dewhirst, *Thomas Willis as a Physician* (University of California, Los Angeles, 1964), and Robert G. Frank, Jnr., *Harvey and the Oxford Physiologists. A Study of Scientific Ideas* (Univ. of California Press, Berkeley and London, 1980), 165–92.
19. Wood, *Athenae Oxonienses*, **II** (n. 8), *Fasti*, 163. I am also indebted for information on the premises and addresses of the Oxford scientists to my research student Carol Brooke's 'Experimental Chemistry in Oxford 1648 – c. 1700. Its Techniques, Theories, and Personnel' (Part II, Chemistry Thesis, Oxford University, 1985). This thesis has probably gone further than any similar piece of research to physically 'pin down' the Oxford community, c. 1648 to 1700, as far as the locations of their laboratories is concerned.
20. Maddison, *Life... of Boyle* (n. 8), 89–132.
21. Franks, *Oxford Physiologists* (n. 18), 116. Charles Webster, 'The discovery of Boyle's Law, and the concept of elasticity of air in the seventeenth century', *Archive of the History of Exact Sciences*, **2** (1965), 441–502. Dr Webster's article is one of the most thorough studies of Boyle's pneumatic work. For his treatment of Roberval, see 448–51.
22. Robert Boyle, *New Experiments Physico-Mechanical, Touching the Spring of the Air, and its Effects (Made, for the most part, with a new Pneumatical Engine)...*, 2nd edn. (Oxford, 1662), 4. *Posthumous Works*, 'Life of... Hooke', iii, states that Greatorex's machine had been 'too gross to perform any great matter'.
23. *Posthumous Works*, 'Life of... Hooke', iv. For the ingenious Christopher Brookes, see C. S. L. Davies, 'The Mathematical Manciple', *Wadham College Gazette* (Wadham College, Oxford, January 2003), 73–4.
24. Boyle, *New Experiments* (n. 22), 8.
25. I express my appreciation to Mr Patrick Rolleston, who has also built a replica Boyle air-pump, and when I gave my 'Discourse' on Hooke to the Royal Institution in November 1994, he generously allowed me to demonstrate it to the audience. Dr Wright kindly loaned several reconstructed pieces on the same evening. See, further, A. Chapman, 'England's Leonardo: Robert Hooke and the Art of Experiment in Restoration England', *Proceedings of the Royal Institution of Great Britain*, **67**, ed. Peter Day (OUP, Royal Institution, 1996), 239–75.
26. Boyle, *New Experiments* (n. 22), 5–6.
27. Boyle, *New Experiments* (n. 22), 10–11, 40–1.
28. Birch, *History of the Royal Society*, **II** (25 Jan. 1665), 10. Boyle, *New Experiments* (n. 22), Experiments 4 and 6: 32.

29. R. Boyle, *Certain Physiological Essays...* (London, 1661), 107–10.
30. I am indebted to my student Michael Osborne for his work on the chemical character of *aurum fulminans, aurum potabile* and their relatives, though from the recipes that came down to us from the Revd John Ward (an Oxford friend of Boyle, Hooke and Willis), it is hard to be certain of the exact gold-chlorine compound—$AUCl_4$, $AUCl_6$, etc.—being produced. And because of the dangerous and highly unstable nature of this gold compound, it was deemed too risky to attempt a modern laboratory preparation. See Michael Osborne, 'The Medical Interests of the Oxford Chemists in the Late Seventeenth Century' (Oxford University Chemistry Part II Thesis, 2002), 41–8, 72–3. J. R. Partington, *A History of Chemistry*, **II** (London, 1961), 308. Also R. J. Frank, Jnr., 'John Aubrey F.R.S., John Lydall, and science in Commonwealth Oxford', *Notes and Records*, **27** (1973), 193–217; for *aurum fulminans*, see 197–8.
31. T. Willis discusses this 'explosion' theory, with relation to muscle action, spasm and convulsion, in *Pathologiae Cerebri et Nervosi Generis Specimen. In quo agitur de Morbis Convulsivis et de Scorbuto* (Oxford, 1667): see Caput I, 'De Spasmis sive Motibus Convulsivis in genere', 3, 7, 9, 14, etc. For an account of Willis and muscular explosion and *aurum fulminans*, see Hansruedi Isler, *Thomas Willis 1621–1675. Doctor and Scientist* (Hafner, New York, London, 1968), 114–22.
32. Boyle, *New Experiments* (n. 22), 48–50. When Boyle did succeed in focussing sunlight on gunpowder, however, he found that it did not combust as rapidly as in air, but that *aurum fulminans*, when dropped on to a heated plate in the airpump receiver, dissolved in a flash: see Partington, *History of Chemistry*, **II** (n. 30), 527; also Franks, *Oxford Physiologists* (n. 18), 253.
33. R. Boyle, *The Sceptical Chymist* (London, 1661; Everyman edn, Dent, London, 1964), First Part, 42–3. Boyle describes Van Helmont's experiment of heating a coal in a sealed glass and noting that it remained unburnt. Boyle then heated material in a sealed earthen vessel to red heat, and while it remained unburnt while still sealed up, it immediately combusted when the air was admitted (p. 43).
34. Hooke, *Micrographia*, 103.
35. Birch, *History of the Royal Society*, **III** (20 Feb. 1679), 465. Also *Posthumous Works*, 'Life of . . . Hooke', xxi.
36. R. Hooke, *Lampas: or Descriptions of some Mechanical Improvements of Lamps & Waterpoises. Together with other Physical and Mechanical Discoveries* (London, 1677), 4–8. Hooke had already described his work on flame in Feb. 1671, in 'An Experiment to prove the substance of a Candle or Lamp is dissolved by the Air, and the greatest part thereof reduced into a Fluid, in the Forme of Air', 14 March 1671-2, Royal Society MS RBC 3, 201–3. Birch, *History of the Royal Society*, **III** (14 March 1671-2), 19.
37. Thomas Sprat, *History of the Royal Society* (London, 1667), 232.
38. Franks, *Oxford Physiologists* (n. 18), 250.
39. 'The Continuation of the Experiments concerning Respiration...', *Phil. Trans.*, **63**, 5 (12 Sept. 1670), 2035–56: 2043. (Experiments continued from *Phil. Trans.*, **62**, 5 (8 Aug. 1670), 2011–31.) Boyle is not mentioned by name, but the researches were his.
40. The imaginative force of the airpump in particular, and the significance of the experimental phenomena which it revealed in mid-seventeenth-century intellectual society are brilliantly analysed in Steven Shapin and Simon Schaffer, *Leviathan and the Airpump: Hobbes, Boyle, and the Experimental Life* (Princeton Univ. Press, 1985).
41. From *The Ballad of Gresham College*, a broadsheet of *c.* 1663, printed with critical notes by Dorothy Stimson, *Isis*, **18** (1932), 103–17. In addition, the higher animals, such as cats, and all manner of birds, reptiles and fishes were placed into the airpump to see how they responded: 'An Account of what hapned [*sic*] to a Carp inclosed in a Vessill of Water of which the Air was pretty well exhausted', R. Hooke, 20 May 1663, Royal Society MS RBC 2, 32. At the same meeting Dr Croon experimented on a tench: RBC 2, 31. Hooke's experiments on breathing and burning, and their dual relationship to a given

volume of air, were attempted in Jan. 1663: 'Mr Hooke made the experiment of shutting up in an oblong glass a burning lamp and a chick; the lamp went out within two minutes, the chick remaining alive, and lively enough', Birch, *History of the Royal Society*, I (28 Jan. 1663), 180.

42. *Posthumous Works*, 'Life of ... Hooke', xiv. Birch, *History of the Royal Society*, II (27 June 1667), 184; (11 and 25 July 1667), 189ff.

43. Birch, *History*, II (23 March 1671), 472.

44. Birch, *History*, III (23 January 1679), 460.

45. Birch, *History*, III (6 March 1679), 468.

46. Birch, *History*, III (6 Feb. 1679), 461; (20 Feb. 1679), 465.

47. Birch, *History*, III (6 March 1679), 468–9.

48. R. Hooke, *A Description of Helioscopes, And some other Instruments* (London, 1676), 26.

49. Wood, *Athenae Oxonienses*, II (n. 8), 549.

50. Hooke, *Micrographia*, 'Preface', unpaginated, sig. *gi* (recto).

51. Hooke, *Micrographia*, 'Preface', sig. *gi* (recto).

52. Hooke, *Micrographia*, 'Preface', sig. *d* (verso).

53. Sprat, *History of the Royal Society* (n. 37), 67.

54. Hooke, *Helioscopes* (n. 48), 'Postscript', 26–30.

Chapter 3

1. Birch, *History of the Royal Society*, I (12 Nov. 1662), 124.

2. Aubrey, *Brief Lives*, 'Life of Wm Harvey', p. 128, for Harvey tutoring the Royal Princes on the battlefield of Edge Hill, 1642. Also 'Life of T. Hobbes', p. 152, for King Charles II's friendship with Hobbes.

3. Francis Bacon, *The New Organon or True Directions concerning the interpretations of Nature*, in *The Works of Francis Bacon*, IV, ed. James Spedding, Robert L. Ellis, and Douglas D. Heath (London, 1860), 'Aphorism XCIX', p. 95. Also Charles Webster, *The Great Instauration. Science, Medicine, and Reform, 1626–1660* (Duckworth, London, 1975), 336–40 for *Novum Organon*.

4. Hooke, *Micrographia*, Dedication 'To the Royal Society'.

5. Horace, *Epistles* 1.1.14. I am indebted to Professor Sir Henry Harris of Christ Church, Oxford, for kindly drawing my attention to the passage in Horace on which the Royal Society's motto was based. As the study of Horace would have been a normal part of a seventeenth-century schoolboy's education, the wider significance of 'Nullius in Verba' would have been immediately recognizable to the early Royal Society Fellows.

6. Hooke, *Diary, 1672–80*, p. 4, for a list of Fellows for 30 November. It is probably true to say that no modern scholar has done more to elucidate the origins and early Fellowship of the Royal Society than Michael Hunter, in *The Royal Society and its Fellows 1660–1700. The Morphology of an Early Scientific Institution* (British Society for the History of Science, 1982); 159–253, for a list of all Fellows elected 1660–1700. Robert Hooke (p. 184) was elected on 3 June 1663 as No. 136. See, further, Hunter, *Establishing the New Science. The Experience of the Early Royal Society*, 9 (Boydell Press, Woodbridge, 1989), 279–338, in particular deals with Hooke, especially the Cutlerian Lectureship.

7. Birch, *History of the Royal Society*, I (28 Nov. 1660), Thomas Sprat, *History of the Royal Society* (London, 1667), 52–61ff.

8. Hooke, *Micrographia*, 'Preface', sig. *g* verso.

9. Hooke to Robert Boyle, 3 July 1663: *The Correspondence of Robert Boyle*, II, 96–8.

10. Birch, *History of the Royal Society*, I (12 Nov. 1662).

11. *Posthumous Works*, 'Life of ... Hooke', x, xi. Also Birch, *History of the Royal Society*, I (2 Nov. 1664), 479, where 'Dr Wilkins related, that Sir John Cutler had declared to him, that he was firm in his resolution to settle upon Mr. Hooke 50L *per annum*, for such employment as the Royal Society should put upon him.'

12. For eventual settlement of the £30 salary, see Birch, *History of the Royal Society*, I (23 Nov. 1664), 490–6, and Birch, *History*, II (11 Jan. 1665), 4, where Hooke is confirmed in his Curatorship and his salary '30L a year *pro tempore*'.

13. Michael Cooper, 'Hooke's Career', in Jim Bennett, Michael Cooper, Michael Hunter, Lisa Jardine, *London's Leonardo. The Life and Work of Robert Hooke* (OUP, 2003), 12.

14. 'Hooke's Career' (n. 13), 20. M. Cooper, *'A more beautiful City.' Robert Hooke and the Rebuilding of London after the Great Fire* (Sutton, Stroud, Gloucestershire, 2003), also discusses Hooke's income, 33–4, 36, 39, 40, etc.

15. *Posthumous Works*, 'Life of . . . Hooke', xxv. Cooper, 'Hooke's Career' (n. 13), 20.

16. Cooper, *'A more beautiful City'* (n. 14), 37–9.

17. Hooke to Boyle, 6 Oct. 1664, in *The Correspondence of Robert Boyle*, II, 342–3.

18. John Ward, *The Lives of the Professors of Gresham College* (London, 1740), 112.

19. Lisa Jardine, *The Curious Life of Robert Hooke* (Harper Collins, London, 2003), 139, argues that evidence suggests that Hooke was allowed to remain in residence in Gresham College after the Fire, surrounded by the temporarily-accommodated City Government.

20. *Posthumous Works*, 'Life of . . . Hooke', xxvii.

21. *The Diary of John Evelyn III, Kalendarium 1650–1672*, ed. E. S. de Beer (OUP, 1955), 7 Aug. 1665: p. 416. *Posthumous Works*, 'Life of . . . Hooke', xi.

22. Pepys, *Diary*, IV (26 July 1663), 246–8. Claire Tomalin, *Samuel Pepys. The Unequalled Self* (Penguin, 2002), 188, etc., for memories of Durdans.

23. Charles Singer and E. Ashworth Underwood, *A Short History of Medicine* (Clarendon Press, Oxford, 1962), 41–7. *Peter* 1:24 'For all flesh is as grass, and all the glory of man as the flower of grass.'

24. Hooke, *Micrographia*, 'Preface', sig. *ai* (recto). See, also, Richard Nichols, *Robert Hooke and the Royal Society* (Book Guild Ltd, Lewes, 1999), 43–8, for Hooke's Curatorship. This book, which looks at most aspects of Hooke's scientific career, draws heavily on the details contained in the *Diary, 1672–80*.

25. *Micrographia*, sig *bi* (recto and verso).

26. Sprat, *History of the Royal Society* (n. 7), 'Epistle Dedicatory'. Also *Genesis* 4:22.

27. Anthony Wood, *Athenae Oxonienses*, II (London, 1721), 'Hooke', 1039. Aubrey, *Brief Lives*, 'Hooke', 157.

28. Galileo Galilei, 'Letter to the Grand Duchess Christina' (1615), in *Discoveries and Opinions of Galileo*, ed. and trans. Stillman Drake (Doubleday, New York, 1957), 175–216.

29. Robert G. Frank, Jnr., *Harvey and the Oxford Physiologists* (Berkeley, California, 1980), 133, 135.

30. Hooke discusses his 'aether' in a variety of places: e.g. *Posthumous Works*, 172, 174, 184. For a modern study, see John Henry, 'Robert Hooke, incongruous mechanist', in *Robert Hooke, New Studies*, ed. Michael Hunter and Simon Schaffer (Boydell Press, Woodbridge, 1989), 157–62.

31. *Micrographia*, 'Preface', sig. *aa*, verso. For studies of Hooke's approach to instrumentation, see Jim Bennett, 'Robert Hooke as Mechanic and Natural Philosopher', *Notes and Records*, **31**, 1 (1980), 33–48. Also Bennett, 'Hooke's Instruments', in *London's Leonardo* (n. 13), 63–104.

32. *Posthumous Works*, 'Life of . . . Hooke', xxvi.

33. R. Hooke, Lecture to the Royal Society, 4 Dec. 1689: R. Soc. MS C.I.P. XX 79.

34. *The Politics of Aristotle*, trans. Ernest Barker (OUP, 1948), ch. IV, 1277b, p. 120.

35. Hooke, *Micrographia*, 'Preface', sig. *di* (recto).

36. Hooke to Boyle, 26 Sept. 1665: *Correspondence of Robert Boyle*, II, 537–8, where Hooke reports the extinction of candles down a deep well in Surrey.

37. *Phil. Trans.*, **2** (3 June 1667), 482–4, for report of inflammable aire coming from the ground near Wigan, Lancashire.

38. Hooke had dissected a viper for the Royal Society in 1664: see Royal Society MS RBC II, 191–3, 'An Account of a Viper by Mr Hooke'. Hooke noted that while not easily visible

with a living snake, in a dead viper the teeth seemed transparent and hollow. Ibid., 194–5, 'An Account of the opening of a Viper by Mr Hook', where he examined the snake's respiratory system; also Birch, *History of the Royal Society*, I (2 Nov. 1664), 480–2. For rattlesnake, see 'An Account of some Experiments on the Effects of the Poison of the Rattle-Snake. By Captain Hall, Communicated by Sir Hans Sloane', *Phil. Trans.*, **35**, 399 (July, Aug., Sept. 1727), 309–15.

39. Richard Townley, 'A description of an instrument for dividing a foot into many thousand parts; and thereby measuring the diameters of the planets to great exactness', *Phil. Trans.*, **2** (1667), 541–4 and plate.

40. Sprat, *History of the Royal Society* (n. 7), 115, 173, 193, 277.

41. Lord Edmond Fitzmaurice, *The Life of Sir William Petty* (London, 1895), 110–11. Petty's vessel consisted of *two* separate keels and narrow wooden hulls, decked together as in a conventional ship, but with a current or 'sluice' of water flowing between them, as in a catamaran. It was 30 tons burthen. Also Birch, *History of the Royal Society*, I (12 Nov. 1662), 124; (26 Nov. 1662), 131; (10 Dec. 1662), 141; (21 Jan. 1663), 183–92; for references to Petty's ship. The last reference, 21 Jan. 1663, contains a series of very detailed reports.

42. Oliver Impey and Arthur MacGregor, *The Origins of Museums: The Cabinet of Curiosities in Sixteenth- and Seventeenth-Century Europe* (Clarendon Press, Oxford, 1985), 147–57, 158–68, 169–78, for Ashmole and the Tradescent family.

43. *Oxford in 1710 from the Travels of Zachirias Conrad von Uffenbach*, ed. W. H. Quarrell and W. J. C. Quarrell (Oxford, 1928), 31 (sub-custos).

44. Claire Tomalin, *Samuel Pepys. The Unequalled Self* (Penguin, 2003), 62–5.

45. *Micrographia*, 107–8.

46. Hooke, *Diary, 1670–80*, 9 April 1673.

47. Birch, *History of the Royal Society*, **II** (30 May 1667), 190. Curiously, the record of the Royal Society for its 30 May meeting says nothing about the Duchess's visit. On the other hand, very little business was recorded on that occasion, and one suspects that the Fellows were occupied in entertaining her. Pepys, *Diary*, 30 May 1667.

48. 'The Ballad of Gresham Colledge', *c.* 1663, ed. Dorothy Stimson, *Isis*, 1932, 101–17.

49. Samuel Butler, *A Satire on The Royal Society. A Fragment*, reprinted in *The Works of the English Poets with Prefaces Biographical and Critical*, **VII**, ed. Samuel Johnson (1779), p. 187, lines 1–4.

50. Butler, *A Satire on the Royal Society* (n. 49), p. 190, lines 87–90.

51. Butler, *The Elephant in the Moon*, in *The Works of the English Poets*, **VII** (n. 49), p. 147, lines 1–4; p. 158, lines 324–30; p. 164, lines 501–4.

52. Jonathan Swift, *Gulliver's Travels* (1726), Part III, 'A Voyage to Laputa...'. Part III, Chapter V, describes the visit to the 'Grand Academy of Lagado', which is very much a skit on the Royal Society. The Revd Stephen Hales' *Vegetable Staticks* (1727), which brought together nearly 30 years of research into plants and gases, and the attempted analyses of organic substances by distillation and other techniques, was published one year after *Gulliver*, though Hales' work was already familiar to the Royal Society, of which he was a Fellow, well before 1727.

53. Adrian Tinniswood, *His Invention So Fertile. A Life of Christopher Wren* (Jonathan Cape, London, 2001), 347.

Chapter 4

1. Robert Hooke, *Diary, 1672–80*, 25 May and 2 June 1676.

2. Thomas Shadwell, *The Virtuoso* (1676), ed. Marjorie Hope Nicholson and David Stuart Rodes (Edward Arnold Ltd, London, 1966). This modern critical edition of Shadwell's text is especially good when dealing with the scientific content of the play.

3. Henry Power, *Experimental Philosophy in Three Books* (Microscopical, Mercurial, and Magnetic) (London, 1663). The Microscopical Observations, which occupied pp. 1–83, dealt with insects—such as the Flea (Obs. 1)—and other objects that would occupy Hooke's attention in *Micrographia* two years later. One can also see how Power's experiments on capillary action might have stimulated Hooke to look at these topics. Yet *Micrographia* was not only a richly and spectacularly illustrated work, which Power's *Experimental Philosophy* was not, but it also injected a new and staggering originality of insight into the topics discussed by Power.

4. Hooke, *Micrographia*, 'Preface', sig. *gg* (verso).

5. Hooke, *Micrographia*, R. T. Gunther's 'Introduction', vii. *Micrographia* was licensed for printing by the Royal Society on 23 Nov. 1664: Birch, *History of the Royal Society*, **I** (1756), 490.

6. Pepys, *Diary*, **VI**, 2 Jan. 1665. This is when Pepys 'bespoke' the book, though he collected his copy from his booksellers on 20 Jan., and sat up reading it until 2 a.m. on the night of 21/22 Jan. 1665.

7. Pepys, *Diary*, **V**, 13 Aug. 1664.

8. Hooke, *Micrographia*, 'Preface', sig. *e* (verso), praises Richard Reeves' telescope lenses, and while Hooke does not specify the maker of his microscope lenses, it is clear from his Diary and other documents that Reeves was his favourite optician.

9. The instrument makers Richard Reeves and John Spong took to calling around to Pepys' house to show him telescopes and other instruments for purchase: Pepys, *Diary*, **VII**, 7 and 9 Aug. 1666, etc. Also, Marjorie Hope Nicolson, *Pepys' Diary and the New Science* (Univ. of Virginia, Charlottesville, 1965), 26–7, etc.

10. *Micrographia*, 'Preface', sig. *aa* (recto).

11. *Micrographia*, 'Preface', sig. *aa* (verso).

12. *Micrographia*, 'Preface', sig. *gg* (recto and verso).

13. *Micrographia*, 'Preface', sig. *gg* (verso).

14. *Micrographia*, 'Preface', sig. *dd* (verso), *e* (recto), *f* (verso), *ff* (recto), etc., for Hooke's description of his microscope. Sig. *f* (verso) describes the size of the tube as 6 or 7 inches in length, with four draw tubes to make it longer, and hence produce a more magnified (yet darker) image. His microscope worked with three lenses, an object glass, eyepiece and 'middle lens' or condenser, though he often preferred to remove this glass and work with the object and eye lenses alone: 'I find generally none more useful than that which is made with two Glasses', sig. *ff* (recto).

15. Hooke realized that microscopes, just like telescopes, gave brighter images with the least possible number of lenses, usually two. In *Micrographia*, 'Preface', sig. *f* (verso), however, he described the making of single lens microscopes using polished globules of melted 'Venice' glass, set on to the end of a stick with wax or mounted within a brass plate; though in spite of the high magnifications they could give, they were nonetheless 'troublesome' to use. Hooke had a great respect for van Leeuwenhoek and corresponded with him over several years. See Hooke, *Lectures and collections made by Robert Hooke, Secretary to the Royal Society... Microscopium. Containing Mr Leeuwenhoek's two letters concerning some late microscopical discoveries...* (London, 1678), 81–104. On p. 99 of *Microscopium*, Hooke claimed that he had made and used a single-lens microscope, with a very short-focus double convex glass 'since I publisht it in the year 1664 in the 20 page Preface to *Micrographia* [sig. *f* verso]: for though some other reasons discouraged me from prosecuting these enquiries, yet I hoped that others might long before this have carried it further'. See, also, Edward G. Ruestow, *The Microscope in the Dutch Republic. The Shaping of Discovery* (CUP, 1996), 6–36, for single-lens microscopes.

16. In spite of his interest in the geometry of light-ray tracing through his microscope optical systems, Hooke preferred to speak of his magnification in empirical, comparative terms, rather than as derived from an optical formula. He would compare the size of a microscopic image as it registered upon the retina of his eye with the fine graduations

engraved upon a precision ruler, as seen with his other, *naked*, eye, and thereby calculate the object's size in terms of tiny fractions of an inch. It is really a rough guide to the linear size of the object, rather than an optical magnification: *Micrographia*, 'Preface', sig. *f* (verso). He also spoke of the problem of illuminating an 'Object less than an hundred part of an inch distant from the Object Glass', *Micrographia*, 'Preface', sig. *ee* (verso); and, very optimistically, 'Microscopes and Telescopes, as they now are, will magnifie an Object about a thousand thousand times bigger than it appears to the naked eye', *Micrographia*, 'Preface', sig. *dd* (verso).

17. *Micrographia*, 'Preface', sig. *dd* (verso).
18. Hooke, *Diary, 1672–80*, 2 Jan. 1676. Also E. G. R. Taylor, *The Mathematical Practitioners of Tudor and Stuart England 1485–1714* (CUP, 1968), 248, Item 285; A. D. C. Simpson, 'Robert Hooke and Practical Optics: Technical Support at a Scientific Frontier', in *Robert Hooke, New Studies*, ed. Michael Hunter and Simon Schaffer (Boydell Press, Woodbridge, 1989), 59, etc., for references to Cock's or Cox's microscopes.
19. *Micrographia*, 'Preface', sig. *e* (recto).
20. *Micrographia*, 'Preface', sig. *ff* (verso).
21. *Micrographia*, Obs. I, p. 1, Scheme 2, fig. 1a.
22. *Micrographia*, Obs. IV, pp. 6–7; Obs. V, pp. 8–10, Scheme 3, fig. 1.
23. *Micrographia*, Obs. VIII, pp. 44–7, Scheme 5.
24. *Micrographia*, Obs. LVII, pp. 216–7.
25. The eyepiece micrometer provided a precise standard whereby the astronomer could measure the angular sizes of objects seen through the telescope: A. Chapman, *Dividing the Circle: The Development of Critical Angular Measurement in Astronomy, 1500–1850* (Praxis-Wiley, London, 1990, 1995), 40–5.
26. Hooke observed and illustrated fungus moulds in *Micrographia*, Obs. XX, pp. 125–31, though it was Antoni van Leeuwenhoek who did the pioneering work in this branch of microscopy a few years after Hooke. See Hooke, *Lectures and Collections... Microscopium*, 1678 (n. 15), where Hooke not only prints two of Leeuwenhoek's recent letters to the Royal Society, pp. 81–9, but adds his own observations of blood, milk globules, sputum, and other substances containing micro-organisms 'and found them much the same with what Mr Leeuwenhoek has declared', pp. 103–4. By the late 1670s, Hooke seems to have been using for these observations both single-lens and compound microscopes that were the work of 'Mr *Christopher Cock* [Cox], in *Long Acre*'. Also Ruestrow, *The Microscope and the Dutch Republic* (n. 15): micro-organisms extensively discussed, e.g. chs. 8 and 9.
27. *Micrographia*, Obs. XII, pp. 81–2.
28. *Micrographia*, Obs. XVI, pp. 101–6.
29. *Micrographia*, p. 104.
30. 'Description of a Well, and Earth in Lancashire [near Wigan] taking Fire by a Candle approached to it', *Phil. Trans.* 2, **26** (3 June 1667), 482–4. Henry Power's *Experimental Philosophy* (n. 3), 'Subterraneous Experiments', 173–81, spoke of three different types of 'Damps' found in mines: 'Common', 'Suffocating' and 'Fiery'. These included pits near Leeds.
31. Power, *Experimental Philosophy* (n. 3), 'Mercurial Experiments', 121–6, speaks of participating in barometric experiments with Richard, John and Charles Townley, and Mr George Kemp. On 27 April 1667, Power and Townley took a barometer to the top of Pendle Hill, Lancashire, and noted that its mercury level stood at 27.4 inches, whereas at the bottom it had been 28.4. Townley also communicated to Hooke an account of a sudden eruption of water from Pendle Hill on 18 Aug. 1669: *Philosophical Experiments*, 32–5. Richard Townley continued to make barometric and other experiments for decades to come, and to submit his results to the Royal Society: 'Prospect of the Weather, Winds and Height of the Mercury in the Barometer, on the first day of the Month; and of the whole Rain in every Month in the year 1703, and the beginning of 1704; Observed at

Towneley [*sic*] in Lancashire, by Ri[chard] Towneley, Esq., and at Upminster in Essex. By the Reverend Mr W. Derham F.R.S.', *Phil. Trans.* 24, **297** (March 1705), 1877–8.

32. Rudolf Virchow enunciated the biological principle *cellula ex cellula* (all cells arise from previous cells) in his *Cellular Pathologie* (1858). See Charles Singer and E. Ashworth Underwood, *A Short History of Medicine* (Oxford, 1962), 635.

33. *Micrographia*, Obs. XVIII, p. 114.

34. *Micrographia*, Obs. XVIII, p. 114. By the 1670s, when Hooke and Leeuwenhoek were examining what we now call micro-organisms, both men became occupied with trying to measure them as fractions of inches or other units. In his letter to Hooke, 5 Oct. 1677, Leeuwenhoek estimated that one drop of pepper-water contained 'more than 1,000,000 living creatures', Hooke, *Lectures and Collections... Microscopium* (n. 15), 81. Then, in another letter, 6 Jan. 1680, Leeuwenhoek calculated that 27,000,000 animal-cules made up the volume of a single sand grain: Leeuwenhoek to Hooke, 6 Jan. 1680, published in *Philosophical Experiments*, 55.

35. *Micrographia*, Obs. XVI, p. 101. In *Lectures and Collections... Microscopium* (n. 15), p. 103, Hooke calculated from observations and estimates through the microscope that 'a hair of my head being by examination found but the 640 part of an inch' across.

36. *Micrographia*, Obs. XXXVIII, 'Of the Structure and Motion of the Wings of Flies', 172–4.

37. Pepys, *Diary*, VII, 7 Aug. 1666.

38. *Micrographia*, Obs. LIX, 'Of multitudes of small Stars discoverable by the Telescope', 141–2.

39. 'Method for making a History of the Weather. By Mr Hook', in Thomas Sprat, *History of the Royal Society* (London, 1667), 173–9. In the critical edition of Sprat's *History*, edited by Jackson J. Cope and Harold W. Jones (Washington Univ. Press, St. Louis, and Routledge and Kegan Paul, London, 1959 and 1966), the archive version of Hooke's 'Method for Making a History of the Weather' is printed in Appendix C, 75–8.

40. 'Of the Invention of the Barometer in the Year 1659', in *Philosophical Experiments*, 2. Though the atmospheric researches of Wren, Boyle and Hooke into the barometer and its sensitivity to changing air pressure were original in their own right, especially in their realization of the connection between air pressure and changes in the weather, it is an exaggeration on the part of Hooke to claim that the instrument had been *invented* in 1659. For as he must have known, physicists had been experimenting with water- and mercury-filled tubes since Torricelli, Viviani, Pascal and Bertie in Italy from around 1644: see *Posthumous Works*, 'Life of... Hooke', vii–viii. Hooke would also have been familiar with the researches of Charles and Richard Townley and their friends in Lanca-shire, who in 1661 had been noting the different barometric pressures at the top and bottom of Pendle Hill, as reported in: Henry Power, *Experimental Philosophy* (n. 31), 104, 121, 126, 127. It is further likely that Hooke was the author of the article 'A Relation of some Mercurial Observations and their Results', *Phil. Trans.* 1, **9** (12 Feb. 1665), 153–9. One suspects that this was so from comments such as 'My Wheel-barometer I could never fill so exactly with Mercury, as to exclude all Air' (p. 155), while he further spoke of being 'imployed' by Boyle and others (p. 153), not to mention references to the barometer described in the 'Preface' to *Micrographia* (p. 154).

41. *Micrographia*, 224–7.

42. It was proposed by the Royal Society on 20 Feb. 1679 that Hooke should undertake baro-metric experiments at the top and bottom of the newly-finished Fire Monument, or 'Piller': Birch, *History of the Royal Society*, **III**, 463–4; also *Philosophical Experiments*, 2ff. Some 323 years later, on the morning of 1 Nov. 2002, I repeated Hooke's experiment on the Monument. It was a blustery day, with low-pressure systems coming and going across London, but when measuring the air pressure with an excellent mercury barometer of French make, of *c.* 1830, I found that the air pressure on top of the Monu-ment was about half an inch lower than at the bottom.

43. *Micrographia*, 'Preface', sig. *c* (verso); Sprat, *History of the Royal Society* (n. 39), 173–9.

44. 'A new Contrivance of Wheel-Barometer, much more easy to be prepared than that which is described in the *Micrography*; imparted by the Author of that Book [Hooke]', *Phil. Trans.* 1, **13** (June 1666), 218–9. Also 'A Description of an Invention, whereby the Divisions of the Barometer may be enlarged in any given proportions...' [Hooke], *Phil. Trans.* 16, **185** (Nov.–Dec. 1686), 241–4.
45. *Micrographia*, Obs. LVII, 'Of a new Property in the Air...', 217–40: see 220.
46. Birch, *History of the Royal Society*, III, 445, 5 Dec. 1678.
47. In a narrative subsequently published in *Philosophical Experiments*, 41, Hooke gives more details of the December 1678 Weather Clock. It was designed to have an eight-day pendulum movement that turned a cylinder, around which a strip of paper had been placed. Small hammers punched the paper to give a barometer reading every 15 minutes. It also recorded wind and rainfall. Alas, there is no evidence that it ever worked successfully.
48. *Posthumous Works*, 'Life of ... Hooke', xxii. Also *Philosophical Experiments*, 49–53.
49. R. Hooke, *De Potentia Restitutiva, or of Spring Explaining the Power of Springing Bodies* (London, 1678), 38–9 (for Edmond Halley's and other scientists' problems with damps and mists on the mountains of St. Helena and Tenerife).
50. Hooke, 'Diary, 1688–93', 26 March 1693.
51. Birch, *History of the Royal Society*, II, 187, 11 July 1667, for Hooke's 'desiring to be excused' from the further experiments on dogs; Hooke to Boyle, 10 Nov. 1664, in *The Correspondence of Robert Boyle*, II, 398–400: Hooke had previously complained of the 'torture of the creature' in these experiments.
52. Thomas Shadwell, *The Virtuoso* (1676), ed. Marjorie Hope Nicolson and David Stuart Rodes (Edward Arnold Ltd, London, 1966), p. 22, line 6.

Chapter 5

1. Michael Hoskin (ed.), *Cambridge Illustrated History of Astronomy* (CUP, 1997), Chs. 4 and 5. René Taton and Curtis Wilson, *Planetary Astronomy from the Renaissance to the Rise of Astrophysics, Part 2A: Tycho Brahe to Newton* (CUP, 1989), give an excellent survey of Renaissance astronomy. A. Chapman, *Gods in the Sky: Astronomy from the Ancients to the Renaissance* (Macmillan, Channel 4 Books, 2001), 249–80.
2. Galileo Galilei, *Siderius Nuncius* ('The Starry Messenger') (1610), in *Discoveries and Opinions of Galileo*, trans. and introduced by Stillman Drake (Doubleday, Anchor Books, 1957), 1–58. The Milky Way had fascinated scholars long before the telescope, and many had considered it to be a part of the stellar sphere made up of smaller stars: Stanley L. Jaki, 'The Milky Way before Galileo', *JHA*, **2/3**, 5 (Oct. 1971), 161–7.
3. '"A World in the Moon": John Wilkins and his Lunar Voyage of 1640', *QJRAS*, **32** (1991), 121–32, for a discussion of Wilkins's ideas and full bibliographical citation of relevant texts. Christiaan Huygens, *Cosmotheoros* (1695), translated into English as *The Celestial Worlds Discover'd, of Conjectures concerning the Inhabitants, Plants, and Productions of the World in the Planets* (London, 1695). Also A. Chapman, 'Christiaan Huygens (1629–1695), Astronomer and Mechanician', *Endeavour* N.S. **19**, 4 (1995), 140–5.
4. Edward Grant, *Planets, Stars, and Orbs. The Medieval Cosmos 1200–1687* (CUP, 1994), 176. This book is one of the best accessible scholarly treatments of medieval cosmology.
5. John Wilkins, *Discovery of a New World ... in the Moon* (1638 and 1640), reprinted in *The Mathematical and Philosophical Works of the Rt. Revd. John Wilkins* (London, 1708). In 'Proposition XIII', pp. 104–13, Wilkins discusses various classical and Christian ideas about the possibility of the souls of the terrestrial dead inhabiting other worlds, but on p. 113 he concludes: '... I know not any Ground upon which to build any probable Opinion ... about *Selenites* or Moon men.'

6. For the earliest telescopes, see Albert Van Helden, 'The Invention of the Telescope', *Transactions of the American Philosophical Society*, **67**, 4 (Philadelphia, 1977), 0–64. Van Helden, 'The Telescope in the Seventeenth Century', *Isis*, **65** (1974), 38–58. Professor Van Helden shows, pp. 47–8 of the latter, how particular object glass focal lengths were obtained at specific dates, as the optical technology advanced. Also H. C. King, *The History of the Telescope* (Griffin, London, 1955), 34–66. Hooke wrote his own aggressively English account of the development of the large astronomical telescope in 'Dr Hook's Answer to some particular Claims of Mons. Cassini' in *Philosophical Experiments*, 388–91; see also 257–67, 269–70, etc., for Hooke's views on telescopic invention.

7. Max Caspar, *Kepler*, trans. C. Doris Hellman (Abelard-Schuman, London and New York, 1959), 123–42, 264–90. Angus Armitage, *John Kepler* (Faber and Faber, London, 1966), 149–51, 179–88. Taton and Wilson, *Planetary Astronom*, **2A** (n. 1), 54–78.

8. A. Chapman, 'Jeremiah Horrocks, the Transit of Venus, and the "New Astronomy" in early seventeenth-century England', *QJRAS*, **31** (1990), 333–57. Also Chapman, 'Jeremy Shakerley (1626–1655?): astronomy, astrology and patronage in Civil War Lancashire', *Transactions of the Historic Society of Lancashire and Cheshire*, **135** (Liverpool, 1985–6), 1–4. Both works reprinted in Chapman, *Astronomical Instruments and their Users: Tycho Brahe to William Lassell* (Variorum Collected Studies, Aldershot, 1996).

9. 'Dr Hook's Answer . . .', in *Philosophical Experiments*, 388–91.

10. *Jeremiae Horrocci Liverpoliensis Angli ex Palinatu Lancastriae Opera Posthuma* (London, 1673). John Wallis was the principal editor of these Horrocks manuscripts for the Royal Society, and his largely biographical 'Epistola Nuncupatoria' in *Opera Posthuma* provides us with several unique insights into Horrocks' life and circumstances. In his own 'Autobiography', written when he was an old man, however, Wallis says nothing about his undergraduate contemporary Horrocks at Emmanuel College, although he lists several tutors whom both men would have known, as well as remarking upon the extra-curricular status of mathematics, both at his school and at Emmanuel; for Wallis privately 'diverted' himself with '*Astronomy and Geography* (as part of *Natural Philosophy*) and to other parts of *Mathematicks*; though at that time, they were scarce looked upon, with us, as Academic Studies': Christopher J. Scriba, 'The Autobiography of John Wallis F.R.S.', *Notes and Records*, **25**, 1 (June 1970), 17–46: see pp. 27–31.

11. Frances Willmoth, *Sir Jonas Moore and Restoration Science* (Boydell Press, Woodbridge, 1993), 24–6. Moore's early career in Lancashire and his Townley associations are discussed in Charles Webster, 'Richard Towneley (1629–1707), the Towneley Group and Seventeenth-Century Science', *Transactions of the Historic Society of Lancashire and Cheshire for 1966*, **118** (Liverpool, 1967), 51–76: see pp. 60–1.

12. 'A Spot in one of the Belts of Jupiter' [Hooke], *Phil. Trans.* 1, **1** (1665), 3.

13. 'Some Observations Lately made at London concerning the Planet Jupiter' [Hooke], *Phil. Trans.* 1, **14** (July 1666), 243–5. Also Hooke, *Cometa, or Remarks about Comets* (London, 1678), Supplement 'The Period of Revolution of Jupiter upon its Axis', 78–80, for later observations of Jupiter's axial rotation by Hooke and G. D. Cassini. *Cometa* was published as an independently-paginated study in *Lectures and Collections made by Robert Hooke* (London, 1678).

14. Hooke, *Diary, 1672–80*, 22 Jan. 1673.

15. Hooke, *Diary, 1672–80*, 26 March 1673.

16. Robert Grant, *A History of Physical Astronomy from the Earliest Times to the Middle of the Nineteenth Century* (London, 1852), 268–9.

17. 'A late Observation of Saturn made by the Same' [Observer, i.e. Hooke], *Phil. Trans.* 1, **14** (July 1666), 246–7. For Peter and William Ball's observations of what they interpreted as a pair of symmetrical rings, see 'Of an Observation, not long since made in England, of Saturn', *Phil. Trans.* 1, **9** (1665), 152–3. The earliest English observations of Saturn, however, were probably those of the young Christopher Wren, when in *De Corpore* he not only mentioned using telescopes of 6, 12, 22 and 35 feet focal length in a succession of

281

Saturn observations that began in 1649, but also described his construction models of Saturn, as a way of trying to make sense of the planet's *ansae*: Albert Van Helden, 'Christopher Wren's *De Corpore Saturni*', *Notes and Records*, **23** (1968), 213–9. See also J. A. Bennett, 'Christopher Wren: Astronomy, Architecture, and the Mathematical Sciences', *JHA*, **6** (1975), 149–84: p. 57, for Wren's Saturn models and telescopes. It was Christiaan Huygens, however, who finally resolved the problem of Saturn's *ansae* when, in 1655, and observing with a new 23-foot-focus telescope, he realized that Saturn was surrounded by a ring around its equatorial plane: Huygens, *Systema Saturni* (The Hague, 1659), reprinted in Huygens's *Oeuvres Complètes*, **XV** (Amsterdam, 1925), 208–353: see p. 299 for rings. Then, in the mid 1670s, G. D. Cassini noted that there was a *division*, which later immortalized his name, in the rings: 'An Extract of Signor Cassini's Letter concerning... a remarkable observation of Saturn...': this Latin letter announcing the ring division discovery, dated Paris, 26 Aug. 1626, was published in *Phil. Trans.* 11, **128** (1676), 689–90.

18. Hooke, *Diary, 1672–80*, 10 April 1673.

19. Hooke, *Diary, 1672–80*, 19 Oct. 1678. I am indebted to Kevin Kilburn of the Manchester Astronomical Society for calculating the aspects of this eclipse for me: from London, Hooke and Aubrey would have witnessed a fine total lunar eclipse.

20. R. Hooke, *Micrographia*, Obs. LX, 'Of the Moon', 242–6.

21. *Micrographia*, Obs. LIX, 'Of multitudes of small Stars discoverable by the Telescope', 242.

22. *Micrographia*, Obs. LX, 243.

23. *Micrographia*, Obs. LIX, 241.

24. *Micrographia*, 'Preface', sig *e* (verso), for Hooke's praise of Reeves. Also Obs. LIX for references to 12- and 36-foot-focus object glasses. It is likely, however, that the 30-foot telescope with which he reported observing the moon in Oct. 1664 (*Micrographia*, 242) was a misprint for that of 36 feet, for while Hooke makes no other references to working with a 30-foot glass, he greatly prized one of 36 feet focal length and used it in several separate research investigations.

25. *Micrographia*, Obs. LIX, 241–2. For further discussion of the performance of his telescopes, see 'Mr Hook's answer to Monsieur Auzout's Considerations...', *Phil. Trans.* 1, **1** (6 March 1665), 64–9.

26. Hooke, *Diary, 1672–80*, 2 Jan. 1676.

27. *Philosophical Experiments*, 390. Several seventeenth-century object glasses survive in various museum collections, and the Royal Society owns three signed by Constantijn Huygens. They have focal lengths of 122, 170 and 210 feet respectively, with diameters of around 8 or 9 inches. On examination, I found the glass remarkably clear, but for a detailed analysis of these lenses, see A. A. Mills and M. L. Jones, 'Three Lenses by Contantine Huygens in the Possession of the Royal Society', *Annals of Science*, **46** (1989), 173–82.

28. References to this monster telescope are tantalizingly vague: 'They are making at Oxford a telescope of 80 foot length to see at once the whole Moon', Samuel Hartlib, *Ephemerides*, August 1655. See Hartlib, *Ephemerides*, Pt IV, item 29/5/46A in *Hartlib Papers Project*, ed. Judith Crawford *et al.*, on CD ROM (Ann Arbor, *c.* 1995). It is doubtful whether European optical technology would have been up to producing an 80-foot-focus lens in 1655, when 25–30 feet seem to have been the limit. Eighty feet would have been more feasible by 1660–5, for lens-making technology was developing very rapidly at this time. One would *not*, moreover, have used an 80-foot lens to see 'at once the whole Moon', as this would have demanded a fairly low magnification, whereas 80-foot or similar focus lenses were designed to give highly magnified views of very small regions of the moon and planets. Also, for big telescope gossip within the Wilkins circle, see Arthur Worsley to Boyle (late 1658 to early 1659) for Wren's 'best telescope in the world'; and Thomas Barlow to Boyle, 13 Sept. 1659, for Dr Wilkins' 'Great Telescope': both in *The Correspondence of Robert Boyle*, **I**, 302 and 370 respectively. Wilkins's 80-foot telescope is also mentioned by Bennett, 'Christopher Wren...' (n. 17), 154.

29. Hooke, 'A Spot... on Jupiter', *Phil. Trans.* 1, **1** (1665), 3. For an excellent account of the telescopes owned by the early Fellows of the Royal Society and their friends, see A. D. C. Simpson, 'Robert Hooke and Practical Optics: Technical Support at a Scientific Frontier', in *Robert Hooke: New Studies*, ed. M. Hunter and S. Schaffer (Boydell Press, Woodbridge, 1989), 33–61. Also Colin A. Ronan and Sir Harold Hartley, 'Sir Paul Neile (1613–1686)', *Notes and Records*, **15** (1960), 159–65; and Angus Armitage, 'William Ball F.R.S. (1627–1690)', ibid., 167–72. William Ball also owned a 12-foot telescope (p. 167).

30. *Micrographia*, 'Preface', sig. *e* (verso) and *ee* (recto). This really began with 'Considerations of Monsieur Auzout upon Mr Hook's New Instrument for Grinding Optick Glasses', *Phil. Trans.* 1, **1** (6 March 1665), 57–63. And Hooke's response 'Mr Hook's answer...' (1665) (n. 25), 64–9, with further rumbles thereafter. Also Auzout to Oldenburg, 2 June 1665, in *Correspondence of Henry Oldenburg II 1663–5*, ed. and trans. A. Rupert Hall and Marie Boas Hall (Univ. of Wisconsin Press, Madison, Milwaukee, London, 1966), letter 380: original French, 410–18, English translation, 419–27.

31. Hooke, *Diary, 1672–80*, 3 Aug. 1672.

32. Hooke, 'Some Observations... Jupiter', *Phil. Trans.*, **1666** (n. 13), 245[misprinted as 145]– 246. I am indebted to Tony Morris of the Mexborough and Swinton Astronomical Society, Yorkshire, for kindly computing the angular size of Jupiter in June 1666, and for then calculating the prime focus image size, and hence the magnification of Hooke's 60-foot telescope. For more detail on Hooke's telescopic astronomy, see A. Chapman, 'Robert Hooke's telescopic observations of solar system bodies', in *Robert Hooke and the English Renaissance*, ed. A. Chapman and Paul Kent (based on the lectures and proceedings of the Oxford Hooke Conference, 2 Oct. 2003) (Gracewing, Leominster, forthcoming, 2004/5), ch. 7.

33. Pepys, *Diary*, **VI**, 1 March 1665. Hooke made several references to the 1664 comet, and thought that it and the comet of 1618 may have been somehow related. He suggested, for instance, that had anyone bothered to do it at the time, it would probably have been possible to follow the comet of 1618 telescopically until it was no more than a round point, as he did with that of 1664: Hooke, *Cometa* (n. 13), 44.

34. Hooke, *Cometa* (n. 13), 9–10, 32–4. Hooke repeated his views that comets were self-luminous bodies in 'A Discourse on the Nature of Comets', delivered to the Royal Society during Michaelmas 1682 when reporting his observations and conclusions upon the comet of 1680: *Posthumous Works*, 168–9.

35. Hooke, *Cometa* (n. 13), 47. Also 'Of Comets and Gravity', in *Posthumous Works*, 166ff.

36. Hooke, *Cometa* (n. 13), 4–5.

37. *Cometa*, 4–5. *Micrographia*, Obs. LVIII, 'Of a new Property in the Air', 237. The angular sizes for the cometary nucleus and coma which Hooke had measured, and published in *Phil. Trans.* and *Cometa* (see n. 36), were also discussed by John Flamsteed in a letter to Sir Jonas Moore, 2 April 1678, wherein Flamsteed says of Hooke 'he professes himselfe to be a Philosopher not an Astronomer and therefore I have no more to say to him but that Astronomy is ill handled when it must be ordered by the Whimsyes of Philosophy': *The Correspondence of John Flamsteed, The First Astronomer Royal, I, 1666–1682*, compiled and edited by Eric G. Forbes and (for Maria Forbes) by Lesley Murdin and Frances Willmoth (Institute of Physics, Bristol and Philadelphia, 1995), Letter 333, pp. 618–20. Flamsteed's belief that Hooke's ingenious researches into 'Philosophy' were but 'Whimsyes' could go some way towards explaining the difficult relationship which clearly existed between the two men. For more on the history of micrometers and the measurement of small angles, see King, *History of the Telescope* (n. 6), 97. Richard Townley, 'A description of an instrument for dividing a foot into many thousand parts, thereby measuring the diameter of planets to great exactness', *Phil. Trans.* 2, **29** (1667), 541–4. [Hooke drew the picture of the micrometer for the plate which accompanied this article.]

38. Jobst Burgi was clockmaker to Tycho's astronomical scientific friend the Landgrave of Kassel, and the Landgrave's Astronomer, Christopher Rothman. See J. L. E. Dreyer, *Tycho Brahe, A Picture of Scientific Life and Work in the Sixteenth Century* (Edinburgh, 1890, reprinted Dover, New York, 1963), 324. Also John Robert Christiansen, *On Tycho's Island. Tycho Brahe and his Assistants 1570–1601* (CUP, 2000), 106–7, etc. Though probably on hearsay evidence, William Derham, *The Artificial Clockmaker, A Treatise of Watch and Clock-work* (London, 1714), 95, claims that 'Justus Borgen' (Burgi) built a pendulum clock for the Emperor Rudolf, and that Tycho Brahe used it to make astronomical observations.

39. Christiaan Huygens' revolutionary clock mechanism was announced to the world in his 15-page illustrated treatise *Horologium* (The Hague, 1658): facsimile and English translation in *Antiquarian Horology* (Dec. 1970), 35–54. Solomon Coster made the first practicable pendulum timepieces. Frederick James Britten, *Britten's Old Clocks and Watches and their Makers*, 9th edn, ed. G. H. Baille, Courtney Ilbert and Cecil Clutton (Bloomsbury, London, 1986), 73–9.

40. William Derham, *The Artificial Clockmaker* (n. 38), 66.

41. Thomas G. Jackson, *Wadham College, Oxford: Its Foundation, Architecture, and History* (Clarendon Press, Oxford, 1893), 158.

42. C. F. E. Beeson, *Clockmaking in Oxfordshire 1400–1850*, with new introduction and index by A. V. Simcock (Museum of the History of Science, Oxford, 1962, 1967; with Simcock, 1989), 64–6.

43. H. T. Pledge, *Science Since 1500* (London, 1966), 70.

44. Derham, *Artificial Clockmaker* (n. 38), 66. A. Chapman, *Dividing the Circle. The Development of Precise Angular Measurement in Astronomy 1500–1850* (Praxis-Wiley, Chichester, 1990, 1995), 89.

45. J. Richer, 'Observationes Astronomiques et Physiques faites en L'Isle de Caienne' (1672), *Mémoires de l'Académie Royale des Sciences* depuis 1666 jusqu'à 1699, **8**, 1 (1729). For Halley at St Helena see Hooke to Newton, 6 Jan. 1680, *The Correspondence of Isaac Newton*, **II**, 310.

46. The only surviving account of Gascoigne's telescopic sights that seems to survive under his own hand is in Gascoigne's letters to William Oughtred, (1) 2 Dec. 1640, (2) undated, but on p. 58 listing an astronomical observation dated 9 Feb. 1641: published in *Correspondence of Scientific Men in the Seventeenth Century*, **I**, ed. S. P. Rigaud (Oxford, 1841), 33–59 inclusive. S. B. Gaythorpe, 'A Galilean Telescope made about 1640, by William Gascoigne, inventor of the filar micrometer', *JBAA*, **39**, 7 (June 1929), 238–41. Also Chapman, *Dividing the Circle* (n. 44), 35–40, and for telescopic sights bibliography, 174.

47. *Micrographia*, Obs. LVIII, 237.

48. R. Hooke, *Some Animadversions on the First Part of Hevelius, his 'Machina Coelestis'* (London, 1674), 8 (Hooke claimed the distance to be $287\frac{1}{3}$ feet). Also *Posthumous Works*, 'Life of . . . Hooke', xv, xix.

49. Johannes Hevelius described and illustrated his instruments and observing methods (based as they were upon a refinement of those of Tycho) in *Machina Coelestis*, **I** (Dantzig, 1673). See also Chapman, *Dividing the Circle* (n. 44), 31–3, 35–40. The telescopic sights controversy became so acrimonious that in 1679 the Royal Society sent the young Edmond Halley to Dantzig to act as arbitrator and make an assessment of the quality of Hevelius's observations. Halley was immensely impressed and, while himself a user of telescopic sights, admitted to the excellence of Hevelius's instruments, and also to the quality of his eyesight and observing technique: Alan Cook, *Edmond Halley, Charting the Heavens and the Seas* (Clarendon Press, Oxford, 1998), 89–100.

50. R. Hooke, *A Description of Helioscopes and some other Instruments* (London, 1676), 9–12: on p. 10, Hooke described experiments with roller divisions made in 1665.

51. Hooke, *Diary, 1672–80*, 23 Sept. 1679: 'At Sir Leoline Jenkins and Childs [Coffee House], discourse of murall quadrant divided by wires and rowles [rollers]'.

52. Hooke, *Animadversions* (n. 48), 55.
53. Hooke, *Animadversions* (n. 48), 54.
54. *Posthumous Works*, 'Life of ... Hooke', xi.
55. William Molyneux to John Flamsteed, 22 Dec. 1685, in *The Correspondence of John Flamsteed, The First Astronomer Royal, II, 1682–1703*, compiled by Eric G. Forbes and (for Maria Forbes) by Lesley Murdin and Frances Willmoth (IOP, Bristol and Philadelphia, 1997), Letter 550, pp. 267–9.
56. Flamsteed to Molyneux, 16 June 1688, in *Correspondence of John Flamsteed*, II (n. 55), Letter 597, 881–3, for mention of Stephen Flamsteed's sudden death: '... an happy *Euthanasia* when he wanted onely 4 dayes of being 70 yeares old compleate'. (Flamsteed, of course, used the word *euthanasia* in the Greek sense of easeful death, rather than to mean deliberate 'mercy killing' as we tend to use it today.) For Flamsteed at Burstow, see John L. Birks, *John Flamsteed. The First Astronomer Royal at Greenwich* (Avon Books, London, 1998), 107–15.
57. Zacharias Conrad von Uffenbach, *London in 1710*, edited from his *Merkwürdige Reisen durch Niedersachsen Holland und Engelland* (Ulm, 1753), by W. H. Quarrell and Margaret Mare (London, 1934), 22. The monies that must have come to Flamsteed, from his Burstow Rectory and from his father's death in 1688 (see n. 56), gave him the resources to commission his famous seven-foot mural arc from Abraham Sharp, at a cost of £120, and while offering no acknowledgement to Hooke, this instrument did incorporate Hooke's screw-edge micrometer: *The Preface to John Flamsteed's 'Historia Coelestis Britannica' or British Catalogue of the Heavens* (1725), edited and introduced by Allan Chapman, based upon a translation by Alison Dione Johnson (National Maritime Museum, Greenwich, Monograph No. 52, 1982), 120.
58. Hooke's involvement in the design of the *original*, 1676, set of instruments for Flamsteed at Greenwich is incontrovertible, especially in the design of the seven-foot-radius sextant, made by Tompion, the limb-screw micrometer and telescopic sights of which came out of Hooke's *Animadversions* (n. 48) of 1674. For Flamsteed's complaint about Hooke taking away the three-foot quadrant and other instruments, see Flamsteed, 'Foul Observations Book', R.G.O. Library MS Fl 1/188, entry 26 Sept. 1679. The Flamsteed MSS are now in Cambridge University Library, with the prefix 'R.G.O. 1'.
59. Hooke, *Animadversions* (n. 48), 68–9; also Plate 22. Hooke's clock-drive for a telescope was realized as a fully-working mechanism by Allan Mills, John Hennessy and Stephen Watson of Leicester University, and demonstrated to the Royal Society Hooke Conference in July 2003. Their paper 'Hooke's design for a driven Equatorial Mounting' was included in the volume of papers presented to the conference delegates, pp. 82–94, and should be published in 2005.
60. 'Dr Hook's Answer [to] ... Cassini', *Philosophical Experiments*, 390: Hooke's 'A 6th thing is the Application of Clock-Work to keep a Glass directed to the Object ...'
61. Flamsteed to Sir Jonas Moore, 16 July 1678, *The Correspondence of John Flamsteed*, I (n. 37), 643–5; Cuthbert on 645.
62. Derek Howse, *Francis Place and the Early History of the Greenwich Observatory* (Science History Publications, New York, 1975), 52, Plate Xb 'Quadrans Muralis Merid.: 10 Pedum Rad', and engraving of quadrant. Also D. Howse, *Greenwich Observatory*, 3, *The Buildings and Instruments* (Taylor and Francis, London, 1975), 17–18.
63. Flamsteed to Townley, 3 July 1673, *Correspondence of John Flamsteed*, I (n. 37), Letter 222, p. 356. It is surprising, considering their closely-connected ideas on the importance of instrumental accuracy, that scarcely any letters survive between Hooke and Flamsteed. Indeed, only three letters passing between the two men appear in the two massive volumes of Flamsteed's *Correspondence* covering the years 1666–1703, and the last of these was a fragment deriving from one of Flamsteed's letters to Sir Jonas Moore: *Correspondence of John Flamsteed*, I (n. 37), 13 Oct. 1674, Letter 193, p. 311, describes Flamsteed's gift of a firkin of Derby ale to Hooke (the Flamsteeds being Derby brewers).

On the other hand, Flamsteed's letters to his regular correspondents, right down to Hooke's death in 1703, often contain references—invariably adverse—to what Hooke has said, published, or done: they clearly did not get on.

64. Hooke, *Diary, 1672–80*, 26 Sept. 1679. Hooke had, according to his Diary, heard of Sir Jonas Moore's death on 27 Aug. 1679; on 2 Sept. he attended his funeral and received a mourning ring.

65. Flamsteed, 'Foul Observations Book' (n. 58), 26 Sept. 1679. Hooke, *Diary, 1672–80*, 26 Sept. 1679, itemized a similar list of instruments removed from Greenwich, viz., 'Iron Screw quadrant, little brasse screw quadrant, wooden quadrant and ring, 3 reflecting rules'. Flamsteed informed Richard Townley of the problems which Hooke's withdrawal of the instruments—'upon pretence that Sir Jonas Moores Executors might put them into the inventory of his goods'—had occasioned him in a letter, 25 Oct. 1679: *Correspondence of John Flamsteed*, **I** (n. 37), Letter 366, pp. 704–5.

66. Hooke, *Diary, 1672–80*, 26 Sept. 1679.

67. A. Chapman, *Dividing the Circle* (n. 44), 54.

68. Aristotle, *De Generatione Animalium*, Bk V, 1, trans. Arthur Platt, in *The Works of Aristotle*, V, ed. J. A. Smith and W. D. Ross (Clarendon Press, Oxford, 1965), 780a–b.

69. R. Hooke, *An Attempt to prove the Motion of the Earth from Observations* (London, 1674), 26.

70. Flamsteed's attempts to measure a parallax using his 87.5-foot-focus 'Well' telescope were clearly modelled on Hooke's 1669 observations made with the 36-foot-focus lens, as described in *An Attempt...* (n. 69). Unfortunately, Flamsteed made no zenith observations that gave him any reliable results. The Well telescope object glass by Borell still exists, and is in the collections of the Science Museum, London. It has a diameter of 9.7 inches and is 0.36 inches thick. See Howse, *Greenwich Observatory*, **3** (n. 62), 58–60. It was depicted by Francis Place in 1676: Howse, *Francis Place* (n. 62), Plate XIIa, 58–61.

71. Hooke, *Attempt to Prove* (n. 69), 26.

72. Hooke, *Attempt to Prove* (n. 69), 22–3.

73. Hooke's ideas on the nature of science and the scientific method were deeply influenced by the writings of Sir Francis Bacon, Robert Boyle and his early mentors at Oxford. Perhaps his own most complete statement on the subject is found in 'A General Scheme, or Idea of the Present State of Natural Philosophy, and how its Defects may be Remedied by a Method Proceeding in the making of Experiments and collecting Observations. Whereby To Compile a Natural History, as the Solid Basis for the Superstructure of True Philosophy [or science]', in *Posthumous Works*, 1–70.

74. James Bradley, 'A Letter to Dr Halley giving Account of a New Discovered Motion of the Fixed Stars', *Phil. Trans.*, **35** (1726), 637–61.

75. R. Hooke, *Helioscopes* (n. 50), 1–6, for technical descriptions.

76. *Helioscopes* (n. 50), 3, 6.

77. *Helioscopes* (n. 50), 4–5.

78. In *Posthumous Works*, 'Life of... Hooke', xxv–xxvi, Hooke tried to 'fold' the optical system of a 120-foot-focus lens by Huygens into a 40-foot tube by means of two reflecting planes.

79. Hooke, *Animadversions* (n. 48), 73, for early reference to Universal Joint, and *Helioscopes* (n. 50), 13, for detailed explanation. Also engraving of Universal Joint in *Helioscopes*, Figs. 10, 11.

80. Sir Isaac Newton, *Opticks* (London, 1704), Query 12–4, pp. 135–6, also speculated as to whether light rays 'falling upon the bottom of the Eye excite vibrations in the *Tunica retina*' and thereby cause 'the sense of seeing' (Qu. 12). And could not these 'vibrations' 'according to their bignesse excite sensations of several Colours'? (Qu. 13). One cannot help wondering from whence these ideas first entered Newton's cognizance: after reading *Micrographia* in 1665, perhaps?

Chapter 6

1. Charles Singer and E. A. Underwood, *A Short History of Medicine* (Clarendon Press, Oxford, 1962), 16–67. Roy Porter, *The Greatest Benefit to Mankind. A Medical History of Humanity from Antiquity to the Present* (Harper Collins, London, 1997), 44–82.

2. William Harvey, *De Motu Cordis et Sanguinis in Animalibus* ('The motion of the heart and blood in living things'), 1628, trans. K. J. Franklin (Oxford, 1957). K. D. Keele, *William Harvey* (Nelson, London, 1965).

3. Robert J. Frank, Jnr., *Harvey and the Oxford Physiologists: A Study of Scientific Ideas* (Univ. of California Press, Berkeley, Los Angeles and London, 1980), 158, etc. Kenneth Dewhurst, *Thomas Willis as a Physician* (Univ. of California, Los Angeles, 1964).

4. Willis describes his discovery in *Cerebri Anatome* (1664): see Samuel Pordage's translation, *The Anatomy of the Brain and the Description and Use of the Nerves in the Remaining Medical Works of that Famous and Renowned Physician Dr Thomas Willis* (London, 1681). Also in facsimile, ed. William Feindel (McGill Univ. Press, Montreal, 1965): see ch. VII, 81–4, 'Of the thinner Meninx or Pia Mater...' for Willis' account of the dissection of a man with a blocked right carotid artery, but whose left carotid had swollen to three times its normal size by way of compensation, 82–3, and yet who suffered no impairment of 'the free exercise of his mind and animal function'. See, further, Hansruedi Isler, *Thomas Willis 1621–1675, Doctor and Scientist* (Hafner, New York and London, 1968), 104–5.

5. Willis, *Cerebri Anatome* (n. 4), 'Preface to the Reader'.

6. *Thomas Willis's Oxford Lectures*, ed. Kenneth Dewhirst (Sandford, Oxford, 1980), make numerous references to regional functions in the cerebral cortex, being what he believed (from subsequent post-mortem evidence) to be physically connected to mental, behavioural, or intellectual characteristics displayed by the anatomical subjects when alive. On p. 141, for instance, he argued that matter found in the brains of patients post mortem had been the cause of severe headaches and, p. 147, that musicality was connected with the cerebellum.

7. Birch, *History of the Royal Society*, **I** (2 and 9 Nov. 1664), 482, 485–6.

8. 'An Account of a Dog dissected. By Mr Hook', in Thomas Sprat, *History of the Royal Society* (London, 1667), 232.

9. Birch, *History of the Royal Society*, **II** (11 July 1667), 187.

10. Hooke to Boyle, 10 Nov. 1664, *The Correspondence of Robert Boyle*, **II**, 399.

11. Pepys, *Diary*, **V** (16 May 1664), 151.

12. Birch, *History of the Royal Society*, **II** (11–18 July 1667), 187–8.

13. Birch, *History*, **II** (10 Oct. 1667), 198. Also 'An Account... of an Experiment made by Mr Hook', *Phil. Trans.* 2, **28** (1667), 539–40.

14. 'An Account of the Rise and Attempts, of a Way to conveigh Liquors immediately into the Mass of Blood', *Phil. Trans.* 1, **7** (Mon. 4 Dec. 1665), 128–30.

15. 'The Method observed in Transfusing the Bloud out of one Animal into another', *Phil. Trans.* 1, **20** (Mon. 17 Dec. 1666), 353–7: 356.

16. 'An Account...' (n. 14), *Phil. Trans.* 1, **7** (4 Dec. 1665), 128–30.

17. 'An Account...' (n. 14), 128–30.

18. Pepys, *Diary*, **VII** (16 Nov. 1666), 373.

19. 'An Account...' (n. 14), 128–30.

20. 'An Account...' (n. 14), 128–30. *Crocus* was an old chemical term for the yellow, red and other coloured powders obtained when iron, copper, antimony and other metals were oxidized: *OED*, 'Crocus', Item 3 'Old Chem.'.

21. Pepys, *Diary*, **VII** (14 Nov. 1666), 371.

22. *Phil. Trans.* 2, **27** (July 1667), 489–504: 501, for 15 June 1667 experiment. This letter from Prof. Jean Denis, describing various transfusion experiments, is rare and only got into a limited number of copies of *Phil. Trans.* 2, **27** (1667), due to its suppression by Henry

Oldenburg in the late stages of editing. A copy exists in the Royal Society library, however, where the Royal Society *Phil. Trans.* Alternative Version 1667 is shelved alongside the official Volume 2, **27** (1667), with a different pagination sequence. For more background on the suppression of Denis' letter to Oldenburg, see Geoffrey Keynes, 'The History of Blood Transfusion, 1628–1914', *The British Journal of Surgery*, **XXXI**, 38 (1943), 38–50: 42.

23. 'An Account of more Tryals of Transfusions . . .' etc., *Phil. Trans.* 2, **28** (21 Oct. 1667), 517–25. Also 'A Relation of some Trials of the Same Operation, lately made in France', *Phil. Trans.* 2, **30** (9 Dec. 1667), 559–64.

24. Pepys, *Diary*, **VIII** (21 Nov. 1667), 543. Birch, *History of the Royal Society*, **II** (21 Nov. 1667), 214–15. One suspects that the ethical issues that arose in the lamb-to-man transfusions were also influenced by *Leviticus* 17:11, 'For the life of the flesh is in the blood'. Could lamb's blood in his veins perhaps destroy Arthur Coga's very humanity, as a being in the image of God? The centrality of blood to life and humanity itself was also expressed in John Woodall, *The Surgeon's Mate* (London, 1617), p. 29, as a caution against excessive blood letting: '. . . in the bloud consisteth the life of man'. For Coga's transfusion on 23 Nov., see Birch, *History*, **II** (28 Nov. 1667), 216.

25. Pepys, *Diary*, **VIII** (30 Nov. 1667), 554.

26. H. Oldenburg to R. Boyle, 17 Dec. 1667, *The Correspondence of Henry Oldenburg*, **IV**, 1667–88, ed. A. R. Rupert Hall and Marie Boas Hall (Univ. of Wisconsin Press, Madison, Milwaukee and London, 1967), 58–60: see 59. Birch, *History of the Royal Society*, **II** (12 Dec. 1667), 225, for second Coga experiment; ibid. (19 Dec. 1667), 227, for Coga's own account of his response to transfusion.

27. 'An Account of an Experiment of *Transfusion*, practised upon a *Man* in *London*', *Phil. Trans.* 2, **30** (9 Dec. 1667), 557–9: 559. But the precise technique for conducting a transfusion was described in a letter in *Phil. Trans.* 2, **28** (21 Oct. 1667): incorrectly paginated pp. 519–623, but in sequence 522–3. Blood transfusion earned its own comic place in Thomas Shadwell's *The Virtuoso* (1676) (ed. Marjorie Hope Nicholson and David Stuart Rodes, Edward Arnold Ltd, London, 1966), Act II Sc. II, p. 5 line 215, where Sir Nicholas Gimcrack, the foolish *Virtuoso*, or scientist, was accused thus: 'you kill'd four or five that I know with your transfusion'.

28. The Anne Greene (sometimes spelt 'Ann Green') episode caused a sensation in Oxford, the revived 22-year-old woman becoming something of a celebrity and a profitable public attraction (financially managed by her father), while the experimental physicians won great kudos from their fortunate detection of her faint breathing. See Robert Plot, *The Natural History of Oxfordshire* (Oxford–London, 1705), 201–3.

29. Plot, *Oxfordshire* (n. 28), 202. The original narrative dealing with Anne Greene's story is *Newes from the Dead. Or a True and Exact Narration of the miraculous deliverance of Anne Greene, Who being Executed at Oxford Decemb. 14 1650, afterwards revived . . . Written by a Scholar of Oxford for the Satisfaction of a friend . . .*, Second Impression (Oxford, 1651). Anonymous, but the Bodleian Library copy in *Wood S16* has 'Rd [Richard] Watkins C.C.C. Ox [Corpus Christi College, Oxford]' handwritten on the title page. This pamphlet was clearly the product of detailed investigation and probably of interviews on the part of the author. He obviously admired both the ingenuity and the humanity of the doctors involved, especially as Anne's execution was widely believed to be a miscarriage of justice in the first place. Her accuser, Sir Thomas Reid, for instance, died suddenly soon after Anne's merciful deliverance. She was also asked, after her recovery, if she had experienced what we would now call a 'near death experience' following her hanging (p. 7). It was found, however, that not only had she no recollection whatsoever of what happened to her soul after hanging, but that she could not even recall the events which had occurred on the fateful morning before her execution when she had been fully conscious. The author suggested that her mind must have been so 'fixed or benumbed with fear' as to obliterate all recollection. He then suggested that such amnesias were

common 'with men that are buzz'd in the head with drink, or transported with madnesse, who, though they seem sensible enough of every present object that moves them, yet after they recover can own but little of what they did or said before' (p. 8). One can understand how Anne Greene's deliverance so captivated the Oxford scholars, replete as it was with the New Science, the psychology of memory and theology.

30. George Valentine Cox, *Recollections of Oxford* (Macmillan, London, 1870), 'Anatomy School Stories', 23. This poem is undated, but sounds as though it comes from the eighteenth and not the seventeenth century: Jack Ketch, for instance, was an eighteenth- and nineteenth-century nickname for the public executioner, after the notorious real live man of that name who beheaded the Duke of Monmouth in 1685. *Newes from the Dead* (n. 29), following the essential medical and historical narrative, prints 24 pages of poems and rhymes composed in Anne's honour by various Oxford scholars. Some are spiritual, others ironical or comical, but none has the balladeering punch of the 'slippery quean'!

31. John Wallis, 'A Relation of Thunder and Lightening, at Oxford', *Phil. Trans.* 1, **13** (1666), 222–6: 222. The death by lightning of 10 May 1666 was in addition investigated and recorded by Anthony Wood: *The Life and Times of Anthony Wood*, **II**, ed. Andrew Clark (Oxford, 1892), 77. The antiquary Wood, moreover, was well placed to get at the facts, being a next-door neighbour to Dr Willis in Merton Street, Oxford. (Willis lived at Beam Hall, Wood at what is now Postmaster's Hall, across the road from Merton College gate.) Wood also recorded the lightning storm and initial news of the death on the river in his Diary: *Life and Times of Anthony à Wood*, ed. Andrew Clark and abridged by Llewelyn Powys (OUP, 1961), 10 May 1666, 156–7.

32. A. Chapman, 'The Scholar, the Thunderbolt, and the Anatomist', *Wadham College Gazette* (Jan. 1993), 59–62.

33. 'A Brief Relation of Some Observations made, and of the best Information [that] could be had concerning the Effect of a Clapp of Thunder and Lightning' [occurred 7 June 1664 at 4 p.m. in Piccadilly], Royal Society MS RB 6, 2:169.

34. Anthony Wood, *Athenae Oxonienses*, **II**, 549, says that Willis moved from Oxford to London in 1666, as does *DNB*. From his own researches into Willis, however, Kenneth Dewhirst says that the London move was in late 1667 or 1668: Dewhirst, *Thomas Willis as a Physician* (n. 3), 13. One might, however, have expected Wood to have known, as he and Willis were neighbours in Merton Street, Oxford; though as Wood's 'Life' of Willis was probably written in the 1680s, one can understand him making a mistake regarding the date of an event which had occurred 20 years earlier and which he had not recorded at the time.

35. See n. 5.

36. Willis saw his neurological researches as inextricably bound up with his Christian faith: in *Cerebri Anatome* (n. 4), 'Epistle Dedicatory' to his friend Gilbert Sheldon, Bishop of London, he 'resolved to unlock the secret places of Man's Mind, and to look into the living breathing Chapel of the Deity', and confute the narrowness of atheism. Likewise, in *The Anatomy*'s 'Preface to the Reader', he hoped to explicate those fleshly structures through which the soul interacts with the body.

37. *Micrographia*, 'Preface', sig. *aa* recto.

38. *Micrographia*, 'Preface', sig. *aa* verso to *c* recto, discusses memory and discernment based upon accurate facts. And as if inspired by Willis' work on the brain, Hooke discusses, sig. *c* recto, the way in which the sense of smell begins with an 'effluvium' transmitted by the nerves of the nose to the brain.

39. *Posthumous Works*, 'Lectures on Light', June 1681, 125.

40. *Posthumous Works*, 'Lectures on Light', Section VII, 138–48.

41. *Posthumous Works*, 'Lectures on Light', Section VII, 141.

42. *Posthumous Works*, 146.

43. *Posthumous Works*, 'Life of . . . Hooke', xix.

44. Hooke, *Diary, 1672–80*, 4 Feb. 1675: 'Dr King read discourse of Muscles. I told them my hypothesis and observations of Round pipes in the flesh of Muscles'. Birch, *History of the Royal Society*, **III** (4 Feb. 1675), 180. Hooke's discussion of muscle structure before the Royal Society at this meeting seemed to follow on from a paper on body tissue given by Dr Edmund King, 179–80. In the *Diary, 1672–80*, 10 Feb. 1678, he also recorded 'Examind Lobster muscule for 1500 of them [fibres, no doubt] in the Length of an inch and the thicknesse of the threads not more than a 2000th part of an inch, whence there must be 4 million in a round inch'.

45. *Posthumous Works*, 'Life of... Hooke', xx. On 25 April 1678, in his paper to the Royal Society, Hooke had said of the structure of muscles: 'the globules of the fibres of the Muscles, which seemed like a necklace of pearl, might be of [the] same fabric as this of bladders [here he demonstrated his idea by means of an experiment] in which might be included a certain portion of air or other very agile matter': Birch, *History of the Royal Society*, **III** (25 April 1678), 402. Did Hooke think of muscle structure as resembling pipes through which a system of 'chain pumps' ('like a necklace of pearl') moved animal spirits? While the classical physicians had spoken of 'animal spirits' in the nervous system, one suspects that Hooke was being influenced by Willis's ideas, whose *Cerebri Anatome* (n. 4) saw a more refined, physical version of these spirits as playing a fundamental role in the working of the cerebral and nervous systems. Also Dewhirst, *Thomas Willis's Oxford Lectures* (n. 6), 153, 154.

46. Exact figures for the mortality of the 1665 Plague of London are impossible to obtain. Since 1603, each of the parishes of the City had been required to submit a 'Bill of Mortality' for *all* deaths to the City authorities, as a barometer of plague deaths. But as a reported plague death demanded the closure of the house of the victim, along with family and servants, the official 'Searchers' were sometimes bribed by families to conceal a plague death. Officially 68,576 people died of the disease, but 100,000 was probably closer to the mark: *The London Encyclopaedia*, ed. Ben Weinreb and Christopher Hibbert (Macmillan, Book Club Associates, London, 1984), article, 'Great Plague', 327.

47. Hooke to Boyle, 8 July 1665, reprinted in John Ward, *The Lives of the Professors of Gresham College* (London, 1740), 174. Also in *The Correspondence of Robert Boyle*, **II** (8 July 1665), 492–4.

48. Hooke, *Diary, 1672–80*, 17 April 1673.

49. Hooke, *Diary, 1672–80*, 16 April 1673.

50. Hooke, *Diary, 1672–80*, 22 Dec. 1672.

51. Graham Martin, 'Prince Rupert and the Surgeons', *History Today* (Dec. 1990), 38–43. Loss of blood causes a secretion of adrenalin which can, to a sinking, post-operative patient, have a beneficial, rousing effect. Such an effect may only be temporary, but if expertly performed, could prevent a patient sinking into a fatal coma. Of course, seventeenth-century medical men knew nothing about adrenalin, but as acute observers they noticed that a 'pick me up' effect often followed moderate blood-letting. I discussed this ancient procedure with a surgeon friend at St Luke's Hospital, Fargo, North Dakota, in Sept. 2002, and he described such blood-letting as 'milking the adrenals'. Of course, if the adrenals were milked too frequently then the post-operative or otherwise sick person died.

52. Martin, 'Prince Rupert' (n. 51), 42. Blood-letting remained in use for over two centuries after Prince Rupert was operated on in 1666, and for pretty much the same reasons: Robert Druitt, *The Surgeon's Vade Mecum* (London, 1878), 24–5. 'Bloodletting' was said to ease the patient, and produced 'paleness of the lips'.

53. John Woodall, *The Surgeon's Mate* (n. 24), 'Of Medicines and their uses', 40–124, gives a good list of the drugs used in seventeenth-century medicine, and how they were supposed to operate. Many were prized for their real or supposed purgative virtues.

54. Hooke, *Diary, 1672–80*: see editor's section 'Taverns and Coffee Houses Mentioned by Hooke', 463–70. There are 154 premises on the list, 65 of which are designated Coffee Houses.

55. Hooke, *Diary, 1672–80*, 25 Oct. 1673. It would be wrong, however, to assume that Hooke was averse to alcohol or did not drink it. His Diary contains many references to drinking wine and beer in moderation with food. But there is only one reference to being drunk or tipsy: 2 Oct. 1673, 'Drank brandy with Captain fudled, drank coffe with Odell'.

56. Aubrey, *Brief Lives*, 'Harvey', 132. Aubrey, who was a great haunter of coffee houses, reminded his friend Antony Wood, somewhat exaggeratedly, that, before they appeared in London and Oxford, 'men knew not how to be acquainted': Aubrey to Wood, 15 June 1680, in Maurice Balme, *Two Antiquaries. A Selection from the Correspondence of John Aubrey and Anthony Wood* (Durham Academic Press, Edinburgh, Cambridge and Durham, USA, 2001), 92. Like all seventeenth-century people, Aubrey was a remedy-collector and self-medicator. Following a surfeit of peaches, for instance, he sent for 'a good lusty Vomit': Aubrey to Wood, 2 Sept. 1694, Balme, *Two Antiquaries*, 148; Wood to Aubrey, 15 Aug. 1694, ibid., 154, contains remedies to clear the head, eyes, etc.

57. Hooke, *Diary, 1672–80*, 22 Dec. 1672.

58. Hooke, *Diary, 1672–80*, 26 Nov. 1674.

59. Hooke, *Diary, 1672–80*, 31 Jan. 1673.

60. Hooke, *Diary, 1672–80*, 23 April 1673.

61. Hooke, *Diary, 1672–80*, 3 Feb. 1674.

62. This market-place approach to all healing in the pre-nineteenth-century world is explored in detail in Roy Porter, *Quacks, Fakers and Charlatans in English Medicine* (Tempus, Stroud, Gloucestershire, 1989, 2000), esp. 31–62.

63. Hooke, *Diary, 1672–80*, 15 July 1676.

64. Hooke, *Diary, 1672–80*, 22 Dec. 1672; 16 Feb. 1673.

65. Hooke, *Diary, 1672–80*, 21 Nov. 1675. On at least one occasion, however, one surmises that a purge failed to take effect, as on 15 Feb. 1677 he recorded, 'Cheated of a shitt. Slept ill'.

66. Hooke, *Diary, 1672–80*, 19 Aug. 1675.

67. Hooke, *Diary, 1672–80*, 15 May 1676. Sweating, of course, was a valued purgative procedure in the physician's armoury. In June 1664, for instance, Hooke's Oxford friend Dr Richard Lower cured a 60-year-old man with a sweat. The patient was being driven mad with pain in his gouty feet, but Lower prescribed a 'repercussive plaister', which drove the 'gout' (a generic name for an acute inflammation) into his heart and stomach! But the man felt immediate relief, and was saved when he broke into a profuse and sustained sweat. Lower to Boyle, 8 June 1664, *The Correspondence of Robert Boyle*, II, 277–82: 277–8 for gout.

68. Hooke, *Diary, 1672–80*, 19 Aug. 1675 for vomiting; 15 May 1676 for socializing when 'Ill with cold'.

69. Hooke, *Diary, 1672–80*, 20 Nov. 1672.

70. Hooke, *Diary, 1672–80*, 19 Nov. 1672.

71. Porter, *Quacks* (n. 62), 136.

72. Hooke, *Diary, 1672–80*, 28 Dec. 1672.

73. Hooke, *Diary, 1672–80*, 25 June 1680.

74. Michael Osborne, 'The Medical Interests of the Oxford Chemists of the Late Seventeenth Century' (unpublished Oxford University Master of Chemistry Thesis, 2002), 41–8, 72–3. Copy on deposit, History Faculty Library, Oxford University. I was Mr Osborne's research supervisor, and believe that this thesis touches upon aspects of the medico-chemical culture of the seventeenth century not covered in the published literature. The original Ward *Diaries* are now in the Folger Shakespeare Library, Washington DC, though a microfilm is held by the Museum of the History of Science, Oxford.

75. Hooke, *Philosophical Experiments*, 'An Account of a Plant, call'd *Bangue*, before the Royal Society 18 Dec. 1689', pp. 210–2. The supplier of the *Bangue* is not specified, though it could have been Hooke's friend Captain Knox, who had spent many years in the Far East. After these seventeenth-century reports, *Cannabis Indica* (if that is what *Bangue*

was) waited another 150 years before attracting medical attention in the West, in the publications of Sir W. B. O'Shaughnessy: see Lorna Ronald, 'An Early Study of Psychoactive Chemistry: a Study of the Cannabinoid Group, 1840–1940' (unpublished Oxford University Master of Chemistry Thesis, 1998), 3, 11.

76. *[George] Herbert, Poems and Prose*, selected by W. H. Auden (Penguin, 1985), 116, verse 6. See verse 4 for the imagery of the tincture.

77. The words 'purge', 'purging' and such ran deep into the medical, social and spiritual literature of the seventeenth century. The *Authorized Version* of the Bible (1611), which would have been familiar to every literate person of Hooke's day, uses them over 30 times in the sense of cleansing badness, while 'purification' and its variants are used over 100 times in the Bible. And as a man familiar with Christ Church, Oxford, and Westminster Abbey, after the Restoration of the Anglican hierarchy in 1660, Hooke would also, no doubt, have been acquainted with the text 'Purge me, O Lord', set to music as an anthem by Thomas Tallis in the sixteenth century.

78. *Micrographia*, 'Preface' sig. *bb* (recto and verso). See p. 41 of present book for full quotation.

79. Hooke, *Diary, 1672–1680*: for Tom Gyles's fatal illness see 8–12 Sept. 1677, and for Grace's recovery, 26 July 1679. In each case, Hooke called out the best physicians of the day.

80. For an excellent earlier study of Hooke's medical practices, see Lucinda McCray Beier, 'Experience and Experiment. Robert Hooke, Illnesses and Medicine', in *Robert Hooke. New Studies*, ed. M. Hunter and S. Schaffer (Boydell Press, Woodbridge, 1989), 235–52. See also Lisa Jardine, 'Hooke the Man: His Diary and His Health', in Jim Bennett, Michael Cooper, Michael Hunter and Lisa Jardine, *London's Leonardo: The Life and Work of Robert Hooke* (OUP, 2003), 163–206: esp. 181–90.

81. Amanda Foreman, *Georgiana, Duchess of Devonshire* (Harper Collins, London, 1999), ch. 19: 299.

Chapter 7

1. *Encyclopaedia Britannica*, XIV edn, vol. 9 (London and New York, 1929), 274, 'Fire Prevention and Extinction'. Also *The London Encyclopaedia*, ed. Ben Weinreb and Christopher Hibbert (Macmillan Book Club Associates, 1984), 324–5, 'Great Fire, 1666'.

2. The legend of King Charles II assisting with the fire-fighting was immortalized in W. Harrison Ainsworth's *Old St. Paul's* (1841), Book 6, Ch. V.

3. For Wallis' geometrical structures for Oxford's Sheldonian Theatre, see Robert Plot, *The Natural History of Oxfordshire* (London, 1705), 277. Also Nehemiah Grew, *Catalogue and Description of the Natural and Artificial Rareities . . . in the Royal Society* (London, 1681), Pt IV. Adrian Tinniswood, *His Invention So Fertile. A Life of Christopher Wren* (Jonathan Cape, London, 2001): Sheldonian Theatre, 102–7; London plan, 150–9.

4. *Posthumous Works*, 'Life of . . . Hooke', xiii.

5. Hooke, *Diary, 1672–80*: 20 Oct. 1673; 14 May 1674; 17 May 1674; 20 Jan. 1675; 23 March 1675.

6. Hooke, *Diary, 1672–80*, 22 Sept. 1672.

7. Alistair Service, *The Architects of London and their Buildings 1066 to the Present Day* (Architectural Press, London, 1979), 9–17, for Inigo Jones. For a technical history of English Renaissance architecture, see Sir Banister Fletcher, *A History of Architecture on the Comparative Method*, 14th edn (Batsford, London, 1948), 766–81. For Pratt, see R. T. Gunther, *The Architecture of Sir Roger Pratt: Charles II's Commissioner for the Rebuilding of London after the Great Fire* (John Johnson for the author, OUP, 1928).

8. Hooke, *Diary, 1672–80*, 16 Sept. 1673, for reference to Jaggard and £1 payment. Also Michael Cooper, *'A More Beautiful City': Robert Hooke and the Rebuilding of London after the Great Fire* (Sutton, Stroud, Gloucestershire, 2003), 141, for charges and fees.

9. Geoffrey Beard, *The Work of Christopher Wren* (John Bartholomew & Son, Edinburgh, 1982), 16–36. Also Service, *Architects of London* (n. 7), 19–29.

10. An early scholarly assessment of Hooke's contribution to the creation of post-Fire London and a list of his buildings is given in Marjorie Isabel Batten (later Webb), 'The Architecture of Dr Robert Hooke F.R.S.', *Walpole Society* (London) 25 (1936–7), 83–113.

11. John Ward, *The Lives of the Professors of Gresham College* (London, 1740), 1–32, 'The Life of Sir Thomas Gresham'. For his Will, 5 July 1575, and intention to found the College, see 19–25. Sir Thomas died suddenly on returning home from The Exchange on 21 Nov. 1579, and Lady Gresham died on 23 Nov. 1596. After her death their mansion became Gresham College.

12. Cooper, *A More Beautiful City* (n. 8), 153, 154, etc. Also M. Cooper, 'Hooke's Career', in J. Bennett, M. Cooper, M. Hunter and L. Jardine, *London's Leonardo. The Life and Work of Robert Hooke* (OUP, 2003), 1–62, especially 28–49.

13. Cooper, *A More Beautiful City* (n. 8), 141.

14. Hooke, *Diary, 1672–80*, 7 July 1674.

15. Hooke, *Diary, 1672–80*, 7 July 1674.

16. Edward (Ned) Ward, *The London Spy* (London, 1698–1703): book form, 1703, ed. Kenneth Fenwick (Folio Society, 1955), Part III, 48.

17. These two statues, the work of Caius Gabriel Cibber, were rescued for posterity when Hooke's 1674 Bedlam was demolished in the early nineteenth century. They are now preserved in the Museum of the Royal Bethlem Hospital, Beckenham, Kent. A later pair, copies of the Cibber originals, are on display in the Later Stuarts Gallery, Museum of London (2004). I wish to thank the Curator of the Bethlem Hospital Museum for his assistance, and also Christine Riding, of Tate Britain, for providing the initial leads in my search for the surviving Bedlam statues.

18. Ward, *London Spy* (n. 16), Pt III, 47.

19. Cooper, *A More Beautiful City* (n. 5), 192–5. Sadly, Hooke's beautiful building had developed alarming structural problems by the late eighteenth century, due to the shallowness of its foundations, and by the early nineteenth century a new Bethlehem Hospital had been built south of the Thames at Southwark (now the site of the Imperial War Museum), and Hooke's 1674 building was demolished.

20. Hooke, *Diary, 1672–80*, 2 Dec. 1672.

21. Hooke, *Diary, 1672–80*, 29 Nov. 1672; also 23 Jan. 1673.

22. Aubrey, *Brief Lives*, 165. Several other Hooke buildings are also listed by Aubrey, ibid.

23. *Philosophical Experiments*, 2. See also p. 67 in the present book.

24. Hooke, *Diary, 1672–80*, 5 Oct. 1677.

25. Derek Howse, *Francis Place and the Early Royal Observatory* (Science History Publication, New York, 1975), Plate IX, 48, for engraving of interior of Octagon Room *c.* 1677, showing long pendulum clocks. Derek Howse, 'The Tompion Clocks at Greenwich, and the dead beat escapement', Pt I, *Antiquarian Horology* 7, **1** (Dec. 1970), 18–34, Part II, *Ant. Horol.* 7, **2** (March 1971), 114–33.

26. Batten, 'The Architecture of Dr Robert Hooke' (n. 10), 97.

27. Hooke, *Diary, 1672–80*, 25 June 1680.

28. Hooke, *Diary, 1672–80*, 27 June 1680.

29. Hooke, *Diary, 1672–80*, 27 June 1680.

30. Hooke, *Diary, 1672–80*, 7–8 May 1678.

31. Hooke, *Diary, 1672–80*, 28 March 1679.

32. Hooke, *Diary, 1672–80*. On p. 409 is reproduced a Hooke drawing for a church for Dr Busby: British Museum Add. MSS 5238 'Dr Hooke's Drawings'. See also *Diary*, 21 April 1679.

33. Documents pertaining to the building of Willen Church by Dr Busby: Westminster Abbey Archives: Westminster Abbey Busby Trustees Box 27.

34. Hooke, *Diary, 1672–80*, 3–5 May 1680, for details of journey to Willen.

35. Hooke, *Diary, 1672–80*, 22 Dec. 1673.
36. *Posthumous Works*, 'Life of ... Hooke', xiii. For Wren and the taking of 'perquisites', see Adrian Tinniswood, *His Invention So Fertile* (n. 3), 207.
37. *Posthumous Works*, 'Life of ... Hooke', xxv.
38. Aubrey, *Brief Lives*, 165. Also 'Hooke's Possessions at his Death: A hitherto unknown Inventory' (based on the researches of Frank Kelsall), in *Robert Hooke. New Studies*, ed. M. Hunter and S. Schaffer (Boydell Press, Woodbridge, 1989), 287-94.
39. Cooper, *A More Beautiful City* (n. 5), 141.
40. Cooper, *A More Beautiful City* (n. 5), 193.
41. M. Cooper, 'Hooke's Career', in Bennett *et al.*, *London's Leonardo* (n. 12), 20.
42. Hooke, *Diary, 1672–80*: for Alvington, 6 Nov. 1675 and 10 Feb. 1676. To get a rough idea of the extent of this estate (albeit without any precise weightings for land values in different parts of the country), it might be remembered that not many years previously Hooke's old Oxford mentor Dr Thomas Willis had purchased for an investment property Burlton Court, Herefordshire, consisting of the manor house and 240 acres of land, for £1,900: Kenneth Dewhurst, *Willis's Oxford Lectures* (Sandford, Oxford, 1980), 15.

Chapter 8

1. David R. Oldroyd, 'Geological Controversy in the Seventeenth Century: "Hooke *vs.* Wallis" and its Aftermath', in *Robert Hooke. New Studies* (Boydell Press, Woodbridge, 1989), 207-33. Ellen Tan Drake, *Restless Genius. Robert Hooke and his Earthly Thoughts* (OUP, 1996). Part I consists of eight chapters, pp. 1–152, examining Hooke's geological ideas and giving bibliography; the remainder of the book reprints, with critical commentary, Hooke's 'Earthquake Discourses', delivered to the Royal Society between 1668 and 1700 (though noting the existence of previous 'Discourses' going back to 23 May 1666, p. 159), and originally published in Hooke's *Posthumous Works* (1705). His earliest lecture on geology to the Royal Society, however, seems to have been delivered as early as 1664, according to a comment in *Posthumous Works*, 439. Dr Tan Drake, who brings a trained geologist's insights into Hooke's 'Discourses', adds dates to them. As she acknowledges on p. 129, however, these dates derive from the 'Earthquake Discourses' chronology established by Rhoda Rappaport, 'Hooke on Earthquakes: Lectures, Strategy and Audience', *BJHS*, **19** (1986), 129-46.
2. *Micrographia*, Obs. XVII, 'Of Petrify'd Wood, and other Petrify'd bodies', 107-12, where he discusses the possible causes of petrifaction of seemingly organic substances.
3. *Posthumous Works*, Lecture or Discourse, 26 May 1697, 439. *Philosophical Experiments*, 'Dr Hooke's Discourses to the Royal Society, in the Beginning of 1697', p. 323.
4. *Philosophical Experiments*, 304-14, 2 Dec. 1696, for Nautilus shell. Also 315-38 for his studies of amber and petrified resins dug up in Germany.
5. Hooke to Boyle, 26 Sept. 1665, *Correspondence of Robert Boyle*, **II**, 537-8.
6. *Posthumous Works*, 292. Tan Drake, *Restless Genius* (n. 1), 159, dates this Discourse to before 15 Sept. 1668. It could well have been delivered to the Royal Society on 27 June 1667, when Hooke described 'a cliff in the Isle of Wight': Birch, *History of the Royal Society*, **II** (27 June 1667), 183.
7. James Barr, 'Why the World was created in 4004 BC: Archbishop Ussher and Biblical Chronology', *Bulletin of the John Rylands Library, Manchester*, **67**, 2 (Spring 1985), 575-608.
8. *Posthumous Works*, 377-84, for Ovid and ancient histories of remote times and changes. Also 394-410 for ancient narratives. Tan Drake, *Restless Genius* (n. 1), also discusses Hooke's awareness of the classical writers and their perceived relevance to geological history (especially Hooke's ideas on the subject during the late 1680s), 85, 86, 102, 109, etc.

9. *Posthumous Works*, 288.
10. *Micrographia*, 109.
11. Thomas Burnet's *A Sacred Theory of the Earth* (London, 1691) was the foremost seven-teenth-century statement of this idea. Thomas Burnet (Master of Charterhouse, and not to be confused with Gilbert, Bishop of Salisbury and historian) argued that Noah's Flood had wiped out the original state of the globe, as a result of human sin, and the present broken state of our planet was the result.
12. *Posthumous Works*, Discourse, 26 May 1697, 439.
13. *Posthumous Works*, 439–40.
14. Edmond Halley, 'A Discourse concerning an Hypothesis of the manner of the Generall Deluge', 12 Dec. 1694, Royal Society MS Register Book Copy, 9, 40–44; eventually published as 'Some Considerations about the Cause of the Universal Deluge', *Phil. Trans.*, **33** (1724), 118–23, and 'Some further thoughts', 123–5, appended to the same. Also A. Chapman, 'Edmond Halley's Use of Historical Evidence in the Advancement of Science' [Royal Society Wilkins Lecture 1994], *Notes and Records*, **48** (1994), 167–91, for Halley's own ideas of a dynamic earth.
15. *Posthumous Works*, 289.
16. Galileo, *Letter to Madame Christina of Lorraine, Grand Duchess of Tuscany* (1615). Galileo did not originate the phrase '... the Holy Ghost is to teach us how to go to heaven, not how heaven goes', but attributes it to an epigram of Cardinal Baronius (1538–1607): see *Discoveries and Opinions of Galileo*, ed. and trans. Stillman Drake (Doubleday, Anchor, New York, 1957), p. 186.
17. *Posthumous Works*, 334–5, 342.
18. *Posthumous Works*: for the Alps once under the sea, 291, 311, etc.; shells on Alps, 324, etc. For Amsterdam digging, 289.
19. *Micrographia*, 109.
20. *Posthumous Works*, 342.
21. *Philosophical Experiments*: part of Hooke's Third Discourse on Amber, 19 May 1697, 332–3.
22. *Posthumous Works*, 284.
23. *Posthumous Works*, 284.
24. *Posthumous Works*, 329ff. This set of Earthquake Discourses Waller (who edited them) originally claimed were written around 1668, and later brought up to date—but left undated—by Hooke. Tan Drake, *Restless Genius* (n. 1), 226, however, believes this series of lectures to have been read before the Royal Society between 8 Dec. 1686 and 19 Jan. 1687. They contain Hooke's theory of fossilization, of the organic origin of fossils, and of how they can be used to understand the ancient condition of the earth. For Hooke's reference to fossils as 'Medals of Nature', see 341.
25. *Posthumous Works*, 342.
26. *Posthumous Works*, 346ff. Tan Drake, *Restless Genius* (n. 1), 246, dates the first of Hooke's Earthquake Discourses to discuss the 'Wandering Poles' idea to 26 Jan. 1687. The idea of polar wandering had clearly been forming in his mind well before, however, for at the end of his previous Discourse, which was probably given in early or mid Jan. 1687, he concluded with the 'teaser': 'Thirdly, Whether the Axis of its [the Earth's] Rotation do change its Situation or Position in respect of its Parts of the Earth; and thence, Whether the Latitudes and Meridional Lines of places do differ in process of time ...' and what are its causes: *Posthumous Works*, 345.
27. *Posthumous Works*, 343. For a full treatment of the Polar Wandering idea, see Tan Drake, *Restless Genius* (n. 1), 87–95.
28. Jean Richer, 'Observationes Astronomiques et Physiques faites en L'Isle de Caienne' (1672), *Mémoires de l'Académie Royale des Sciences depuis 1666 jusqu'à 1699*, **8**, 1 (1729), 88. Also A. Chapman, *Dividing the Circle: The Development of Critical Angular Measurement in Astronomy, 1500–1850* (Praxis–Wiley, Chichester, 1990, 1995), 102–5.

29. Hooke to Newton, 6 Jan. 1680, *The Correspondence of Isaac Newton, II, 1676–1687*, ed. H. W. Turnbull, FRS (CUP, for the Royal Society, 1960), Letter 239, pp. 309–10. The French Cayenne and English St Helena findings are in *Posthumous Works*, 349.

30. *Posthumous Works*, 349 (for Siam reference). Also Newton, *Principia Mathematica* (1687), Book III, 20:IV for table of pendulum swings: see, for instance, Newton, *The Principia. Mathematical Principles of Natural Philosophy*, trans. I. B. Cohen, Anne Whitman and J. Burdenz (Univ. of California Press, Berkeley, Los Angeles and London, 1999), 826–9. In addition to the different rates at which pendulums of the same length swung in different parts of the earth, dispute regarding the true shape of the globe (was it an oblate or a prolate sphere?) came out of a set of careful surveys of France, based on a meridian running through the Paris Observatory, between 1669 and 1718, by Jean Picard, G. D. Cassini and others. See A. Chapman, *Dividing the Circle. The Development of Critical Angular Measurement in Astronomy, 1500–1850* (Praxis-Wiley, Chichester, 1990, 1995), 102–5.

31. *Posthumous Works*, 349ff.

32. *Posthumous Works*, 436 for elephant remains in Germany, and 438–9 for English finds. (These Discourses seem to have been delivered by Hooke in the mid 1690s.)

33. *Philosophical Experiments*, 323 (1697).

34. *Philosophical Experiments*, 334.

35. *Philosophical Experiments*, 322.

36. *Posthumous Works*, 291, suggests 'that there have been many other Species of Creatures in former Ages, of which we can find none at present; and that 'tis not unlikely also but that there may be divers new kinds now, which have not been from the beginning'. Hooke is not suggesting species evolution here, but rather the extinction of some species and perhaps the special creation of others. And through his knowledge of Ovid, Hesiod and other classical poets who spoke of *metamorphoses* (such as those cited on 8 March 1693, p. 402) and his familiarity with the Roman natural historian Pliny, Hooke had come up with no coherent mechanism by which living species could change (as opposed to a generalized interest in the subject). See also Tan Drake, *Restless Genius* (n. 1), 96–100, for Hooke and evolution.

37. *Posthumous Works*, 291.

38. *Posthumous Works*, 435.

39. Barr, 'Why the World was created in 4004 BC' (n. 7). Also Julius Africanus, Sextus, in *The Oxford Dictionary of the Christian Church*, 3rd edn, ed. J. L. Cross and E. A. Livingstone (OUP, 1997), p. 913, for 5500 BC date for Creation. Edmond Halley drew attention to the *Septuagint* (Jewish Greek) and original Hebrew chronologies for the age of the earth in his 'A Short Account of the Saltness of the Ocean, and of several Lakes that emit no Rivers; with a Proposal, by help thereof, to discover the Age of the World', *Phil. Trans.* 29, **334** (1715), 296–7.

40. *Posthumous Works*, 372–3. In 1680, of course, no one could yet read the ancient Egyptian, Phoenician and other ancient languages directly: their historical narratives could only be approached through what survived in Greek or Latin commentators, such as Plato, Herodotus, Ovid, Hanno, Plutarch, etc., and from Hebrew accounts.

41. *Posthumous Works*, 395.

42. *Posthumous Works*, 343: in 1687, Hooke was telling the Royal Society that he had 'about ten or twelve years since' '. . . indeavour'd to show that the form of the Earth was probably somewhat flatter towards the Poles than towards the Equinoctial [Equator], since which I have met with some Observations that do seem to make a probability of my Conjecture and Hypothesis'. Hooke is clearly claiming a priority here for the realization that the earth is oblate, or flattened at the poles, long before Newton wrote *Principia*. Also *Principia*, Book III, Prop. 20, Problem 4 (n. 30).

43. Cited in Tan Drake (n. 1), 91. For a detailed analysis of Hooke's and Wallis' dispute about polar wandering, and an anciently remodelled earth, see D. R. Oldroyd, 'Geological Controversy in the Seventeenth Century' (n. 1), 207–33.

44. Wallis to Halley, 4 March 1687, cited in Oldroyd, 'Geological Controversy in the Seventeenth Century' (n. 1), 210–12: 212, para. VIII. See, also, A. J. Turner, 'Hooke's Theory of the Earth's Axial Displacement: Some Contemporary Opinions', *BJHS*, **7** (1974), 166–70.

45. *Posthumous Works*, 416.

46. Hooke is right regarding *Pyramidographia* (London, 1646), for, in spite of the thoroughness of his survey of the Great Pyramid of Giza, John Greaves does not record the astronomical or magnetic orientations of the monument. *Posthumous Works*, 353.

47. Wallis to Halley, 4 March 1687, in Oldroyd, 'Geological Controversy in the Seventeenth Century' (n. 1), 211, para. V. Edmond Halley also had his doubts about the Wandering Poles idea, arguing that cities that had been the locations of careful accurate observations over several millennia showed no changes of polar elevation, as they should have done over 2,000 or more years: Halley, 'An Account of some Observations made at Nuremberg by Mr P. Wurtzelbaur, showing that the Latitude of that Place continued without sensible alteration for 200 years past...', *Phil. Trans.* 16, **190** (misprinted 1678, but really and sequentially 1687): see 405 for Alexandrian latitude from Greek to modern times.

48. Hooke, 'Ansr to Dr Wallis & Ways to find ye Meridian', reprinted in Oldroyd, 'Geological Controversy' (n. 1), 213–18: 216 for Copernicus and Galileo.

49. Hooke, *Diary, 1672–80*, 12 Jan. 1676 and 9 Oct. 1674.

50. *Posthumous Works*, 418.

51. *Posthumous Works*, 420.

52. *Posthumous Works*, 420, for explosion of Hackney gunpowder mills. Hooke does not give a date for the explosion.

53. *Posthumous Works*, 416, headed '23 July 1690'.

54. *Posthumous Works*, 421.

55. *Posthumous Works*, 424, headed '30 July 1690'.

56. Charles Daubeny, *A Description of Volcanoes Active and Extinct* (London, 1826), 357–8, for assumed chemical action of volcanoes; 370ff. for volcanoes near to the sea.

57. *Philosophical Experiments*, 273–4.

58. *Posthumous Works*, 'Life of... Hooke', xxv. *The Life and Times of Anthony à Wood*, ed. Andrew Clark and abridged by Llewelyn Powys (OUP, 1961), 8 Sept. 1692.

59. *Life and Times of... Wood* (n. 58), 24 Sept. 1693.

60. Hooke, 'Ansr. to Dr Wallis', in Oldroyd, 'Geological Controversy' (n. 1), 217.

61. Tan Drake, *Restless Genius* (n. 1), 217, argues that as early as his *Lampas* (1677), 208, Hooke had been accusing Henry Oldenburg of passing on his, Hooke's, geological ideas to Steno. But one suspects that Hooke was being over-sensitive here, for it was Oldenburg's job, as Secretary of the Royal Society, to correspond with the learned men of Europe and pass on original ideas.

62. Halley, 'An Estimate of the Quantity of Vapour raised out of the Sea by the Warmth of the Sun...', *Phil. Trans.* 16, **189** (1687), 366–70. Halley, 'An Account of the Circulation of the watry [*sic*] Vapours of the Sea, and the Cause of Springs, presented to the Royal Society', *Phil. Trans.* 16, **192** (1691), 468–73.

63. Halley had been in correspondence with Hooke since Halley had been on St Helena in the late 1670s, as Hooke intimates in his letter to Newton, 6 Jan. 1680, wherein he discusses Halley's pendulum clock anomalies (n. 29). And as Clerk to the Royal Society after 1686, Halley would have worked closely with Hooke. Halley delivered his ideas on Noah's Flood in a lecture to the Royal Society: 'A Discourse, concerning an Hypothesis of the Manner of the Generall Deluge', undated, but 'laid before R. Soc.' 12 Dec. 1694: Royal Society MS Register Book Copy 9, 40–4. It was not published in *Phil. Trans.* 33, **383** (1724), 118–23. See also A. Chapman, 'Edmond Halley's Use of Historical Evidence in the Advancement of Science' [Royal Society Wilkins Lecture 1994], *Notes and Records* 48, **2** (1994), 167–91: 178ff. for Deluge.

64. Halley, 'Saltness of the Seas...' (n. 39), 293–300. Halley had first proposed cometary impacts as agents of change on the archaic, pre-Adamite globe in 'An Account of Some Observation lately made at Nuremberg' (n. 47), 405–6. He developed them further in his 'General Deluge' paper (n. 63). It was, however, William Whiston who took up the idea of comets influencing the earth, either by impact or gravitational drag, or even by supplying the water—from their vaporous tails—for Noah's and previous deluges: Whiston, *A New Theory of the Earth, From its Original, to the Consumation of all Things, Wherein the Creation of the World in Six Days. The Universal Deluge...* (London, 1696), 34–5; 300–2, etc. for cometary contact.

65. Halley, 'Latitude of Nuremberg' (n. 47), 405.

66. Halley, 'Generall Deluge' (n. 63), 41; *Phil. Trans.* 33, **383** (1724), 119. The source of the vast quantity of water necessary to flood the entire globe fascinated both traditional theologians and scientists alike. Thomas Burnet's orthodox, *A Sacred Theory of the Earth* (n. 11), Bk I, Ch. II, p. 30, wrestled with the same problem.

67. Whiston, *New Theory of the Earth* (n. 64), 301–3.

68. It is a popular modern misconception, fostered by evangelical materialists such as Thomas Henry Huxley in the late nineteenth century and Richard Dawkins in our own time, that before Darwin everyone believed in the simple literal truth of *Genesis*, and to doubt it provoked damning ecclesiastical censure. It is, however, a view that can only be sustained by deliberately ignoring the abundant historical record to the contrary, and by turning a blind eye to the Church's own academic debates about exactly what Scripture is intended to teach us. St Augustine, for instance, in his critical commentary *De Genesi ad Litteram*, around AD 410, was aware that the flat earth and 'tabernacle' or tent-like heavens of the ancient Jewish Scripture did not accord with the Greek astronomical knowledge of a spherical earth and cosmos with which he was familiar. Indeed, St Augustine set out ways in which the spiritual truths embodied in *Genesis*—God's creation of the universe and man in God's image—could be reconciled with the more complex physical knowledge of nature accepted by educated men of his own time. Medieval theologian-scientists such as Albertus Magnus and Cardinal Nicholas of Cusa further developed those ideas about Scripture and science. Galileo (who in his own spiritual life lived and died a devout Catholic) argued similarly in his *Letter to the Grand Duchess Christina* (1615), citing St Augustine and Cardinal Baronius as authorities. John Wilkins, Hooke's Oxford mentor and later Bishop of Chester, reasoned likewise in *A Discourse concerning a New Planet, tending to prove, That, ('tis probable) our Earth is one of the Planets* (1638), Prop. V, 183–90: 'That divers learned Men have fallen into great Absurdities, whilst they have looked for Sects of Philosophy [or science] in the Words of Scripture', and, p. 184, 'so that there is not a Demonstration in Geometry, or Rule in Arithmetick; not a Mystery in any Trade, but it may be found out in the *Pentateuch*': in *The Mathematical and Philosophical Works of the Right Reverend John Wilkins* (London, 1708). And in 1836, the Revd Dr William Buckland, Canon of Christ Church and Regius Reader in Geology at Oxford University, also argued that an archaic earth, which had gone through many geological changes, could be reconciled with the Creation narrative in *Genesis*: Wm. Buckland, *Geology and Mineralogy considered with reference to Natural Theology* (London, 1836), 7–33: see esp. 18–19.

69. Hooke, *Diary, 1672–80*, 20 Nov. 1673, for Hooke's purchase of a work by Scaliger.

Chapter 9

1. Hooke, List of Inventions, containing 106 'inventions', Royal Society MS, Classified Papers XX, No. 54; undated, but sequentially *c.* 1667. In *Philosophical Experiments* (1726), 293, Hooke describes a portable Camera Obscura or 'picture box'.

2. Wood, *Athenae Oxonienses*, **II** (London, 1721), 'Hooke', 1030. Aubrey, *Brief Lives*, 'Hobbes', 157: 'Mr Robert Hooke loved him, but was never but once in his company.'

3. *Micrographia*, 'Preface', sig. *a* recto.
4. *Micrographia*, 'Preface', sig. *g* recto.
5. *Micrographia*, Obs. VI, 'Of small Glass Canes', 10–32.
6. *Micrographia*, 'Preface', sig. *c* verso.
7. *Micrographia*, 'Preface', sig. *c* verso and *cc* recto.
8. *Micrographia*, 'Preface', sig. *c* verso.
9. 'A new Contrivance of Wheel-Barometer, much more easy to be prepared than that which is described in the *Micrography*; imparted by the Author of that Book [Hooke]', *Phil. Trans.* 1, **13** (June 1666), 218–9.
10. Thomas Sprat, *History of the Royal Society* (London, 1667), 173–9.
11. Hooke, *Philosophical Experiments and Observations* 41 for Weather Clock. See, also, Chapter 4 of present book, p. 68.
12. *Micrographia*, 'Preface', sig. *dd* verso.
13. *Micrographia*, 'Preface', sig. *e* verso, *ee* recto. Also 'Considerations of Monsieur Auzout upon Mr Hook's New Instrument for Grinding Optick-Glasses', *Phil. Trans.* 1, **4** (June 1665), 55–63. Auzout also wrote to Henry Oldenburg criticizing the Hooke lens-making machine in *Micrographia*: Auzout to Oldenburg, 2 June 1665, in *The Correspondence of Henry Oldenburg, II, 1663–1665*, ed. A. R. Hall and M. Boas Hall (Univ. of Wisconsin Press, Madison, Milwaukee and London, 1966), Letter 380: French 410–8, English trans. 419–27.
14. 'Considerations of Monsieur Auzout' (n. 13), 55–8.
15. 'Monsieur Auzout's Judgment touching the Apertures of Object-Glasses, and their Proportions, in respect of the several Lengths of Telescopes', *Phil. Trans.* 1, **4** (June 1665), 55–6.
16. 'Considerations of Monsieur Auzout' (n. 13), 1. See also 'An Accompt of the improvement of Optick Glasses', *Phil. Trans.* 1, **1** (6 March 1665), 2–3. An account of Signor Giuseppi Campani's object glasses, problems of chromatic aberration ('Rain-bow Colours'), and planetary observations.
17. The life on other worlds issue fascinated seventeenth-century scientists: 'Considerations of Monsieur Auzout' (n. 13), 61. Also Auzout to Oldenburg, 2 June 1665, *Correspondence of Henry Oldenburg II* (n. 13), French 424, English 417.
18. Christiaan Huygens, *Cosmotheoros, The Celestial Worlds Discovered; or conjectures concerning the inhabitants, plants and productions of the worlds in the planets* (London, 1698). Even as late as 1835, indeed, the American and part of the British public were willing to believe that Sir John Herschel had discovered advanced civilized creatures on the Moon: Richard Adams Locke, 'The Great Astronomical Discoveries Lately made by Sir John Herschel at the Cape of Good Hope', appearing originally in the New York *Sun*, 1835, and reproduced in *The Man in the Moon*, ed. Faith K. Pizor and T. Allen Camp (Sedgewick and Jackson, London, 1971), 190–216.
19. 'Mr Hook's answer to Monsieur Auzout's Considerations, in a Letter to the Publisher of these *Transactions*', *Phil. Trans.* 1, **4** (1665), 64–9.
20. 'Mr Hooke's answer' (n. 19), 65–6.
21. 'Mr Hooke's answer' (n. 19), 65–6.
22. 'Mr Hooke's answer' (n. 19), 65–6, and *Micrographia*, Obs. LIX, 'Of Multitudes of small Stars discoverable by the Telescope', 242.
23. A. Chapman, 'Christiaan Huygens (1629–1695), Astronomer and Mechanician', *Endeavour* N.S. vol. 19, no. 4 (1995), 140–5. For Cassini Division discovery: 'An Extract of Signor Cassini's Letter Concerning... with a remarkable Observation of Saturn, made by the same', Latin letter, Paris, 26 Aug. 1676; *Phil. Trans.* 11, **128** (1676), 689–90. 'Accompt of the improvement of Optick-Glasses' (n. 16), 2–3.
24. 'Accompt of the improvement of Optick-Glasses' (n. 16), 3.
25. *Posthumous Works*, 'Life of ... Hooke', xiii; this date, 28 Feb. 1674, is a little ambiguous, as the reference to his 28 Feb. reflecting telescope experiments is two paragraphs below the date 1674.

26. Hooke, *Diary, 1672–80*, 5 Aug. 1672.
27. Hooke, *Diary, 1672–80*, 28 Feb. 1674.
28. 'Blebbs' and 'Veins' were bubbles and striations in optical glass: used in Auzout—Hooke correspondence about telescope lenses, *Phil. Trans.* 1, **4** (1665), 57–69 (see n. 13).
29. Hooke, *Diary, 1672–80*, 6 and 23 Aug. 1672.
30. Hooke, *Diary, 1672–80*, 27 Jan. 1673.
31. Hooke, *Diary, 1672–80*, 3 Oct. 1672.
32. Hooke, *Diary, 1672–80*, 3 Oct. 1672.
33. The first person to develop a viable reflecting telescope was John Hadley, who described his six-inch-diameter speculum mirror of around six feet focus mounted on an adjustable altazimuth stand. It was found superior in image quality and brightness to long refractors, and formed the prototype for the Newton reflectors of Sir William Herschel and subsequent astronomers: John Hadley, 'An Account of a Catadioptrick Telescope, made by John Hadley F.R.S. With the Description of a Machine contriv'd by him for the applying of it to use', *Phil. Trans.* **32** (1723), 303–12.
34. *The Diary of John Evelyn*, ed. E. S. de Beer, III, *Kalendarium 1650–1672* (OUP, 1955), 7 Aug. 1665, 416.
35. The provenance of the 'Hooke-type' clock-wheel dividing and cutting engine is far from clear, with some modern scholars, such as Michael Wright, being doubtful whether the original invention was Hooke's at all. William Derham, *The Artificial Clock-Maker*, 4th edn (London, 1759), 'Preface', X, mentions amongst various other devices ' . . . the Invention of *Cutting Engines* (which was Dr Hooke's) . . .' Derham, after all, had known the elderly Hooke and edited some of his unpublished papers in *Philosophical Experiments and Observations* (1726). Without citing his source, Thomas Reid, in his *A Treatise on Clock and Watchmaking, Theoretical and Practical* (Edinburgh, 1826), 284, dates Hooke's invention to 1655! For a surviving 'Hooke-type' machine, preserved in the Science Museum, London, see K. R. Gilbert, *The Machine Tools Collection: Catalogue of Exhibits with Historical Introduction* (HMSO, London, 1966), 74, Item 170, 'Wheel-cutting Machine, seventeenth century'. See Plate 18.
36. Hooke, *Diary, 1672–80*, 16 Aug. 1672.
37. F. A. Bailey and T. C. Barber, 'The Seventeenth-Century Origins of Watchmaking in West Lancashire', in *Liverpool and Merseyside*, ed. J. R. Harris (London, 1969), 1–15.
38. Many of the documented and anecdotal aspects of Horrocks' early life in Toxteth were drawn together by John E. Bailey in the nineteenth century: 'Jeremiah Horrocks and William Crabtree, Observers of the Transit of Venus, 24 November 1639', *The Palatine Note-Book*, **II** (Manchester, 1882), 253–66: esp. 254–5, for Horrocks' background. Also Bailey, 'The Writings of Jeremiah Horrox [*sic*] and William Crabtree', *The Palatine Note-Book* III (Manchester 1883), 17–22.
39. Hooke, *Diary, 1672–80*, 18 March 1673.
40. Hooke, *Diary, 1672–80*, 20 March 1673. 'Dividing Plates' were large workshop tools, either circles or quadrants, divided up into degrees and minutes. Hooke, in *Some Animdaversions of the First Part of Hevelius, his 'Machina Coelestis'* (London, 1674), 14, mentions using 'a very large quadrantal dividing plate of 10 feet radius' to graduate a quadrant.
41. Hooke, *Diary, 1672–80*, 2 May 1675.
42. See Gilbert, *The Machine Tools Collection* (n. 35), 70, Plate 18. A wheel-dividing engine is also described in the manuscript notebook, covering the years 1692–1727, that *may* have been the notebook of Samuel Watson of Coventry, now in the Collection of the Worshipful Company of Clockmakers: Guildhall Library, London, MS 6619/1, 17–20 and p. 51 verso.
43. A. Chapman, *Dividing the Circle. A History of Critical Angular Measurement in Astronomy, 1500–1850* (Praxis-Wiley, Chichester, 1990, 1995), 123–7.
44. Hooke, *Diary, 1672–80*, 31 Jan. 1673.

45. Hooke, *Philosophical Experiments*, 225–48. Hooke had described this device and the importance of ocean depth sounding in *Phil. Trans.* 1, **9** (12 Feb. 1665), 147–9, in 'An Appendix to the Directions for Seamen, bound for long Voyages', and in *Phil. Trans.* 2, **24** (8 April 1667), 433–4, 'For Observations and Experiments to be made by Masters of Ships, Pilots, and other fit Persons in their Sea-Voyages'.

46. Hooke, *Philosophical Experiments*, 293.

47. Hooke, *Lampas; or, Description of some Mechanical Improvements of Lamps and Waterpoises. Together with some other Physical and Mechanical Discoveries* (London, 1677), 4–8, for flame structures.

48. Hooke, *De Potentia Restitutiva, or, of Spring. Explaining the Power of Springing Bodies* (London, 1678): pp. 1 and 5 give keys to the Spring anagrams originally set out in the 'Appendix' of his *A Description of Helioscopes. And Some other Instruments* (London, 1676), 31–2. Item No. 3 in this Appendix, 31, read *ceiiinosssttuu*, while Item 9, 32, read *cdeiinnoopsssttu* [this seems to have contained a typographical error, for it was re-set in *De Potentia* as *cediinnoopsssttu*]. In *De Potentia* (1678), Hooke provided the keys: p. 1, *ceiiinossttuu* = '*Ut tensio sic Vis*; That is, The Power of any Spring is in the same proportion with the Tension thereof'; and p. 5, '*cediinnoopsssttu*, namely *Ut Pondus Sic tensio*', or the weight hung from a spring is proportionate to the Spring's tension. This was the idea that lay at the heart of the 'new sort of Philosophical Scales' (*Description of Helioscopes*, p. 32) or Spring Balance. It also related directly to the physics of his hairspring balance in watches.

49. *Posthumous Works*, 'Life of . . . Hooke', xxi.

50. John Harris's *Lexicon Technicum* (London, 1704), under 'Engine', does indeed mention the vacuum-induced steam suction Engine of Thomas Savary, FRS, of 1698, though there is no mention of Hooke.

51. L. T. C. Rolt, *Thomas Newcomen. The Prehistory of the Steam Engine* (David and Charles, Dawlish, Macdonald, London, 1963), 49.

52. *Philosophical Experiments*, 388–91. Hooke can sometimes appear xenophobic in regard to discoveries claimed by foreigners where he believed himself to possess a priority. For with the exception of Newton, most of his perceived rivals were either foreigners living abroad—Huygens, Hevelius, Cassini, Steno—or else foreigners who had become English residents, such as Henry Oldenburg. I would suggest, however, that this is not a generic xenophobia as such, but rather an awareness that living around Europe were a dozen or so very clever men interested in the same subjects as himself, and often enjoying powerful backers—such as King Louis XIV of France—whereas he, Hooke, had none. Even so, his suspicion is aimed at specific individuals and circumstances, not at foreigners in general. He had, for instance, acknowledged Cassini's discoveries of spots on Jupiter, being 'obliged to him for the perfecting the Theory [of Jupiter's rotation] as we are for many other rare Discoveries and excellent improvements in Astronomy' (*Cometa* (London, 1678), 78), and one of his regular friends during the 1670s was Theodore Haak, the elderly German expatriate with whom, according to his Diary entries, he often played chess.

53. *Philosophical Experiments*, 390.

54. *Philosophical Experiments*, 390.

55. For early equatorial mount and primary sources, see Chapman, *Dividing the Circle* (n. 43), under 'Equatorial Mount'.

56. Francis Bacon, *Sylva Sylvarum, or a Natural History* (1627) *in Ten Centuries*, consisting of 1,000 queries and experiments, organized across 10 books of 100 each: *The Works of Francis Bacon*, ed. James Spedding, Robert Leslie Ellis and Douglas Devon Heath, in Bacon's *Philosophical Works*, **II**, Part II (London, 1876). Also Edward Somerset (Marquis of Worcester), *A Century . . . of Inventions* (London, 1663). Though not arranged in centuries, the devices described by John Wilkins in *Mathematical Magick* (London, 1648) were immensely influential in focusing attention on the power and usefulness

of invention. A. Chapman, ' "A World in the Moon": John Wilkins and his Lunar Voyage of 1640', *QJRAS*, **32** (1991), 121–32. For seventeenth-century flying references, see p. 127 and ref. 25.

57. Aubrey, *Brief Lives*, 'Hooke', 166.
58. *Posthumous Works*, 'Life of . . . Hooke', xx.

Chapter 10

1. Aubrey, *Brief Lives*, 265.
2. *Posthumous Works*, 'Life of . . . Hooke', iv.
3. Ibid., iv. It is hard to imagine that Hooke's ideas on flight and 'Flying Chariots' had not been influenced by the writings of John Wilkins: *Mathematical Magick: or the Wonders that may be perform'd by Mechanical Geometry* (1648); see *The Mathematical and Philosophical Works of the Right Reverend John Wilkins* (London, 1708): *Mathematical Magick*, Propositions VII and VIII, 116–29, which dealt with flight and flying machines.
4. Thomas Hobbes, *Leviathan* (London, 1651), 'The Introduction', where at the very beginning of his treatise, Hobbes states: 'For seeing life is but a motion of Limbs . . . For what is the *Heart*, but a *Spring*; and the *Nerves*, but so many *Strings*; and the *Joynts*, but so many *Wheeles*, giving motion to the Whole Body, such as was intended by the Artificer?' One can see many aspects of this approach to mechanism implicit in the researches of Robert Hooke.
5. Hooke, *Diary, 1672–80*, 4 Oct. 1674.
6. Ibid., 11 Feb. 1675.
7. *Posthumous Works*, 'Life of . . . Hooke', iv.
8. *Micrographia*, Obs. XXXVIII, 'Of the Structure and motion of the Wings of Flies', 172–4. (See also present book, Chapter 4, pp. 64–5.) Also Obs. XLII, 'Of a Blue Fly', 182–5.
9. *Micrographia*, 174.
10. Leonardo da Vinci was also fascinated by flight, and made detailed studies of flying creatures — predominantly of birds, for lacking a microscope he could not see much detail in insects — as a preliminary to devising flying machines: *The Notebooks of Leonardo da Vinci*, I, ed. Edward MacCurdy (Reprint Society, London, 1954), 383–466: 'Flight', 367–76: 'Flying Machines'.
11. *Micrographia*, 173. Hooke also described the relation between vibration and music in 'A Curious Dissertation concerning the Causes of the Power & Effect of Musick. By the late famous Dr Robert Hooke', undated paper posthumously communicated by Dr William Derham, 14 Dec. 1727, Royal Society MS RBC 13.3. In this paper, Hooke argued that an awareness of the vibrations of music precedes an awareness of language, as babies respond to music (p. 6).
12. Pepys, *Diary*, **VII** (8 Aug. 1666), 239–41.
13. Hooke, *De Potentia Restitutiva, or of Spring. Explaining the Power of Springing Bodies* (London, 1678), 23.
14. Hooke, *De Potentia Restitutiva* (n. 13), 5. *Posthumous Works*, 'Life of . . . Hooke', vi. Also Michael Wright, 'Robert Hooke's Longitude Timekeeper', in *Robert Hooke. New Studies*, ed. M. Hunter and S. Schaffer (Boydell Press, Woodbridge, 1989), 63–118.
15. Hooke, *A Description of Helioscopes and some other Instruments* (London, 1676), 32, item 9. Also *De Potentia Restitutiva* (n. 13), 5.
16. William Derham, *The Artificial Clockmaker. A Treatise of Watch and Clock-work* (London, 1714), ch. VII, 95, for Burgi.
17. *Posthumous Works*, 'Life of . . . Hooke', iv. Christiaan Huygens, *Horologium* (The Hague, 1658): facsimile reprint and English translation in *Antiquarian Horology* (Dec. 1970), 35–54.
18. William Cuningham, *The Cosmographical Glasse* (London, 1559), fol. 109–10, for finding the longitude by a clock. Also David Wates, *The Art of Navigation in England in Elizabethan*

and early Stuart Times, Modern Maritime Classics Reprint 2 (National Maritime Museum, Greenwich, 1978), 58.

19. *Philosophical Experiments*, 4–6. For a detailed subsequent account of these Huygens-type sea clocks, see Edmund Stone's *Supplement* to his translation of Nicholas Bion's *The Construction and Principal Uses of Mathematical Instruments* (London, 1758), 310–1.

20. *Philosophical Experiments*, 4–5. Also 'A Narrative concerning the Success of Pendulum-Watches at Sea for Longitudes', *Phil. Trans.* 1, **1** (March 1665), 13–5. This narrative mentions the initiative of Lord Kinkardine, and the command of Major Holmes (the rank then held by Sir Robert Holmes), but nothing whatsoever about Hooke. Also Robert Ollard, *Man of War: Sir Robert Holmes and the Restoration Navy* (1969; Phoenix, London, 2001), 84, for Holmes' 1663 voyage in *HMS Reserve*.

21. Hooke, *Helioscopes* (n. 15), 28. Hooke's substantial 'Postscript' to his Cutlerian Lecture on solar telescopes is a detailed justification for and historical narrative of his prior claim for the invention of the hairspring balance for watches, which he hoped would be used successfully to find the longitude at sea.

22. Derham, *Artificial Clockmaker* (n. 16), 66.

23. *Posthumous Works*, iv.

24. Fundamental to a scholarly understanding of Hooke's work on Spring, and its relation to his wider scientific method as a self-conscious Baconian, are Marie Boas Hesse, 'Hooke's Vibration Theory and the Isochrony of Springs', *Isis*, **57** (1966), 433–41, and M. B. Hesse, 'Hooke's Philosophical Algebra', ibid., 67–83.

25. Christiaan Huygens announced his spiral hairspring invention for watches to the Royal Academy at Paris, *Journal des Scavans*, in 1674, and to the Royal Society (of which he was also a member), in an anagram which translated as 'axis circuli mobilis affixus in centro volutae ferreae' (i.e. 'the moveable axle of a little wheel to which is fixed the centre of a spiral iron spring') on 30 Jan. 1675. The ball was really set rolling at the Royal Society, however, on 18 Feb. 1675, when Henry Oldenburg read a letter from Huygens in Paris, outlining the above, and dated 20 Feb. 1675 (New Style, or 10 Feb. Old Style, as was used in England) 'concerning a pocket watch which he affirmed to go as fast as a pendulum...' But 'Mr Hooke said, that divers years ago he had such an invention; and that watches had been made according to the same; to which he appealed to the Journal-books of the *History of the* [Royal] *Society*, and to several members of it': Birch, *History of the Royal Society*, **III** (18 Feb. 1675), 190. Hooke had, indeed, showed pendulum watches to the Society in August 1666 (see Birch, **II**, 8 Aug. 1666: 108, and 29 Aug. 1666: 112, amongst others), but whether these watches contained straight, zigzag, or the physically superior *Volute* or spiral springs of the sort described by Huygens in 1674 is the real point at issue. A full account, complete with an engraved illustration, of the Huygens spring balance was published as 'An Extract of the French *Journal des Scavans* concerning a New Invention of Monsieur Christiaan Zulichem, of very exact and porta-tive Watches', *Phil. Trans.* 10, **112** (25 March 1675), 272–3. One should remember, how-ever, that Hooke was not the only person claiming a new spring balance in the Spring of 1675, for Gottfried Leibniz also wrote to the Royal Society with his own mechanism (which he claimed was different from Huygens') in *Phil. Trans.* 10, **113** (25 April 1675), 285–8.

26. The counterblast against Hooke—almost certainly written by Henry Oldenburg—came in the unsigned book review, 'A Description of HELIOSCOPES, and some other Instru-ments, made by Robert Hooke... F.R.S....', *Phil. Trans.* 10, **118** (1675), 440–2. In addition to chiding Hooke for his claims and conduct, the reviewer also stated of Hooke's earlier spring balance watches 'that none of those Watches succeeded, nor that any thing was done since to mend the... invention, and to render it useful, that we know of, until Monsieur Hugens, who is also Member of the Royal Society as well as he is of the *Royal Academy* at Paris, sent hither a letter dated *Januar*. 30. 1674/75', 440. Hooke's rejoinder to Oldenburg came in the 'Postscript' to his Cutlerian Lecture *Lampas; or a*

Description of some Mechanical Improvements of Lamps and Waterpoises, together with some other Physical and Mechanical Discoveries (London, 1677): see 43–4, 'A New Principle for Watches', where he asserted that he had shown his mechanism to the Royal Society 'some ten or twelve years since' (1665–7), and hence claimed a clear priority over Huygens. The chronology of the watch balance controversy and the exchanges with Oldenburg were also covered by Richard Waller, *Posthumous Works*, vi–vii.

27. Hooke, *De Potentia Restitutiva* (n. 13), 5.
28. Hooke, *Description of Helioscopes* (n. 15), 27.
29. Ibid., 29.
30. Derham, *Artificial Clockmaker* (n. 16), 103.
31. Hooke, *Diary, 1672–80*, 12 May 1675. In Hooke, *Description of Helioscopes* (n. 15), 'Postscript', 26–9, he makes his claim for the spring balance invention.
32. Hooke, *Diary, 1672–80*, 8 March 1675.
33. Ibid., March 1675 [see present book, p. 179, for author's drawing of Hooke's *Diary* sketch]. Hooke, *De Potentia Restitutiva* (n. 13), 5. *Posthumous Works*, 'Life of ... Hooke', vi–vii, for discussion of watch balance. On 25 Feb. 1675, moreover, Hooke had shown the Royal Society 'an invention' which he claimed could find the longitude to 15 arc minutes. 'James Shaen promised, that he would procure for him [Hooke, presumably] either a thousand pounds sterling in a sum, or an hundred and fifty pounds per annum'; but there is no record of Hooke's device winning this prize: Birch, *History of the Royal Society*, **III** (25 Feb. 1675), 191.
34. Hooke, *Diary, 1672–80*, 1 Jan. 1675; *Description of Helioscopes* (n. 15), 23.
35. Hooke, *Diary, 1672–80*, 13 June 1675. Very clearly, horological and vibratory experimentation was dominating much of Hooke's scientific thinking at this time, both in response to the Huygens invention, and also towards the improvement of his own earlier designs, for, likewise on 13 June 1675, he recorded: 'One day this last week I revived my old contrivance for Pocket watch by cutting the ... in two and inserting the halfs joynd by two side pieces'. He also provided a sketch of this double spring balance. This seems to have been the device that fitted Dean Tillotson's watch.
36. Ibid., 12 Aug. 1677.
37. Michael Wright, 'Robert Hooke's Longitude Timekeeper', in *Robert Hooke. New Studies*, ed. M. Hunter and S. Schaffer (Boydell Press, Woodbridge, 1989), 63–118. In this splendid and illuminating paper, Michael Wright publishes a transcription of the undated Hooke manuscript in Trinity College, Cambridge, MS O. 11a1, a document that provides the basis for his analysis of Hooke's clockwork. Michael Wright further illuminates his analysis with engineering drawings of his interpretation of Hooke's balance wheel, spring and escapement mechanisms. He has also built a working replica of the mechanism.
38. Derham, *Artificial Clockmaker*, 1714 (n. 16), ch. VIII, 'Of the Invention of those Pocket Watches, commonly called Pendulum Watches', 99.
39. The best scholarly study of Harrison and his horology is Humphrey Quill, *John Harrison, the Man who found Longitude* (John Baker, London, 1966). For extensive treatment of Harrison's 'Watch', H4, see 71–166.
40. *Posthumous Works*, 'Life of ... Hooke', x. 'G. sol Re Ut' was not a particular pitch or note, but part of the 'Guidonian' system of tuning in stringed instruments. Dr David Skinner, Organist of Magdalen College, Oxford, informs me that Hooke's figure of 272 vibrations per second for a musical pitch suggests that he was talking about a plucked wire producing a note of middle range.
41. *Posthumous Works*, 'Life of ... Hooke', xxiii. These rotating wheels, which it was hoped would emit standard physical pitches and sounds, were all part of Hooke's attempt (a concern which he shared with Huygens) to understand the physics of sound. The best modern scholarly treatment of Hooke's interest in sound is in Penelope Gouk, *Music, Science, and Natural Magic in Seventeenth-Century England* (Yale Univ. Press, New

Haven and London, 1999), ch. 6, 'Robert Hooke, Natural Magician and Experimental Philosopher', 193–223: 207–9ff.

42. Hooke, *Diary, 1672–80*, 15 Jan. 1676. This was another of those 'Saturday Club' meetings at Sir Christopher Wren's house, where Hooke and his friends discussed everything under the sun. On this occasion, Hooke described a number of his experiments and ideas about sound—which he saw as the product of impacts and blows—harmonics and pitch.
43. Hooke, 'A Curious Dissertation' (n. 11).
44. Ibid., 6–7.
45. Ibid., 10.

Chapter 11

1. *Posthumous Works*, 'Lectures of Light', Section I, p. 72, delivered early 1680. Light: 'it was the very first thing in the World to which he Almighty Creator gave his *Fiat*, when he made the World, *fiat Lux*, Let there be Light' (*Genesis* 1:3). Also Section II, Feb. 1680: 97, for Divine nature of light. Hooke also discusses the Divine origin of light in *Cometa, or Remarks about Comets* (London, 1678), 15.
2. *Micrographia*, Obs. IX, 'Of the Colours observable in Muscovy Glass [mica], and other thin Bodies', 47–67; Obs. IX is one of Hooke's great virtuosic creations, showing the way in which his scientific imagination could begin with observations, and then move on to a series of brilliant inductions which explore how nature might well work. Also Obs. X, 'Of Metalline, and other real Colours', 67–79.
3. *Micrographia*, Obs. X, 67–9: one really senses Hooke's delight in colours from the rich and evocative language that he uses.
4. *Posthumous Works*, 79.
5. *Micrographia*, Obs. IX (n. 2), 55. The motion origin of light is returned to several times in the Royal Society Lectures on Light in the early 1680s: *Posthumous Works*, 113, 121, etc.; while these motions are also suggested as the cause of the pulse or wave origin of light, p. 121. Hooke was in addition fascinated by phosphorescence and other things that glowed in the dark.
6. Hooke, *Cometa* (n. 1), 15. *Posthumous Works*, 'A Discourse of the Nature of Comets', after Michaelmas (Autumn) 1682, 166ff.
7. *Posthumous Works*, 71, 77.
8. *Posthumous Works*, 78–9.
9. *Micrographia*, Obs. IX, 54, speaks of Cartesian 'turbinated Globules' and light. In *Posthumous Works*, 71–6, Hooke reviews ancient and modern theories of light.
10. The primary and secondary qualities of objects and the way in which they were thought to govern our perceptions had fascinated seventeenth-century philosophers and scientists, but a definitive treatment and influential explanation are to be found in Robert Hooke's Christ Church contemporary John Locke, *An Essay Concerning Human Understanding* (1690), Book II, Chapter VIII, 'Some further Considerations Concerning our Simple Ideas'. Book II also develops Locke's wider theory of knowledge as experience-based, and looks at sense-impression, perception and memory.
11. *Micrographia*, Obs. IX, 67.
12. *Micrographia*, Obs. IV, 'Of fine Waled Silk or Taffety', 6–7, and Obs. X, 'Of Watered Silk or Stuff', 8–10.
13. *Micrographia*, Obs. IX, 48–9.
14. *Micrographia*, Obs. IX, 50. For colours on surface of heated metals—'vitreous laminae'—see 53.
15. *Micrographia*, Obs. IX, 64–5. He also advanced this 'puls' model of light at the Saturday Club, which met at Sir Christopher Wren's house: Hooke, *Diary, 1672–80*, 1 Jan. 1676. In

this Diary entry, Hooke even sketches how he believes these waves move: this might well be the first surviving drawing of a wave action in the history of physics.

16. *Micrographia*, Obs. IX, 57–9. For Hooke's original drawing, see Scheme VI, fig. 6.
17. *Micrographia*, Obs. X, 67.
18. *Micrographia*, Obs. X, 74.
19. *Micrographia*, Obs. X, 74–5.
20. *Micrographia*, Obs. LVIII, 'Of a new Property in the Air, and several other transparant Mediums nam'd Inflection, whereby very many considerable Phaenomena are attempted to be solv'd, and divers other uses are hinted', 218.
21. In *Micrographia*, Obs. LVIII, 217–40, Hooke explores a wide variety of optical phenomena, and this, I would suggest, constitutes, along with Obs. IV, IX and X, the first truly original piece of optical research conducted in the European scientific Renaissance. His idea of 'Inflection', or 'multiplicate refraction', opened up a whole new way of looking at optical phenomena. He also described new experiments using his Inflection model of progressive light bending in a Lecture to the Royal Society: Birch, *History of the Royal Society*, **III** (18 March 1675), 194–5.
22. Hooke saw Inflection as occurring because the air became progressively dense the closer one approached the earth's surface, in accordance with what we could now call a pressure gradient. His own barometric experiments, going back to at least August 1661 (*Micrographia*, Obs. LVIII, 225), and the work of Richard and Charles Townley (see present book, pp. 68, 278), convinced him that air also becomes a progressively denser *optical* (and increasingly refractive) medium as one descends from space to the earth. John Wallis' work on the fractional division of curves and infinite numbers is published in his *De Sectionibus Conicis* (1655) and *Arithmetica Infinitorum* (1656). See Christopher J. Scriba, 'John Wallis', in *Dictionary of Scientific Biography*, **13**, ed. Charles Coulton Gillespie (Scribner's, New York, 1980), 146–54. Dr Jacqueline Steddall of Queen's College, Oxford, is the contemporary historian of mathematics who has done the most original research into Wallis.
23. *Micrographia*, Obs. LXVIII, 220. On the same page, 220, Hooke describes his attempts to create a transparent fluid mass, getting denser towards the bottom—fresh water, salt water, alcohol, etc.—so that he could observe the geometry of a pencil of light passing through it from above.
24. *Posthumous Works*, 'Lectures on Light', Section V, 'Read June, 1681'. He also returns at the beginning of this lecture, 220, to two of his favourite optical themes: the pulse or wave model of light; and his sense of awe that light has the power to radiate in all directions from a single source, and travel vast distances, seemingly instantaneously.
25. Isaac Newton, 'A Letter of Mr Isaac Newton . . .; containing his New Theory about Light and Colors . . .', *Phil. Trans.* 6, **80** (19 Feb. 1672), 3075–87: see 3075.
26. Hooke, *Diary, 1672–80*, 1 Jan. 1676.
27. Newton, 'New Theory' (n. 25), 3084.
28. Newton, 'New Theory' (n. 25), 3078. See Hooke, *Micrographia*, Obs. IX, 54, for 'Experimentum Crucis'.
29. *Micrographia*, Obs. LVIII, 234.

Chapter 12

1. *Posthumous Works*, 'Of Comets and Gravity', 175.
2. Hooke, 'Of Gravity', Royal Society Lecture, 21 March 1665/6, Royal Society MS RBC 2:223. See also Michael Nauenberg's 'Robert Hooke's Seminal Contribution to Orbital Dynamics', a paper presented to the Royal Society Hooke Conference, July 2003; included in the circulated Conference Proceedings Papers, as yet unpublished.
3. Birch, *History of the Royal Society*, **II** (2–3 May 1666), 90–2.

4. Birch, *History of the Royal Society*, **IV** (28 April 1686), 479, for Dr Vincent's presentation of Newton's *Principia* manuscript to the Royal Society.
5. The idea of a 'Sphere of Magnetick Virtue', or attraction, around a magnetic body, such as a magnetized sphere or 'terrella', was first proposed by William Gilbert in *De Magnete* (1600): see *On the Loadstone and Magnetic Bookes on the Great Magnet of the Earth*, trans. P. Fleury Mottelay, in *Great Books of the Western World*, No. 26 (Encyclopaedia Britannica, Chicago, 1952, 1990). Book II, Ch. 27, p. 51, discusses the 'Orbis Virtutis', while Book III, 111–7, discusses magnetism and the earth's rotation. John Wilkins, *The Discovery of a World... in the Moon* (1638–40), in *The Mathematical and Philosophical Works of the Right Reverend John Wilkins* (London, 1708), Prop. XIV, 'That 'tis possible for some of our Posterity, to find out a Conveyance to this other World; and if there be Inhabitants there, to have Commerce with them', 113–35. On pp. 120–1, Wilkins takes up the problem of weight as a function of location, so that big birds find it easier to fly the further they rise from the earth.
6. *Micrographia*, Obs. LX, 'Of the Moon', 244, where Hooke compared lunar 'pits' or craters to the Icelandic and Canary Island volcanoes on earth; 245, where he argues that the moon's spherical shape suggests that it has a power of gravity. The idea of gravity being the central pulling force that made planetary bodies spherical was also further discussed in *Posthumous Works*, 'Of Comets and Gravity', 166, 178.
7. *Posthumous Works*, 'Of Comets and Gravity', 118.
8. Hooke, 'Of the Difference of Gravity by removing the body further from the Surface of the Earth', 24 Dec. 1662, Royal Society MS RBC 1, 288–91: 289 for Bacon reference. Summary in Birch, *History of the Royal Society*, **I** (24 Dec. 1662), 163–5. The Lord Verulam (Sir Francis Bacon) reference relates to a point raised in Bacon's *Sylva Sylvarum, or a Natural History in Ten Centuries* (1627): 10 'Centuries' being 10 books, each containing 100 experiments or queries, or 1,000 in all. In 'Century' 1, Experiment 33, Bacon suggested that research should be conducted to test the belief that a lump of rock or ore which needed six men to move it on the earth's surface needed only two down a deep mine, as the gravity, levity or attraction of the rocks above would help to pull it upwards. This conjecture was seen as very important in the wider business of understanding weight and gravity in the early and mid-seventeenth century. It is why experiments in mines and deep wells were believed to be necessary, as well as experiments conducted in tall buildings such as churches. John Wilkins too had discussed these underground experiments in his *Discovery of a World... in the Moon* (n. 5), Prop. XIV, 120–1, as part of his discourse on flight. Henry Power, *Experimental Philosophy* (London, 1664), Book III, 'Subterraneous Experiments', 175–81, Exp. 4, 177, also reported that a one-pound brass weight weighed an ounce less at the bottom of a 68-yard-deep pit than it did at the top. He calculated this by means of a delicate pair of balances. One can thereby understand the context of Hooke's early gravity experiments, and why, at this stage in the infancy of gravitational physics in 1662, laboratory experiments were seen as a more fruitful line of inquiry than the analysis of planetary dynamics.
9. Birch, *History of the Royal Society*, **I** (24 Dec. 1662), 163.
10. See notes 2 and 8.
11. *Posthumous Works*, 'Of Comets and Gravity', 182.
12. See Chapter 5 of present book, p. 85.
13. Hooke to Newton, 6 Jan. 1680, Letter 235, in *The Correspondence of Isaac Newton*, **II**, 309–10.
14. A. Chapman, 'The Pit and the Pendulum: G. B. Airy and the Determination of Gravity', *Antiquarian Horology* (Autumn 1993), 70–8.
15. *Posthumous Works*, 'Life of... Hooke', xiv. Also Birch, *History of the Royal Society*, **II** (9 March 1671), 471.
16. *Posthumous Works*, 'Life of... Hooke', xxi.
17. *Posthumous Works*, 'Of Comets and Gravity', 185.
18. *Posthumous Works*, 'Of Comets and Gravity', 159, where he mentions the comet seen on 16 Aug. 1682. Also, for levitating forces, see 168, 181, etc. Flamsteed's assistant at

Greenwich had seen the comet for the first time on 15 Aug. 1682: John Russell Hind, *The Comets: A Descriptive Treatise upon those Bodies* (London, 1852), 36–7. See, further, Donald K. Yeomans, *Comets, A Chronological History of Observation, Science, Myth, and Folklore* (Wiley, New York and Chichester, 1991), Chapter 6.

19. *Posthumous Works*, 'Of Comets and Gravity', 168.
20. *Posthumous Works*, 'Of Comets and Gravity', 168, items 2, 3 and 4.
21. Hooke, *An Attempt to Prove the Motion of the Earth from Observation* (London, 1674), 28.
22. Hooke, *Attempt* (n. 21), 27–8. See Birch, *History of the Royal Society*, **II** (23 May 1666), 90–2, for Hooke's description of a body inflected into a curve under the influence of an attracting power.
23. Hooke to Newton, 6 Jan. 1680, Letter 239, *The Correspondence of Isaac Newton*, **II**, 310.
24. Hooke to Newton, 24 Nov. 1679, Letter 235, ibid., 297–8.
25. Ibid., 297.
26. Ibid., 297.
27. Newton to Hooke, 28 Nov. 1679, Letter 236, ibid., 300–3.
28. Ibid., 301.
29. Hooke to Newton, 9 Dec. 1679, Letter 237, ibid., 304–6.
30. Newton to Halley, 27 May 1686, Letter 286, ibid., 433.
31. Hooke to Newton, 2 Jan. 1680, Letter 239, ibid., 309–10.
32. Hooke to Newton, 17 Jan. 1680, Letter 240, ibid., 313.
33. Flamsteed to Richard Townley, 15 Dec. 1680, Letter 389, *The Correspondence of John Flamsteed, The First Astronomer Royal, I, 1666–82*, compiled and edited by Eric G. Forbes, Lesley Murdin and Frances Willmoth (IOP, Bristol and Philadelphia, 1995), 747–8: 'The errand of this is onely to tell you that according as I foresaw and praedicted that it Would the late Comet which was seene before sun rise appears againe after Sun set' . . . having passed behind the sun. No previous letters containing this prediction seem to survive, but it is plain that in this and in his subsequent letters to other astronomers in late 1680 and early 1681, he felt that his clearly well-known prediction of the comet's return had been vindicated. Flamsteed reiterated his prediction that the 1680 comet would pass around the sun and reappear in a draft of his *Historia Coelestis Britannica*, which I restored to the English translation of the text: *The 'Preface' to John Flamsteed's 'Historia Coelestis Britannica' 1725*, ed. Allan Chapman, based on a translation by Alison Dione Johnson, National Maritime Museum Monograph 52 (National Maritime Museum, Greenwich, 1982), 160. See also Yeomans, *Comets* (n. 18), 99–100.
34. Flamsteed's main observations at this period were being made with the seven-foot-radius equatorial sextant, with telescopic sights and micrometers, based on Hooke's design, described in the present book, pp. 89–93. Also *Flamsteed's 'Historia Coelestis Britannica'* (n. 33), 113–6.
35. It had probably been Halley who first grasped the seemingly elliptical, rather than parabolic, orbit of the comet of 1682, from analyses of his own and John Flamsteed's observations of that body. This led Halley to equate the comet of 1682 [Halley's Comet] with the comets of 1607 and 1631: Halley to Newton, 28 Nov. 1695, Letter 532, *The Correspondence of Isaac Newton*, **IV**, 171–2. Also Halley to Newton, 21 Oct. 1695, Letter 536, 182–3.
36. Newton deals with comets at the end of *Principia*, Book III, Propositions 40 (Theorem 20) to 42. In this section, Newton builds upon the premises that comets move around the sun in orbits which are conic sections, the foci of which lie in the sun. He was, however, at this stage, 1685–6, still thinking of them as moving in *parabolic* and not *elliptical* orbits: Isaac Newton, *The Principia, Mathematical Principles of Natural Philosophy*, trans. I. B. Cohen and Anne Whitman, assisted by Julia Burdenz (Univ. of California, Berkeley, Los Angeles and London, 1999), 895–944. In 1687, when *Principia* was published, much of the data for the 1680 and 1682 comets had not yet been fully analysed and correct orbits extracted, but a periodicity of some comets seemed likely, especially as

those of 1680 and 1682 seemed to swing around the sun and fly off back into space once again. Halley's masterpiece, *Astronomiae cometicae synopsis. A synopsis of the astronomy of comets* (London, 1705) established the periodicity of the 1682 comet (Halley's).

37. Halley to Newton, 29 June 1686, Letter 289, *The Correspondence of Isaac Newton*, II, 441–4.
38. Ibid., 442.
39. Ibid., 442.
40. Birch, *History of the Royal Society*, III (28 April 1686), 479–80. Also Birch, *History*, III (19 May 1686), 484; ibid., 30 June 1686, 491, for *Principia* licensed and £50 from Halley.
41. Halley to Newton, 22 May 1686, Letter 285, *Correspondence of Isaac Newton*, II, 431.
42. Halley to Newton, 29 June 1686, Letter 289, ibid., 443.
43. Halley to Newton, 22 May 1686, Letter 285, ibid., 431.
44. Ibid., 431.
45. Ibid., 431.
46. See Newton to Halley, 27 May 1686, Letter 286, 443; 20 June 1686, Letter 288, 435; 14 July 1686, Letter 290, 444; 27 July 1868, Letter 291, 446.
47. This theme runs, with increasing fury, through Newton's letters to Halley from May to July 1686 (n. 46).
48. Hooke to Newton, 9 Dec. 1686, Letter 288, *The Correspondence of Isaac Newton*, II, 438.
49. Newton to Halley, 20 June 1686, Letter 288, ibid., 438.
50. Ibid., 439.
51. Ibid., 439. Newton's reference to Hooke as a plagiarist and a light-weight comes in a furious, anti-Hooke tirade, taking the form of a four-page (*Correspondence*, II, 437–40) postscript to the Newton to Halley letter of 20 June 1686, Letter 288.
52. Newton to Hooke, 28 Nov. 1679, Letter 236, ibid., 303.
53. Hooke to Newton, 9 Dec. 1679, Letter 238, ibid., 306.
54. Newton to Hooke, 13 Dec. 1679, Letter 238, ibid., 308.
55. Hooke to Newton, 17 Jan. 1680, Letter 240, ibid., 313.
56. Ibid., 313.
57. Richard S. Westfall, *The Life of Isaac Newton* (CUP, 1993), 181.
58. Flamsteed, *Historia Coelestis Britannica* (n. 33), 161–2. In my edition of Flamsteed's *Historia* (n. 33), I restore in italics those sections written by Flamsteed, and which still survive in manuscript, in which he describes his own unjust and high-handed treatment from Newton, but which the editors of his original 1725 edition chose to omit.
59. Flamsteed, *Historia* (n. 33), 164.
60. Newton to Hooke, 28 Nov. 1679, Letter 236, *The Correspondence of Isaac Newton*, II, 302. Hooke to Newton, 6 Jan. 1680, Letter 239, 310. Indeed, the Hooke–Newton correspondence between 24 Nov. 1679 and 17 Jan. 1680 contains many references to gravity experiments.
61. Aubrey, *Brief Lives*, 166–7.

Chapter 13

1. *Posthumous Works*, 'Life of … Hooke', xxiv.
2. The myth of the Roman Catholic Church suppressing science was already firmly established in England by 1689 when Hooke gave a history of science lecture to the Royal Society on 4 Dec. 1689: Royal Society MS C1.P, **XX**, 79. In particular, he spoke of 'Ignorant Monks' and the 'Darknesse of those times … [and] … blacker designs of those who ruled the commonwealth of Learning'. Hooke did, however, admire Pope Sylvester II (Pope 999–1003) for his interests in astronomy, mechanics and horology. The intellectual and mathematical writings of medieval Catholic Europe were simply not read by north European Protestants by this time, and the Middle Ages became a byword for ignorance and repression. It was, of course, an opinion that drew heavily

upon the Galileo affair of 1616–32, Galileo being the arch-hero of Wilkins, Hooke and Co. Copernicanism and the works of Galileo, nonetheless, posed problems for Calvinist Protestants in the Netherlands, who initially tended, on points of natural phenomena, to take their Bibles more literally than their Catholic counterparts. The reconciliation of science and Scripture was a significant component in the Dutch Reformation in particular: see Rienk Vermji's excellent *The Calvinist Copernicans: The Reception of the New Astronomy in the Dutch Republic 1575–1750* (Koninklijke Nederlandse Akademie, van Wetenschappen, Amsterdam, 2002).

3. Hooke, *Diary, 1672–80*, 31 Dec. 1676.
4. Pepys, *Diary*, **VI** (15 Feb. 1665), 135–7.
5. Pepys, *Diary*, **VI** (21 Jan. 1665), 19–20. Marjorie Hope Nicolson, *Pepys' Diary and the New Science* (Univ. of Virginia Press, Charlottesville, Virginia, 1965), captures Pepys' excitement at the New Science: see pp. 24–5, for instance, for his fascination with the microscope.
6. Pepys, *Diary*, **VI** (1 May 1665), 94–5.
7. Hooke, *Diary, 1672–80*, 19 Dec. 1676.
8. Ibid., 15 Dec. 1676.
9. Ian Cuddington, 'A Sociable Addiction: Tea, Coffee, Cocoa and the Chemistry of the Xanthines' (unpublished M.Chem. degree, Chemistry Part II Thesis, 2002), 2–15. Edward Robinson, *The Early English Coffee House*, 1893 (Dolphin, Christ Church, Hants, 1972), 71ff.
10. Hooke, *Diary, 1672–80*, 463–70 (Robinson and Adams edition, 1935), 'Taverns and Coffee Houses mentioned by Hooke'.
11. Ibid., 1 Jan. 1676.
12. Ibid., 2 Jan. 1676.
13. Also Hooke, 'Diary, 1688–93', 4 March 1689, 26 Oct. 1689, 2 Nov. 1689, 24 Nov. 1689, for discussion of telescopes, lenses and prices. The optician John Yarwell mentioned, 2 Nov. 1689.
14. Hooke, *Diary, 1672–80*, 28 Feb. 1674.
15. Pepys, *Diary*, **VIII** (16 Feb. 1667), 63–6.
16. *Micrographia*, Obs. XVII, 'Of Petrify'd Wood…', 108. Hooke was given fossilized material 'by the most accomplish'd Mr *Evelin*, my highly honour'd friend'. See also, for a life of Evelyn, John Bowle, *John Evelyn and his World. A Biography* (Routledge and Kegan Paul, 1981), 105–12 for Royal Society; Beatrice Saunders, *John Evelyn and his Times* (Pergamon, Oxford, 1970).
17. *The Diary of John Evelyn*, **III**, *Kalendarium 1650–1672*, ed. E. S. de Beer (Oxford, Clarendon Press, 1955), 7 Aug. 1665, 416.
18. *Micrographia*, 'Preface', sig. *gg* recto.
19. Adrian Tinniswood, *His Invention So Fertile. A Life of Christopher Wren* (Jonathan Cape, London), 111.
20. Attempts to use the 202-foot hollow cylinder of the Monument as a zenith sector to measure a stellar parallax — as Hooke had done with his 36-foot instrument in Gresham College in the late 1660s, and Flamsteed with his 'Well Telescope' a decade later — came to nothing. One major reason was the unavailability of a lens of sufficiently long focal length. Even so, the Monument does have a clear vertical passage-way, extending from the ceiling of its stone-vaulted basement to the very top, though the top is now covered in. I was allowed to examine the basement and other parts of the Monument in November 2002. The Monument was, however, used for air-pressure and barometer experiments: Birch, *History of the Royal Society*, **III** (20 Feb. 1679), 463–4. See also present book, p. 67.
21. Hooke, *Diary, 1672–80*, 9 March 1676.
22. Ibid., 24 Sept. 1675.
23. Ibid., 5 June 1675, for reference to arches. 26 Sept. 1675: 'Riddle of arch, *pendet continuum flexile, sic stabit grund Rigidum*' ('the curved line hangs, and thus the foundation [or pillars]

will stand firm'). The word *grund* is problematical, and my wife Rachel (a classicist) says that Hooke probably derived it from the medieval Latin *groundo* ('to lay a foundation') or *groundarium* ('foundation'). Also Hooke, *A Description of Helioscopes, And some other Instruments* (London, 1676), 31–2, where he concludes his Cutlerian Lecture with a list of 10 inventions and discoveries which he plans to announce, and for which he is currently publishing succinct 'tasters', often in anagrammatic form. No. 3 was the anagram for his famous Law of Spring, while No. 2 was 'The true Mathematical and Mechanical form of all manner of Arches for Building, with true butment necessary to each of them. A Problem which no Architectonick Writer hath ever yet attempted, much less performed. *abcccddee…*' (anagram).

24. Hooke, *Diary, 1672–80*, 22 Oct. 1675.
25. Ibid., 18 Aug. 1677.
26. Ibid., 13 Oct. 1677.
27. Ibid., 6 Oct. 1675.
28. Robert Boyle's Will, in R. E. W. Maddison, *Life of the Honourable Robert Boyle* (Taylor and Francis, London, 1969), 261.
29. *The Correspondence of Robert Boyle*, **II**, records many examples of Hooke in London writing to Boyle in Oxford, Stalbridge or Ireland. These letters are always friendly in tone, and sometimes conclude with Hooke requesting Boyle to pass on his greetings to mutual friends, as when, Hooke to Boyle, 15 Aug. 1665, 512–3, he wishes to be remembered to Mr and Mrs Crosse (Boyle's old landlord in the High Street, Oxford), and to Dr Richard Lower.
30. *The Correspondence of Robert Boyle*, **II**, Hooke to Boyle, 6 Oct. 1664, 342–4.
31. Hooke, *Diary, 1672–80*, 15 Nov. 1673.
32. Ibid., 3 Oct. 1672.
33. Ibid., 17 Oct. 1677.
34. Ibid., 11 Jan. 1673.
35. Ibid., 31 Jan. 1673.
36. Maurice Balme, *Two Antiquaries. A Selection from the Correspondence of John Aubrey and Anthony Wood* (Durham Academic Press: Edinburgh, Cambridge, Durham, USA, 2001), 10. Aubrey was elected FRS in 1663, and this became a central feature of his life thereafter, knowing as he did most of the Fellows.
37. Aubrey, *Brief Lives*, 'The Life and Times of John Aubrey', xxxviii.
38. John Aubrey to Anthony Wood, 15 June 1680, in Balme, *Two Antiquaries* (n. 36), 92.
39. Hooke, *Diary, 1672–80*, 14 Oct. 1673.
40. Aubrey to Wood, 26 June 1679, in Balme, *Two Antiquaries* (n. 36), 77.
41. Aubrey to Wood, 23 Oct. 1688, ibid., 115.
42. Aubrey to Wood, 27 Oct. 1687, ibid., 113–4.
43. Aubrey to John Evelyn, 10 May 1692, in *Aubrey on Education. A hitherto unpublished manuscript by the author of 'Brief Lives'*, ed. J. E. Stephens (Routledge and Kegan Paul, London and Boston, 1972), 15.
44. Aubrey to Wood, 'March the last [31st?]' 1674, in Balme, *Two Antiquaries* (n. 36), 56.
45. Aubrey to Wood, 20 Jan. 1690, ibid., 133.
46. *Aubrey on Education* (n. 43), 94, 164.
47. Ibid., 68.
48. Ibid., 86. For a full modern scholarly treatment of Aubrey, see Michael Hunter, *John Aubrey and the Realm of Learning* (Duckworth, London, 1975).
49. Aubrey to Wood, 24 April 1690, in Balme, *Two Antiquaries* (n. 36), 126.
50. Hooke, *Diary, 1672–80*, 3 Oct. 1674.
51. Ibid., 12 Aug. 1676.
52. Ibid., 2 Oct. 1675.
53. Ibid., 20 Jan. 1674. The 'Pitz' in question was probably the bookseller Moses Pit or Pitt.

54. Hooke, 'Diary, 1688–93', 6 June 1689. If the *Tempest* in question was Shakespeare's, he had seen it at the theatre 15 years previously: *Diary, 1672–80,* 20 June 1674.
55. Ibid., 25 April 1689.
56. Hooke, *Diary, 1672–80,* 22 Sept. 1672.
57. Ibid., 31 Dec. 1672.
58. Ibid., 14 June 1674.
59. Aubrey, *Brief Lives,* 'The Life and Times of John Aubrey', is full of tales of ghosts, spirits and fairies: see xxv, etc. Aubrey collected these tales and wrote them down, although he believed that the Civil Wars, along with printing, had driven out the ghosts: 'the divine art of Printing and Gunpowder have frighted away Robin-goodfellow and the Fayries', xxix, and 'no Suffimen is a great fugator [i.e. putter to flight] of Phantosmes than Gunpowder', xxxii. To Aubrey, the Civil Wars had been utterly pivotal in English history, ushering in not only new political and religious arrangements (which this conservative Wiltshire country gentleman and lover of the 'Olden times' detested), but also science and coffee houses, and driving out ghosts and fairies!
60. See Richard Nichols, *The Diaries of Robert Hooke, The Leonardo of London, 1635–1703* (Book Guild Ltd, Lewes, 1994), 29–34, for a list of Hooke 'Diary' entries relating to eating.
61. Hooke, *Diary, 1672–80,* 31 Aug. 1680.
62. Ibid., 4 May 1673. Robinson and Adams, editors of Hooke's *Diary, 1672–80,* describe 'Scaramuches' as a puppet show, in their footnote, 42. This is probably incorrect. 'Scaramouch' was a stock character in Italian *Commedia dell' Arte* comic theatre — along with Pierrot, Columbine, Arleqino, etc. — which was all the rage at the time. King Louis XIV of France patronized these companies at Court, and one particular French or Italian actor came to specialize in the Scaramouch — a cheeky, mischief-maker — role. As the Duke of York's (the future King James II) own Court, at York House, was famed for its love of all things French, it is not unlikely that Hooke had been one of an invited audience to see 'Scaramouch' himself perform with a visiting *Commedia dell' Arte* company over from France. Hooke's presence in the audience of such a high-status gathering is a clear indication of his gentlemanly social standing. (I am indebted to my friend Nigel Frith, actor, writer and theatre historian, for information about Scaramouch.) It is possible, however, from his Diary that on 12 April 1673 Hooke may indeed have attended a puppet show, from the single word 'Punchanellos' in his entry for that day. As he spent part of the day around Scotland Yard and the Strand, this could well have been some Court entertainment in nearby Whitehall Palace. Such shows, the origin of 'Punch and Judy', were often puppet versions of *Commedia dell' Arte* stories.
63. Hooke, *Diary, 1672–80,* 1 Sept. 1679.
64. Ibid., 12 Aug. 1675.
65. Ibid., 9 Dec. 1678. Hooke's fear of women sometimes broke through into his dreams, as when his Diary records, 23 March 1674, 'Dremt of viragoes and other strange phenomena.'
66. Ibid., 8 May 1673, 20 Jan. 1674, etc.
67. Ibid., 25 Nov. 1674.
68. Ibid., 20 June 1678.
69. Ibid., 1 Aug. 1678: 'I viewd the house and gave her a certificat of what I thought of her wall'.
70. Ibid., 25 June 1678.
71. Ibid., 5 July 1678.
72. Ibid., 12 Sept. 1678.
73. Ibid., 4 Sept. 1675.
74. Ibid., 28 Sept. 1678.
75. Ibid., 2 July 1679; 18 Oct. 1679.
76. Ibid., 16 Aug. 1677.

77. Ibid., 11 Aug. 1673.
78. Aubrey, *Brief Lives*, 'Hooke', 165.
79. Claire Tomalin, *Samuel Pepys. The Unequalled Self* (Penguin, 2003), 207–14.
80. Aubrey, *Brief Lives*, 'Harvey', 131. See Old Testament, *I Kings*, 1:3–4, for the very old King David and Abishag the Shunnamite, the fair damsel of Israel, who was brought in to keep him chastely warm in bed. Aubrey, alas, failed to record the name of the girl who similarly kept the aged Dr Harvey warm, and who benefited under his Will.
81. Hooke, *Diary, 1672–80*, 14 Sept. 1672.
82. Ibid., 26 Jan. 1673.
83. Ibid., 28 Oct. 1672.
84. Ibid., 13–14 Aug. 1673.
85. Ibid., 14 Aug. 1673.
86. Ibid., 15 Aug. 1673.
87. Ibid., 22 Nov. 1673.
88. Ibid., 12 Dec. 1673.
89. Rob Martin, 'The Scientist, the Grocer, the Governor, and Grace', Isle of Wight History Centre, 2002: website file, // A:\hookeweb\sggg.htm. See p. 9 for reference to Mary and Nell Young and the Isle of Wight; ibid., p. 10, for Jane Young as a Hooke family servant on the Isle of Wight. This website is based on a detailed examination of Isle of Wight records by local historians.
90. Bridget Taylor's cousinship with Nell Young is made clear in Hooke, *Diary, 1672–80*, 29 Aug. 1673: 'Nell's Cousin came out of country calld Bridget Taylor'. Also Martin, 'The Scientist...' (n. 89), p. 9.
91. Hooke, *Diary, 1672–80*, 16 Oct. 1673, 'Hird Dol Lord'.
92. Ibid., 27 Dec. 1673. On 28 March 1674, 'at Mrs Storys 40 sh[illings] for Dols wages'. Had relations between Hooke and Doll so broken down that she had moved out unpaid and Hooke needed to pass on her wages via a third party?
93. Ibid., 20 April 1674.
94. Ibid., 17 Aug. 1678.
95. Hooke, 'Diary, 1688–93', Sun. 2 April 1693.
96. Ibid., Sun. 23 April 1693. See n. 93 above.
97. Ibid., Sun. 6 Oct. 1689.
98. Ibid., Sat. 12 March 1693.
99. Ibid., Sun. 21 May 1693.
100. Hooke, *Diary, 1672–80*, 3 April 1674.
101. Ibid., 6 June 1674.
102. Ibid., 20 June 1674; 6 July 1674. Could the 'Cock' which Hooke lost have been the break-age of a lens by Christopher Cock — whom Hooke sometimes referred to as 'Cock'?
103. Ibid., 16 July 1674.
104. Ibid., 30 Sept. 1674.
105. 'Grace ye daughter of Mr. John Hooke, born ye 2nd of May 1660': Newport, Isle of Wight (NPT/REG/COM/4), reproduced in 'The Hooke Family Tree (including the Giles Family Tree)', on Isle of Wight History Centre 'hookeWEB HOMEpage', electronic file // A:hookeweb\tree.htm.
106. Hooke, *Diary, 1672–80*, 13 Sept. 1672.
107. Ibid., 12 Jan. 1679.
108. Ibid., 4 June 1676.
109. Ibid., 20 July 1677.
110. Ibid., 30 June 1677. Whatever might have happened between Grace and Pettis in the cellar, she was back in bed with her uncle Robert the following day, 1 July 1677, 'Tu Grace ⋈.'
111. Ibid., 20 July 1677.
112. Ibid., 15 Dec. 1676.

113. Ibid., 31 Oct. 1677.
114. Ibid., 3 May 1678.
115. Ibid., 7 June 1678.
116. Ibid., 1 March 1678.
117. 'The Life of John Hooke', pp. 1–4, Isle of Wight History Centre 'hookeWEB HOME-page', file // A:\hookeweb\john.htm.
118. Hooke, *Diary, 1672–80*, 21 Dec. 1672; 20 Dec. 1673.
119. Ibid., 10 Nov. 1672.
120. Ibid., 6 Aug. 1675.
121. Ibid., 10 Feb. 1676.
122. Ibid., 2 March 1678.
123. Ibid., 2 March 1678.
124. In Newport, I.o.W. Corporation Books (NBC/45/16b, 8 May 1678). I have not examined this record personally, but cite it from Rob Martin (2000), Isle of Wight History Centre 'hookeWEB HOMEpage'. See 'The John Hooke Tragedy', and 'The Scientist, The Grocer, the Governor, and Grace' (n. 89), p. 10. Quotations in this website are cited from seven primary sources. See also Isle of Wight History Centre 'hookeWEB HOME-page', 'Newport Corporation and the Suicide of John Hooke', from Newport, I.o.W. Borough Convocation Book (NBC/45/16b) at http://freespace.virgin.net/ric.martin/vectis/hookweb/corp.htm, p. 2, 8 July 1678.
125. 'Newport Corporation and the Suicide of John Hooke', Newport I.o.W. Borough Convocation Book (NBC/45/16b), at (n. 124) p. 2, 8 July 1678, for Mrs Elizabeth Hooke (John's widow) and her £10 p.a. pension.
126. Hooke, *Diary, 1672–80*, 19 and 29 April 1678; 11 May 1678.
127. Ibid., 29 April 1678.
128. 'The Scientist, the Grocer, the Governor, and Grace' (n. 89), 10–1. Also Richard Ollard, *Man of War: Sir Robert Holmes and the Restoration Navy* (1969; Phoenix, London, 2001), 188: the birth of Mary Holmes is mentioned, 1678, but no mother is specified. Lisa Jardine, *The Curious Life of Robert Hooke, the Man who Measured London* (Harper-Collins, London, 2003), 257, suggests the possibility of Grace Hooke's pregnancy masquerading as 'measles' in Hooke's Diary. But as I argue on pp. 237–8, I find it hard to accept that a man whose private Diary is so sexually explicit as Hooke's should have been so coy about an illegitimate pregnancy as to call it 'measles', and show no further concern.
129. Hooke, *Diary, 1672–80*, 7 Aug. 1677. I am indebted to Monica Mears for conversation and correspondence, in July 2003, regarding the possible paternity of the woman born in 1678 who became acknowledged as Mary Holmes.
130. Hooke, *Diary, 1672–80*, 26 Feb. 1678, 'Mr Young told me of Grace sick of the Measles'. The identity of this Mr Young is unclear. In the 'Index' of Hooke's *Diary, 1672–80*, Robinson and Adams (eds.) describe him as being 'of Plymouth'. One wonders how extensive the Young family was, and what was their relationship to the Hookes: employees, servants, or even distant relatives?
131. Modern demographic researchers have shown—from parish registers and such—that seventeenth-century births out of wedlock were less common than popularly supposed, though they were more likely to take place in anonymous large communities, such as London, than in the full glare of legal and religious opprobrium back at home. Yet St Margaret's, Westminster, in spite of the large transient and anonymous population in seventeenth-century Westminster, had a bastardy rate of only 2.5%; above the national average, but still low: see Peter Laslett, *The World we have Lost* (Methuen, London, 1965), 135. Tom Brown, the facetious comic writer in the *Lacedemonian Mercury* (prob. Feb. 1692), makes play of the fact that women 'at the other end of Town' (Westminster or the City) from Wapping not infrequently give birth only six, five or even two months after marriage! Brown is almost certainly alluding to the well-known fact that women went to the anonymity of London to have socially embarrassing births:

for Brown's text, see Philip Pinkus, *Grub Street Stripped Bare* (Constable, London, 1968), 89–90.

132. Hooke, *Diary, 1672–80*, 7 June 1678.
133. Ibid., 27 July 1678.
134. Ibid., 2 Nov. 1679.
135. Ibid., 13 Dec. 1676.
136. Ibid., 7 July 1675, 'Sister Hooke hither'. On 10 Dec. 1676, '. . . turkeys sent from S[ister] Hooke last Thursday'. Had the carrier wagon taken three days from Portsmouth — depending on the winter crossing of the Solent — to London?
137. Ibid., 7 June 1678.
138. Ibid., 29 Sept. 1674.
139. Ibid., 25 Nov. 1676.
140. Hooke, 'Diary, 1688–93', 1 July 1693, mentions 'N. Young'. This was probably his old servant-cum-mistress, Nell, from 1672–3, and who, as a married woman and mother, had become a firm friend of Robert, as this Diary shows (see present book, pp. 231–2). There was clearly a community of Hooke-connected Youngs in London and on the Isle of Wight, whose presence is to be found in surviving documents. The 'Index' to Robinson and Adams' *Diary, 1672–80*, for instance, lists one Nicholas Young, a master mason, plus a Mr Young of Plymouth, a Mrs Young and Young the barber. Were they related?
141. Hooke, *Diary, 1672–80*, 12 July 1675.
142. Ibid., 8 Sept. 1677.
143. Ibid., 9 Sept. 1677.
144. Sir Edmund King (1629–1709), MD, had known Hooke since the beginning of the Royal Society and was one of the leading animal experimenters of the day. His career, moreover, had been a remarkable one; he started out as a surgeon's apprentice and ended up an FRCS, Lambeth MD, Knight and Royal Physician. One cannot but wonder, however, how far the copious blood-letting and similar aggressive treatments that King applied to his Majesty, following the onset of what was probably a mild stroke, in Feb. 1685, may have hastened Charles II's end.
145. Hooke, *Diary, 1672–80*, 26 July 1679. Hooke consulted Dr Mapletoft and Mr Whitchurch about Grace's illness. He also wrote to her mother to keep the now widowed Elizabeth Hooke informed about her daughter's condition.
146. Ibid., 12 Sept. 1677.
147. Ibid., 13 Sept. 1677, 'Mrs Kedges in Silver street to acquaint Hanna Gyles of Toms Death'. Was Mrs Kedges another of Hooke's lines of communication with family on the Isle of Wight? Then, 25 Oct. 1677, Hooke 'Received a letter of Gratefullnesse from R.[obert] Gyles [Tom's father] for kindness to poor Tom Gyles'.
148. *Posthumous Works*, 'Life of . . . Hooke', xxiv.
149. Aubrey, *Brief Lives*, 'Boyle', 36.
150. Hooke, *Diary, 1772–80*. The day after receiving the shock of Brother John Hooke's death (2 March 1678), Hooke was 'ill all day', '*miserere mei deus*', 3 March 1678.
151. Ibid., 14 March 1675.
152. Ibid., 16 May 1675.
153. Ibid., 4 May 1673.
154. Ibid., 12 March 1678.
155. Ibid., 9 Sept. 1678.
156. Ibid., 7 Sept. 1678.
157. Ibid., 26 Dec. 1674.
158. *Posthumous Works*, 'Life of . . . Hooke', xxviii.
159. Ibid., xxviii.
160. Ibid., xxv.
161. Hooke, 'Diary, 1688–93', 19 March 1693. These cryptic references to churches on Sundays are not without ambiguity. R. T. Gunther, the editor of the 'Diary, 1688–93', suggests that

the letter 'M' refers to Hooke's servant Martha. Even so, the specific mention of churches such as St Helen's, St Peter Poor and Westminster indicates that Hooke himself was developing some Sunday acquaintance with them. He also seems to have taken the Sacrament at St Peter Poor on 7 July 1689: 'To St. Peters Rd. Sacramt of Perry'. Then between Sunday 11 Dec. 1692 and the sudden termination of his Diary on Tuesday 8 Aug. 1693, Hooke makes 18 references to Church on Sunday mornings. Had an interest in matters spiritual which had lain dormant since his Oxford days been re-kindled?

162. Andrew Marvell (1621–78), 'To his Coy Mistress', line 22. See *The Metaphysical Poets*, ed. Helen Gardner (Penguin, 1972), 250–5. I do not know if Hooke knew this poem.
163. *Posthumous Works*, 'Life of ... Hooke', xxviii.
164. Aubrey, *Brief Lives*, 'Hobbes', 157.
165. For Pepys' dangerous and unintended involvement in the Popish Plot of 1679, see Tomalin, *Samuel Pepys* (n. 79), 314–26.
166. Hooke, *Diary, 1672–80*, 28 Oct. 1678.
167. Ibid., 4 Dec. 1678.
168. Ibid., 19 Dec. 1678.
169. Ibid., 31 Oct. 1678.
170. Ibid., 3 and 5 Dec. 1678.
171. Ibid., 5 Nov. 1673.
172. Hooke, 'Diary, 1688–93', 11 Dec. 1688.
173. Queen Mary, styled Mary II (1662–94), was the daughter of James, Duke of York—later King James II—and Anne Hyde. Mary's maternal grandfather, Edward Hyde, Lord Clarendon, and her grandmother were staunch Protestants, as was her mother Anne, who had done well in the Commonwealth under Oliver Cromwell, but who switched allegiance to the Stuarts in 1660. Mary was brought up and—in spite of her father's Catholic conversion in 1672—remained a devout Protestant to her dying day. William (1650–1702), however, was English on his mother's side, his mother being Mary, daughter of King Charles I and Henrietta Maria of England. His father was William II of Orange. He was Protestant in religion, and anti-French in both his Dutch and English policies. He had married Mary in 1677, and in 1689 inaugurated the rule of William and Mary.
174. Hooke, 'Diary, 1688–93', 5 Nov. 1688.
175. Ibid., 20 Nov. 1688.
176. Ibid., 20 Nov. 1688, 'At Jon[athan's Coffee House] talk of ships revolt'.
177. Hooke, *Diary, 1672–80*, 6 Aug. 1675; 10 Feb. 1676.
178. Jim Bennett, Michael Cooper, Michael Hunter and Lisa Jardine, *London's Leonardo. The Life and Work of Robert Hooke* (OUP, 2003); see Cooper, 'Hooke's Career', 45–8, for Hooke's earnings by the mid 1670s.
179. Richard S. Westfall, *Never at Rest. A Biography of Isaac Newton* (CUP, 1980). For break-down, see 533–54. Also Frank E. Manuel, *A Portrait of Isaac Newton* (Belknap Press of Harvard Univ. Press, Cambridge, Mass., 1968), 'The Ape of Newton: Fatio de Duillier', 191–212, and 'The Black Year, 1693', 213–25.
180. R. Hooke, 'Discourse ... concerning an Experiment of the Penetration of two liquors', 3 July 1689, Royal Society MS CI. P, **XX**, 78, 1, recto and verso.

Chapter 14

1. Thomas Willis, *The Anatomy of the Brain*, trans. Samuel Pordage, from *Cerebri Anatome* (1664), in *The Remaining Medical Works of That Famous and Renowned Physician Dr Thomas Willis* (London, 1681): see 'Postscript' for biography of Willis. Pordage claimed that as the book was being printed, Nov. 1675, news arrived that Dr Willis had died,

suddenly and unexpectedly, from a cough that developed into pneumonia and pleurisy. He died peacefully, after receiving Holy Communion, aged 57.

2. Aubrey was staying briefly in Oxford, on a journey between London and Draycott, when he died: 'John Aubrey' (Richard Garrett), *Dictionary of National Biography*.

3. Ellen Tan Drake, *Restless Genius. Robert Hooke and his Earthly Thoughts* (OUP, 1996), No. 27, 10 Jan. 1700. Based on the dating of Hooke's 'Earthquake Discourses' established in Rhoda Rappaport, 'Hooke on Earthquakes: Lectures, Strategy, and Audience', *BJHS*, **19** (1986), 129–46.

4. *Posthumous Works*, 'Life of . . . Hooke', xxvi.

5. Ibid., xxvi. One wonders how original this demonstration could have been by this date, for, as we saw on p. 77, Wren had been interested in Saturn since the late 1640s, and had also made models of the body of the planet as a way of trying to make sense of Saturn's *ansae*, or rings. See Albert Van Helden, 'Christopher Wren's *De Corpore Saturni*', *Notes and Records*, **23** (1968), 213–9: 215–6 for models.

6. *Posthumous Works*, 'Life of . . . Hooke', xxvi.

7. Jim Bennett, Michael Cooper, Michael Hunter and Lisa Jardine, *London's Leonardo. The Life and Work of Robert Hooke* (OUP, 2003): Cooper, 'Hooke's Career', 20.

8. *Posthumous Works*, 'Life of . . . Hooke', xxiv, xxvii.

9. Ibid., xxvi.

10. Ibid., xxvi.

11. These included Dr Richard Whittington of Oriel College, Oxford, and former Coroner for Birmingham; Dr John Lester, a Walsall physician of extensive diagnostic experience; and the late Professor John Potter, Director of the Radcliffe Medical School, Oxford, and a Fellow of Wadham College.

12. *Posthumous Works*, 'Life of . . . Hooke', xxvi–xxvii, for Waller's description of Hooke's final syndrome of illnesses and accidents.

13. Ibid., xxvi.

14. Private communication: Dr Richard Whittington, 'Notes on the Illnesses of Robert Hooke', to the author, Dec. 2003.

15. *Posthumous Works*, 'Life of . . . Hooke', xxvi.

16. Ibid., xxvii.

17. Hooke, *Diary, 1672–80*, 31 Dec. 1672: 'Dremt of a medicine of garlick and the night before I drempt of riding and eating cream with Cap. Grant'. Richard Nichols, *The Diaries of Robert Hooke, The Leonard of London, 1635–1703* (Book Guild, 1994), 29–34, for references to food and drink.

18. *Posthumous Works*, 'Life of . . . Hooke', xxvi–xxvii.

19. Robert Knox, *An Historical Relation of Ceylon together With somewhat Concerning Severall Remarkable passages of my life that hath hapned [sic] since my Deliverance out of my Captivity* (London, 1681; Glasgow, 1911), 382–3. Also, for manuscript, see Bodleian Library, Oxford, MS Rawlinson Qc 15, fol. 71.

20. *Posthumous Works*, 'Life of . . . Hooke', xxvi.

21. Ibid., xxvii.

22. Godfrey Copley to Thomas Kirke, 2 March 1703; original cited by Lisa Jardine, *The Curious Life of Robert Hooke, The Man Who Measured London* (Harper Collins, London, 2003), 298.

23. Hooke, *Diary, 1672–80*, 4 Jan. 1673.

24. Ibid., 26 Dec. 1674.

25. Jardine, *The Curious Life of Robert Hooke* (n. 22), 315.

26. See n. 19 above.

27. Reproduced photographically in Jardine, *The Curious Life of Robert Hooke* (n. 22), final colour plate, facing p. 295. The Will is dated 25 Feb. 1703.

28. *Posthumous Works*, 'Life of . . . Hooke', xxvii.

29. Thomas Kirke (Jun.) to Thomas Kirke (Sen.), at Cookridge, near Leeds, 9 March 1702/3: in *Publications of Thoresby Society Miscellanea*, xxviii (Leeds, 1928), 457.

30. Knox, *An Historical Relation* (n. 19), 382. Knox gives the date 2 March 1702/3, 'This Night about 11 or 12 of the Clocke'. One presumes that if Hooke died on 3 March, it must have been not much after midnight.
31. Ibid., 383.
32. Kirke to Kirke (n. 29), 457. Also 'Hooke's Possessions at his Death: A Hitherto Unknown Inventory': this document was discovered in the Public Record Office by Frank Kelsall. In *Robert Hooke, New Studies*, ed. Michael Hunter and Simon Schaffer (Boydell Press, Woodbridge, 1989), 287–94: see 290.
33. *Posthumous Works*, 'Life of ... Hooke', xxvi.
34. See 'Hooke's Possessions at Death' (n. 32), 292–4, for Exton's Inventory.
35. All the above detail in Will Inventory (n. 32). For pistols, see Hooke, *Diary, 1672–80*, 3 and 5 Dec. 1678.
36. Hooke, *Diary, 1672–80*, 20 April 1674.
37. 'Hooke's Possessions at Death' (n. 32), 294.
38. Kirke to Kirke (n. 29), 457.
39. Jardine, *The Curious Life of Robert Hooke* (n. 22), 307.
40. *Posthumous Works*, 'Life of ... Hooke', xxvii.
41. 'Hooke's Possessions at Death' (n. 32), 294, for signatures.
42. Kirke to Kirke (n. 29), 457.

Appendix

1. Lisa Jardine, *The Curious Life of Robert Hooke, the Man who Measured London* (Harper Collins, London, 2003), 17–9. See also colour reproduction of the picture alleged by Professor Jardine to be Hooke, facing p. 118. In its top right-hand corner is the inscription 'I. Ray'.
2. Jardine, *The Curious Life of Robert Hooke*, 18.
3. Hooke, *Diary, 1672-80*, 23 March 1674.
4. Ibid., 4 Feb. 1674.
5. Ibid., 'Paid periwig wooman 10 sh[illings]', 16 Feb. 1674; 'Paid the periwig woman at Gresham College Gate 2s. 6d in full', 18 May 1674.

LIST OF WORKS ABBREVIATED IN THE NOTES

Robert Hooke

Diary, 1672–80
The Diary of Robert Hooke, MA, MD, FRS, 1672-1680, ed. Henry W. Robinson and Walter Adams (Wykeham Publications, Taylor and Francis, London, 1935, 1968).

'Diary, 1688–93'
'The Diary of Robert Hooke: Part I, November 1688 to March 1690; Part II, December 1692 to August 1693', in Robert T. Gunther (ed.), *Early Science in Oxford*, **X** (Oxford, 1935).

Micrographia
Micrographia, or some Physiological Descriptions of Minute Bodies Made by Magnifying Glasses with Observations and Inquiries thereupon (London, 1665).

Philosophical Experiments
Philosophical Experiments and Observations of the late Eminent Dr Robert Hooke, ed. William Derham (London, 1726).

Posthumous Works
The Posthumous Works of Robert Hooke, ed. Richard Waller (London, 1705).

Other works

Aubrey, *Brief Lives*
John Aubrey, *Aubrey's Brief Lives*, edited from the original manuscripts with an introduction by Oliver Lawson Dick (Secker and Warburg, London, 1949, 1975).

Birch, *History of the Royal Society*
Thomas Birch, *History of the Royal Society of London*, **I–IV** (London, 1756-7). (Birch's *History* is a published version of the Royal Society Manuscript Register covering the first 30 years of the Society's history.)

The Correspondence of Isaac Newton
The Correspondence of Isaac Newton, **II** (ed. H. W. Turnbull, FRS, 1960), **IV** (ed. J. F. Scott, 1967) (CUP, for the Royal Society).

The Correspondence of Robert Boyle
The Correspondence of Robert Boyle, I, 1636-1661, ed.
The Correspondence of Robert Boyle II, 1662-1665, ed. Michael Hunter, Antonio Clericuzio and Lawrence M. Principe (Pickering and Chatto, London, 2001).

319

Pepys, *Diary*
The Diary of Samuel Pepys, **I–IX**, ed. R. Latham and W. Matthews (G. Bell and Sons, London, 1970–6).

Journals
BJHS
British Journal for the History of Science

JBAA
Journal of the British Astronomical Association

JHA
Journal for the History of Astronomy

Notes and Records
Notes and Records of the Royal Society

Phil. Trans.
Philosophical Transactions of the Royal Society

QJRAS
Quarterly Journal of the Royal Astronomical Society

INDEX

Please note page references in *italics* refer to endnotes

321